18.00

WITHDR/
The U.

D1389707

'TOCK
verpool

Chemical
Thermodynamics

Chemical
Thermodynamics

M. L. McGlashan

ACADEMIC PRESS
A Subsidiary of Harcourt Brace Jovanovich, Publishers
London   New York   Toronto   Sydney   San Francisco

ACADEMIC PRESS INC. (LONDON) LIMITED
24/28 Oval Road
London NW1 7DX
(Registered Office)
(Registered number 5985 14)

US edition published by
ACADEMIC PRESS INC.
111 Fifth Avenue
New York
New York 10003

© 1979 Academic Press Inc. (London) Ltd.
British Library Cataloguing in Publication Data
McGlashan, Maxwell Len
  Chemical thermodynamics.
  1. Thermodynamics
  2. Chemical reactions
  I. Title
  541'.369   QD504   79-40919
  ISBN 0-12-482650-4

Printed in Great Britain by
J. W. Arrowsmith Ltd., Bristol, BS3 2NT

Preface

I believe, and shall repeat on p. 111, that 'Thermodynamics is an experimental science, and not a branch of metaphysics. It consists of a collection of equations, and also some inequalities, which inter-relate certain kinds of measurable physical quantities. In any thermodynamic equation every quantity is independently measurable. What can such an equation "tell one" about one's system or process? Or, in other words, what can we learn from such an equation about the microscopic explanation of a macroscopic change? Nothing whatever. What then is the use of thermodynamic equations? They are useful because some quantities are easier to measure than others.' If every equilibrium property of every kind of system had been measured with high accuracy then thermodynamics would be useless (though it would still be beautiful). Happily for those of us who profess the subject, the almost infinite variety of systems and of properties, the limitations on the accuracy of even the most skilled and careful experimentalist, and the unwillingness of the scientist to undertake a measurement when he can calculate the quantity he needs from someone else's measurements of quite different quantities, all ensure that it will remain both beautiful and extremely useful.

This book was written for chemists. I have excluded electromagnetic, gravitational, and relativistic effects, have said little about transitions of higher order than the first, and have given only an introductory account of the modern thermodynamics of critical points. I have deliberately relegated thermodynamic inequalities to a minor place and so have divided the second law into an equality (chapter 5) having consequences that permeate the whole of the rest of the book, and an inequality having only a few consequences all confined to chapter 7. Heat engines and Carnot's cycle are dismissed, each in a brief section in chapter 7, as consequences of little interest to chemists. On the other hand, I have digressed from classical thermodynamics in two chapters: in chapter 14 on Boltzmann's distribution law which has far-reaching consequences for chemists and without which it is impossible adequately to discuss the 'third law'; and in chapter 19 on galvanic cells so as to reveal the extra-thermodynamic dependence of the formula for the electromotive force of a galvanic cell on Onsager's reciprocal relations, further discussion of which I decided, rather sadly, to omit.

Among the few best accounts of thermodynamics, none seems to me to give among its necessary formalism any proper emphasis to experimental methods. The great charm of thermodynamics to me is that it can at best be both rigorous in the formulation of its algebra and rigorous in its experimental demands. At one moment the thermodynamicist is discussing some trying analysis (say, the finer points of a Washburn correction) and at the next moment some trying experimental point (say, what kind of solder to use for a particular closure). Nor does it seem to me possible to appreciate thermodynamics unless it is continually related to what one actually measures in the laboratory; the reader of this book, as of any other on thermodynamics, should constantly be asking the question: 'Yes, but how exactly do you measure it (and not some vaguely related quantity)?'. In this book I have tried to marry the theoretical and experimental aspects of the subject.

I have used none but SI units. In the naming of quantities and choice of symbols I have followed the International Organization for Standardization's ISO 31 (Parts 0 to 13) except in one instance: my adoption of the symbol Φ for the quantity I call the Giauque function (after the great Nobel laureate in this field).

The first few chapters were drafted about twelve years ago at the instigation of four exceptional second-year undergraduates at the University of Exeter: Dr Kiki Warr, Dr Stephen Sykes, Mr David Fitzsimons, and the late Mr Graham Waller. Dr James Watson (through a provocative question after a lecture), Dr Kenneth Marsh, and Professor Gerhard Schneider (especially on chapter 17) each offered me valuable advice. Dr Selby Angus willingly did some computations for me. Professor Norman Rydon provided the material for problem 18.1, and the excuse to acknowledge here the ten happy years I spent with him and his colleagues at the University of Exeter. Among my present colleagues Professors Allan Maccoll and Charles Vernon each read parts of the typescript and gave me wise advice. Mr John Cresswell made all the drawings (except the few reproduced from elsewhere) with great willingness, skill, and accuracy. I warmly thank all of these. But the book would still not have appeared had it not been for the happy conjunction in my laboratory of Dr Michael Ewing, Dr James Hugill (until he left in September 1978 to work in Amsterdam), and Mr Leonard Toczylkin. From January 1978 until the last proofs had been corrected in July 1979 they came regularly to my room and spent many hours there. They would arrive armed with my latest chapter, with their formidable understanding, and with a library that usually included both kinds of Fowler. Looking most intimidating they would then systematically criticize everything from my punctuation and grammar to the matter itself. When they had finished with that chapter, they would goad me into writing the next. The book is immensely better because of their work and might have been better still had I not sometimes been pig-headed. I am more grateful to them than I can easily say.

That Miss Joan Fujimoto of Academic Press expertly guided the book into publication gives me the excuse to thank her most warmly not only for that but also for more than eleven years' happy cooperation between Publisher and Editor of the Journal of Chemical Thermodynamics. J. W. Arrowsmith Ltd. of Bristol gave me proofs so free of mistakes and so thoughtfully and beautifully printed that they were a joy to read.

Finally I acknowledge my great debt to the late Professor Edward Guggenheim from whom in altogether thirteen years' happy association at the University of

Reading I learnt how to think about thermodynamics, and also to Professor Hugh Parton who first aroused my interest in the subject at Canterbury University College in New Zealand just thirty-five years ago.

M. L. McGLASHAN

University College London
July 1979

Values of some fundamental constants†

gas constant	R	$8.31441 \text{ J K}^{-1} \text{ mol}^{-1}$
Avogadro constant	L	$6.022045 \times 10^{23} \text{ mol}^{-1}$
Boltzmann constant	$k = R/L$	$1.380662 \times 10^{-23} \text{ J K}^{-1}$
charge of a proton	e	$1.6021892 \times 10^{-19} \text{ s A}$
Faraday constant	$F = Le$	$9.648456 \times 10^{4} \text{ s A mol}^{-1}$
Planck constant	h	$6.626176 \times 10^{-34} \text{ m}^2 \text{ kg s}^{-1}$
speed of light in a vacuum	c	$2.997924580 \times 10^{8} \text{ m s}^{-1}$
permittivity of a vacuum	ε_0	$8.85418782 \times 10^{-12} \text{ m}^{-3} \text{ kg}^{-1} \text{ s}^4 \text{ A}^2$

† *Recommended Consistent Values of the Fundamental Physical Constants, 1973. CODATA Bulletin* **1974**, no. 11. (Obtainable from ICSU CODATA Central Office, 51 Boulevard de Montmorency, 75016 Paris, France.)

Molar masses of selected elements‡

	$M/\text{g mol}^{-1}$		$M/\text{g mol}^{-1}$		$M/\text{g mol}^{-1}$		$M/\text{g mol}^{-1}$
H	1.0079	F	18.9984	S	32.06	Kr	83.80
He	4.0026	Ne	20.179	Cl	35.453	Sn	118.69
C	12.011	Na	22.9898	Ar	39.948	I	126.9045
N	14.0067	Si	28.0855	K	39.0983	Xe	131.30
O	15.9994	P	30.9738	Br	79.904	Gd	157.25

‡ *Atomic Weights of the Elements 1977. Pure and Appl. Chem.* **1979**, 51, 409.

General references

To save space, some frequently quoted books will be abbreviated as follows.

ETc (I): *Experimental Thermochemistry.* Rossini, F. D.: Editor. For the International Union of Pure and Applied Chemistry (IUPAC). Interscience: New York. **1956**.
ETc (II): *Experimental Thermochemistry, Volume II.* Skinner, H. A.: Editor. For IUPAC. Interscience: New York. **1962**.

ETd(I): *Experimental Thermodynamics, Volume I, Calorimetry of Non-Reacting Systems.* McCullough, J. P.; Scott, D. W.: Editors. For IUPAC. Butterworths: London. **1968.**

ETd(II): *Experimental Thermodynamics, Volume II, Experimental Thermodynamics of Non-Reacting Fluids.* Le Neindre, B.; Vodar, B.: Editors. For IUPAC. Butterworths: London. **1975**.

ECTd(I): *Experimental Chemical Thermodynamics, Volume 1, Combustion Calorimetry.* Sunner, S.; Månsson, M.: Editors. For IUPAC. Pergamon: Oxford. **1979**.

SPRCT(I): *Specialist Periodical Reports, Chemical Thermodynamics, Volume 1.* McGlashan, M. L.: Senior Reporter. Chemical Society: London. **1973**.

SPRCT(II): *Specialist Periodical Reports, Chemical Thermodynamics, Volume 2.* McGlashan, M. L.: Senior Reporter. Chemical Society: London. **1978**.

Table of contents

Chapter 1

Thermometers and calorimeters

Chapter 2

Composition, and change of composition, of a phase

Chapter 3

Practical thermometry

Chapter 4

Practical calorimetry

Chapter 5

Thermodynamics of a phase

Chapter 6

Change of state of a phase of fixed composition

Chapter 7

Thermodynamic inequalities and their consequences

Chapter 8

Thermodynamics of a heterogeneous system

Chapter 9

Phase equilibria for pure substances

Chapter 10

Dependence of thermodynamic functions on composition

Chapter 11

Standard thermodynamic functions

Chapter 12

Gases and gaseous mixtures

Chapter 13

The principle of corresponding states for fluids

Chapter 14

Digression on Boltzmann's distribution law

Chapter 15

Nernst's heat theorem

Chapter 16

Liquid mixtures

Chapter 17

Fluid mixtures

Chapter 21

Thermodynamics of fluid surfaces

Appendices

Chapter 1

Thermometers and calorimeters

§ 1.1   Introduction and scope

First we must say exactly what we shall mean by several special words, such as *system*, *state*, and *phase*. We shall then remind the reader about the quantity called *work* which we take over unchanged from the older sciences of mechanics and electricity and magnetism. Next we shall introduce the concepts of an *adiabatic enclosure* and of *thermal equilibrium*. These will lead us to the *zeroth law* of thermodynamics, or the recognition of a new property of a system called its *temperature*, and to the devising of instruments called *thermometers* for the measurement of temperature. We shall then introduce the special temperature called the *perfect-gas temperature* and the *gas thermometer* used for its measurement. Practical thermometry will be discussed in chapter 3. Next we shall introduce the quantity called *thermodynamic energy* and the *first law* of thermodynamics. This will lead us, towards the end of the chapter, to the devising of instruments called *calorimeters* for the measurement of changes in thermodynamic energy, or of changes in the closely related quantity called *enthalpy*. Practical calorimetry will be discussed in chapter 4.

In §§ 1.12 and 1.13 the treatment is a little abstract. It will be safe for the reader new to the subject to skim these, though not to skip them altogether.

§ 1.2   System

By a *system* we mean any part of the real world that we choose to study. Everything that is not part of the chosen system is described as belonging to its *surroundings*.

That is simple enough, but the beginner in thermodynamics often makes difficulties for himself by failing to say exactly what his chosen system is before trying to discuss it.

As an example let us consider a closed vessel containing water present partly as liquid and partly as gas. We might choose as our system just the liquid water, or just the gaseous water, or the liquid water plus the gaseous water. One of those choices might turn out to be the most convenient for the discussion of a particular problem, but any of them is a possible choice. The important thing is to be clear about which choice is made.

When the system is chosen in any of those three ways the containing vessel is regarded as part of the surroundings. There is, however, no reason why we should not choose to include the vessel in the system, and in the discussion of calorimetric experiments, for example, it is often convenient to do so.

Similarly, if the vessel is connected to the atmosphere by means of a U-tube containing mercury, or if it also contains a stirrer or electric heating element, then we may treat the mercury, or the stirrer, or the heating element, either as part of the system or as part of the surroundings whichever is the more convenient for our purpose.

A *macroscopic* system is one of tangible size. Thus a drop of water is a macroscopic system but a molecule of water is not.

§ 1.3. State of a macroscopic system

For a complete description of a macroscopic system we must specify not only the substances of which it is composed and where the bounding surfaces are that divide the system from its surroundings, but also its *state*.

The state of a system is completely defined by the values of its properties, that is to say by the values of the measurable physical quantities belonging to the system. Must we then define all the properties of a system in order completely to define its state? Happily that is not necessary. Only a certain number of the properties can be varied independently. It is therefore sufficient to measure the values of that number of independent properties. That number depends on the nature of the system, but for the systems with which we shall have to deal it will rarely exceed five and will often be less. For any system the sufficient number of independent properties can be chosen in many different ways, but for a given system the *number* of them is always the same. Sometimes it is convenient to make one choice and sometimes another; there are no right or wrong choices but only convenient or less convenient ones. Once chosen, the independent properties are called the *independent variables* for the system and all other properties are then *dependent variables*. It is most important always to be clear about which properties are being treated for the time being as the independent variables.

The reader might now expect to be told what the number of independent variables for any particular kind of system actually is. He will have to be content, however, with the answer: 'As many as are found by experiment to be necessary'. Later we shall find a simple general rule, called the *phase rule*, for calculating that number for systems that are in *equilibrium*.

§ 1.4   Extensive and intensive properties

There are two kinds of properties: *extensive* and *intensive*.

Imagine a system divided into two or more parts. Then an *extensive property* is one for which the value of the property of the system as a whole is the sum of the values of the property for each of the parts. It should be obvious that *mass, volume,* and *amount of substance* (for which see § 2.2) are all extensive properties. So are the *energy, enthalpy, Gibbs function, Helmholtz function,* and *entropy,* with which we shall soon be concerned.

Again imagine a system divided into two or more parts. Then an *intensive property* is one that can have, though it need not have, the same values for two or more parts as for the system as a whole. It should be obvious that *pressure* is such an intensive property. *Temperature* is another. It should also be obvious that the quotient of any two extensive properties is an intensive property. Thus *density* (mass divided by volume), *mole fraction* (amount of one substance divided by the sum of the amounts of all the substances), *specific volume* (volume divided by mass) and other *specific* quantities, and *molar volume* (volume divided by amount of substance) and other *molar* quantities, are all intensive properties. So are the *chemical potential* and the *absolute activity*, with which we shall soon be concerned.

§ 1.5 Thermodynamic properties

Thermodynamic properties or *thermodynamic quantities* are those like temperature, pressure, volume, amount of substance, energy, entropy, and chemical potential, that do not depend on the rate at which something happens. Thus electric current, which is the rate at which electric charge flows through a conductor, is not a thermodynamic quantity. Nor is thermal conductivity, which is the rate at which energy flows through a substance under the influence of a temperature gradient.

§ 1.6 Phase

If the intensive properties of a system are uniform, that is to say if the system has throughout itself the same density and the same temperature and the same pressure and the same electric permittivity and so on, then we describe the system as *homogeneous* and call it a *phase*.

Any system consists of one or more phases. If it consists of more than one phase it is described as *heterogeneous*.

In a tall enough system that is otherwise homogeneous some intensive properties such as the density will vary sensibly from the top to the bottom because of the earth's gravitational field. Whenever such variation is important the system must be regarded as a vertical stack of separate phases. In nearly all systems of interest to chemists such complications can safely be ignored.

Another complication arises when we consider the interface between two phases, where over a finite distance normal to the interface the values of some of the intensive properties, such as the density, change continuously from their values in one phase to their values in the other phase. We shall deal with such surface phases in chapter 21.

§ 1.7 Open and closed phases

An *open phase* is one of variable material content, that is to say one that can exchange material either with other phases present in the system under consideration or with the surroundings of the system. A *closed phase* is one that cannot exchange material with other phases or with the surroundings.

§ 1.8   Open and closed systems

An *open system* is one that can exchange material with its surroundings. A *closed system* is one that cannot exchange material with its surroundings.

We shall be concerned in this book only with closed systems, but these will often consist of two or more open phases.

As an example consider a closed vessel containing a uniform gas. If we choose as our system the gas then our system is a closed system consisting of one closed phase. As a second example consider a closed vessel containing water present partly as liquid and partly as gas. If we choose as our system all the water whether liquid or gas then our system is a closed system consisting of two open phases. If instead we were to make the (usually inconvenient) choice of just the liquid water as our system then it would be an open system consisting of one open phase.

§ 1.9   Work

The physical quantity called the *work* done on a system is taken over unchanged from electromechanics, by which we mean mechanics plus electricity and magnetism. There are several ways of causing work to flow from the surroundings of a system into the system itself. Three of these ways are illustrated in figure 1.1. Let the system be a fluid, whether of one phase or more, enclosed in a box having walls that are supposed rigid except for the movable piston A. A stirrer B and an electric heating element D

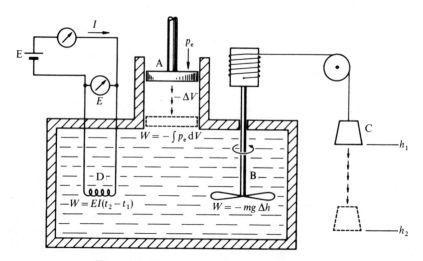

Figure 1.1. Three ways of doing work on a system.

are immersed in the fluid. Work can then be done on the system (i) by arranging for some external force to push in the piston, (ii) by arranging a system of pulleys so that a falling body C rotates the stirrer B, or (iii) by arranging for an external source E of electric potential difference to drive an electric current through the heating element D. We shall now discuss each of these in turn.

Let $\mathcal{A}$ be the area of the piston and p_e the external pressure (force divided by area) acting on the piston when it moves inwards by an infinitesimal distance ds. Then in the absence of friction the work W done on the system when the piston moves from a distance s_1 (measured from some reference mark) to a distance s_2 is given by

$$W = \int_{s_1}^{s_2} p_e \mathcal{A} \, ds, \tag{1.9.1}$$

which, in view of the relation $\mathcal{A} \, ds = -dV$ where V is the volume *of the system*, can be rewritten in the form:

$$W = -\int_{V_1}^{V_2} p_e \, dV. \tag{1.9.2}$$

If $V_2 < V_1$, that is to say if the piston is pushed inwards, then the work W done *on* the system is positive. If $V_2 > V_1$, which would be the case if the pressure of the fluid were greater than the applied external pressure, then W would be negative; we might then say that positive work $-W$ was done on the surroundings *by* the system. We shall always use the symbol W to denote work done on the system.

If, and only if, the pressure is constant throughout the compression then equation (2) can be simplified to give

$$W = -p_e \int_{V_1}^{V_2} dV = -p_e(V_2 - V_1) = -p_e \, \Delta V.\dagger \tag{1.9.3}$$

We now turn to the stirrer B. Let m be the mass of the body C attached by means of pulleys (supposed frictionless) to the stirrer B. Then if the body falls from a height h_1 to a height h_2 above some reference mark, the work W done on the system is the positive quantity:

$$W = -\int_{h_1}^{h_2} mg \, dh, \tag{1.9.4}$$

where g is the local acceleration of free fall. Since m is, and g usually may be

---

† The operator Δ is used here, and will be used throughout this book, to denote an increment: $\Delta X \stackrel{\text{def}}{=} (X_2 - X_1)$ is the excess of the value of any physical quantity X in a state labelled '2' over the value of X in a state labelled '1'. When we wish to specify a ΔX more closely we may attach subscripts and superscripts to the operator. For example we may write $\Delta_1^2 X$ for $(X_2 - X_1)$. Similarly we may write $\Delta_1^g X$ for $(X^g - X^l)$ where the state symbols g and l refer respectively to gas and liquid, and $\Delta_{\text{mix}} X$ for $(X^{\alpha + \beta} - X^\alpha - X^\beta)$ where the subscript $_{\text{mix}}$ indicates mixing of the separate phases α and β (such as two miscible liquids) to give the mixed phase $(\alpha + \beta)$.

assumed to be, independent of h, equation (4) can be simplified to

$$W = -mg(h_2 - h_1) = -mg \, \Delta h.$$ (1.9.5)

Finally we turn to the electric heating element D. Let Q be the electric charge which flows through the electric heater (electric 'worker' would be a better name) when the electric potential difference is E. Then in the absence of friction (resistances of the leads) the work W done on the system is given by

$$W = \int E \, dQ.$$ (1.9.6)

If, and only if, the external circuit is arranged so that the electric potential difference E and the electric current I are both constant from the time t_1 at which the electric circuit is closed until the time t_2 at which it is opened again, then equation (6) can be simplified to give

$$W = EI \int_{t_1}^{t_2} dt = EI(t_2 - t_1) = EI \, \Delta t.$$ (1.9.7)

That completes our discussion of the three ways illustrated in figure 1.1 of doing work on a system. The only other formula we shall need in this book is that for the work done when an interfacial layer separating two fluid phases is increased in area. A simple example would be the work needed to stretch a soap film. We postpone our discussion of that kind of work until we deal with surfaces in chapter 21.

There are still other ways of doing work on a system, with which we shall not deal because they lie outside the scope of this book. They include the work done on a solid (for example a wire or a piece of rubber) when it is stretched and the work done on a body by varying an electromagnetic field. In every case, however, the work can be written as a sum of terms having the form $x \, dX$ where x is an intensive 'force' quantity and X is an extensive 'displacement' quantity.

Before leaving the subject of work we recall that none of the results in this section depends on any thermodynamic argument. In each case the result has been taken over directly from the earlier sciences of hydrostatics, mechanics, and electricity and magnetism. We shall henceforth take these results for granted. We shall also take it for granted that the reader understands how to measure the various quantities involved: pressure, volume, mass, length, electric potential difference, electric current, and time.

§ 1.10 Adiabatic enclosure. Thermally insulated system

It is known from experience that it *can* be the case, though it is not usually the case, that the state of a closed system remains unchanged unless work is done on it.†
Whenever that is the case the system is said to be *adiabatically enclosed* or *thermally insulated.*

—————————

† This statement might well be called the '−1th law of thermodynamics'.

A stoppered Dewar flask is a good approximation to an adiabatic enclosure. If a system has been enclosed in a Dewar flask for long enough to ensure that anything, such as a chemical reaction, that was happening has stopped happening, then the state of the system will remain unchanged unless work is done on it in one of the ways discussed in § 1.9.

A system that is not enclosed by adiabatic walls is said to have *diathermic* or *thermally conducting* walls.

§ 1.11　Isolated system

If a system is adiabatically enclosed and if we also exclude the possibility of doing work on it, then the system is said to be isolated. If a system has been isolated for long enough to ensure that anything that was happening has stopped happening, then the state of the system will remain unchanged.

§ 1.12　Thermal equilibrium

Consider an adiabatically enclosed system consisting of two sub-systems A and B separated by an adiabatic wall as in figure 1.2(a). Let $\mathscr{S}'_A$ denote a set of independent variables defining the state of the sub-system A and $\mathscr{S}'_B$ a set defining the state of B. Now let us replace the adiabatic wall that separates A and B by a diathermic wall as in figure 1.2(b). Then in general the states of both systems will change. When the change is complete let the new states be $\mathscr{S}_A$ and $\mathscr{S}_B$.

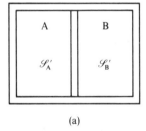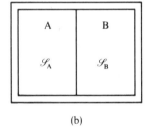

(a)　　　　　　　　　　　　　　　　　(b)

Figure 1.2. Thermal equilibrium. Double lines denote adiabatic walls and a single line a diathermic wall.

The sub-system A is now said to be in *thermal equilibrium* with the sub-system B. Alternatively the composite system (A + B) consisting of the two sub-systems A and B is said to be in *internal thermal equilibrium*.

Whereas the states $\mathscr{S}'_A$ and $\mathscr{S}'_B$ were independent (either could be changed without affecting the other), the states $\mathscr{S}_A$ and $\mathscr{S}_B$ of two systems that are in thermal equilibrium are not independent. We may therefore write

$$f(\mathscr{S}_A, \mathscr{S}_B) = 0, \quad \text{(thermal equilibrium)}, \tag{1.12.1}$$

which we read as 'Whenever two systems A and B are in thermal equilibrium there exists a relation between the set of independent variables $\mathscr{S}_A$ that defines the state of A and the set $\mathscr{S}_B$ that defines the state of B'. Equation (1) might seem rather abstract but we shall make good use of it in the next section.

§ 1.13 The zeroth law. Temperature

We are now ready to introduce the notion of *temperature*.

It is a result of experience that if a system A is in thermal equilibrium with another system B, and if the system B is in thermal equilibrium with a third system C, then the system C is also in thermal equilibrium with the system A. This statement is called the *zeroth law of thermodynamics* and we shall shortly see how it implies the existence of *temperature* as one of the intensive properties of a system. Before we do that, however, let us make sure that we are quite clear about the meaning of the zeroth law and about how it could be verified experimentally. Consider an adiabatically enclosed system consisting of the three sub-systems A, B, and C, the pair A and B and the pair B and C being separated by diathermic walls but the pair C and A being separated by an adiabatic wall, as in figure 1.3(a). Let $\mathscr{S}_A$ denote a set of independent variables defining the state of the system A, $\mathscr{S}_B$ a set defining the state of B, and $\mathscr{S}_C$ a set defining the state of C. Now let us replace the adiabatic wall that separates C and A by a diathermic wall as in figure 1.3(b). Then the zeroth law asserts that C and A were *already* in thermal equilibrium in spite of the adiabatic wall so that replacing the adiabatic wall by a diathermic wall causes no change in the state of any of the systems.

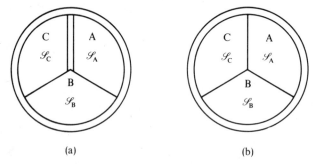

(a)   (b)

Figure 1.3. Zeroth law of thermodynamics. Double lines denote adiabatic walls and single lines diathermic walls.

Thus thermal equilibrium of A with B and of B with C *implies* thermal equilibrium of C with A. We saw in the previous section that thermal equilibrium of A with B requires the existence of a relation having the form:

$$f(\mathscr{S}_A, \mathscr{S}_B) = 0 .\qquad(1.13.1)$$

Similarly thermal equilibrium of B with C requires the existence of a relation having the form:

$$f(\mathscr{S}_B, \mathscr{S}_C) = 0 .\qquad(1.13.2)$$

According to the zeroth law the existence of the two relations (1) and (2) implies the existence of a third relation:

$$f(\mathscr{S}_C, \mathscr{S}_A) = 0 \,. \tag{1.13.3}$$

The simultaneous existence of relations (1), (2), and (3) is possible only if the functions $f(\mathscr{S}_A, \mathscr{S}_B)$, $f(\mathscr{S}_B, \mathscr{S}_C)$, and $f(\mathscr{S}_C, \mathscr{S}_A)$ have the forms:

$$\left.\begin{aligned}
f(\mathscr{S}_A, \mathscr{S}_B) &= \theta(\mathscr{S}_A) - \theta(\mathscr{S}_B) \,, \\
f(\mathscr{S}_B, \mathscr{S}_C) &= \theta(\mathscr{S}_B) - \theta(\mathscr{S}_C) \,, \\
f(\mathscr{S}_C, \mathscr{S}_A) &= \theta(\mathscr{S}_C) - \theta(\mathscr{S}_A) \,.
\end{aligned}\right\} \tag{1.13.4}$$

Since moreover each of the functions in equations (1) to (3) is equal to zero it follows from equations (4) that

$$\theta(\mathscr{S}_A) = \theta(\mathscr{S}_B) = \theta(\mathscr{S}_C) \,. \tag{1.13.5}$$

Expressed in words equation (5) asserts that there exists an intensive property which is a function of the state of any system and which has the following important characteristic: when two or more systems are in thermal equilibrium they have identical values of that property. We call that property temperature and denote it by θ.

Let us now summarize what we have learnt about temperature. One of the intensive properties of a system that is in internal thermal equilibrium is its temperature. Two or more systems that are in thermal equilibrium with one another have identical temperatures.

Throughout this section and the previous section we have tacitly assumed that the systems A, B, and C were each in internal thermal equilibrium, that is to say (i) that no adiabatic partitions were present inside any of the systems, and (ii) that if an adiabatic partition had recently been removed from inside any of the systems, or if any of the systems had recently been tampered with in any other way, that sufficient time had been allowed for the system to achieve internal thermal equilibrium.

Throughout the rest of this book we shall deal only with systems containing no internal adiabatic partitions. Thus any of our systems will either (i) be in internal thermal equilibrium and so have a uniform temperature, or (ii) be changing towards thermal equilibrium as the result of a non-uniform temperature. An example of (ii) would be a metal bar, one end of which had recently been near a flame. The state of such a system cannot be defined. Such a system cannot be described in terms of thermodynamic equations. Indeed we shall find that the only thermodynamic conclusion we can arrive at for such a system is that the hot parts will become cooler and the cold parts will become warmer until the whole system is in internal thermal equilibrium, that is to say until the whole system has a uniform temperature. Except for one place in chapter 7 where we shall deduce that conclusion from one of our axioms, we shall use the word *system* only for a system which is in internal thermal equilibrium and which therefore has a uniform temperature.

§ 1.14 Example

The previous two sections having been rather abstract, it might help now to give an illustrative example. Let us suppose that each of our systems A, B, and C consists of a pure gas, not necessarily the same gas in each, and that the walls separating the systems are impermeable to the gases. The complete description of a pure gas is found by experiment to need a specification of three independent properties (in the absence of external fields) in addition to a specification of the nature of the gas, that is to say whether it is oxygen or nitrogen or water vapour or whatever. One at least of these must be an extensive property because if we chose three intensive properties such as the pressure, the density, and the refractive index, we should be left with no knowledge of the size of the system. Let us choose the amount of substance n,† the volume V, and the pressure p.

We begin by considering the two systems A and B of § 1.12. When the two systems are in thermal equilibrium equation (1.12.1) asserts the existence of a relation of the form:

$$f(n_A, V_A, p_A, n_B, V_B, p_B) = 0 . \tag{1.14.1}$$

We can solve equation (1) for any one of the variables, say p_B, and obtain

$$p_B = F(n_A, V_A, p_A, n_B, V_B) . \tag{1.14.2}$$

The pressure p_B is now a dependent variable. When the two systems A and B are in thermal equilibrium the composite system (A + B) has only five and not six independent variables. Expressed in words, equation (2) asserts that when a specified amount n_A of a gas A occupying a specified volume V_A at a specified pressure p_A is in thermal equilibrium with a specified amount n_B of a gas B occupying a specified volume V_B, then the pressure p_B of the gas B cannot be arbitrarily specified but is fixed by the conditions already specified. The only thing we can do about the pressure p_B is to measure it.

Equation (1) might for example have the explicit form:

$$p_A V_A / n_A - p_B V_B / n_B = 0 , \tag{1.14.3}$$

and as a matter of fact does have this form if the gases are *perfect*.

We turn now to the three systems A, B, and C discussed at the beginning of the present section. According to the zeroth law, if A and B are in thermal equilibrium so that

$$f(n_A, V_A, p_A, n_B, V_B, p_B) = 0 , \tag{1.14.4}$$

and if B and C are in thermal equilibrium so that

$$f(n_B, V_B, p_B, n_C, V_C, p_C) = 0 , \tag{1.14.5}$$

then it follows that C and A are in thermal equilibrium so that

$$f(n_C, V_C, p_C, n_A, V_A, p_A) = 0 . \tag{1.14.6}$$

† For amount of substance see § 2.2.

The simultaneous existence of these three relations is possible only if the functions have the forms:

$$\left.\begin{aligned}
f(n_A, V_A, p_A, n_B, V_B, p_B) &= \theta(n_A, V_A, p_A) - \theta(n_B, V_B, p_B), \\
f(n_B, V_B, p_B, n_C, V_C, p_C) &= \theta(n_B, V_B, p_B) - \theta(n_C, V_C, p_C), \\
f(n_C, V_C, p_C, n_A, V_A, p_A) &= \theta(n_C, V_C, p_C) - \theta(n_A, V_A, p_A).
\end{aligned}\right\} \quad (1.14.7)$$

It follows from (7) with (4), (5), and (6) that

$$\theta(n_A, V_A, p_A) = \theta(n_B, V_B, p_B) = \theta(n_C, V_C, p_C), \qquad (1.14.8)$$

or, in words, that there exists an intensive quantity which is a function of the state of the gas and which has the same value in two or more gaseous systems that are in mutual thermal equilibrium.

Equation (8) might for example have the explicit form:

$$p_A V_A / n_A = p_B V_B / n_B = p_C V_C / n_C, \qquad (1.14.9)$$

and as a matter of fact does have that form if the gases are *perfect*. The quantity pV/n is thus a measure of the temperature of a perfect gas.

§ 1.15 Measurement of temperature. Thermometry

We shall now devise a *thermometer*. We choose from many possible examples a length of thin wire buried in the system whose temperature θ_{system} we wish to measure, the ends of the wire being connected by leads of low electric resistance to an external electric-resistance meter. The electric resistivity ρ of a wire made from a specified metal depends on the temperature θ and on the pressure p. We suppose that the pressure on the wire is kept constant, or alternatively for small variations in pressure that we may ignore the dependence of ρ on p.† For a wire of given length and cross section we may then write

$$\theta_{wire} = f(R_{wire}), \qquad (1.15.1)$$

where θ_{wire} is the temperature of the wire and R_{wire} is its electric resistance, and where we may choose any form we like for the function $f(R_{wire})$. (For example we might make the simple choice $\theta_{wire} = rR_{wire}$ where r is a constant to which we may assign any dimension and any numerical value we choose.) Provided that the wire is in thermal equilibrium with the system we then have $\theta_{system} = \theta_{wire}$ so that

$$\theta_{system} = f(R_{wire}), \qquad (1.15.2)$$

and so we can measure the temperature of the system.

Three sources of error, apart from the possible effect of variations of pressure, must be foreseen in the measurement summarized in equation (2).

---

† As a matter of fact the dependence of ρ on p is sometimes far from negligible in precise resistance thermometry. We accordingly choose a metal for which $|(\partial\rho/\partial\theta)_p|$ is as large as possible while $|(\partial\rho/\partial p)_\theta|$ is as small as possible. (The inverse choice is appropriate for resistance manometry.)

If the temperature of the system is changing at an appreciable rate then the thermometer must follow that temperature without appreciable time lag. For that reason the thermometer must be in the closest possible thermal contact with the system, must have the highest possible thermal conductance, and must have the lowest possible heat capacity (see § 1.22).†

If the unavoidable leads from the thermometer to the external meters are good thermal conductors, the temperature of the thermometer will differ from that of the system in the same sense as the temperature of the surroundings differs from that of the system. If the leads are not good electrical conductors their electric resistance will make an uncomfortably large contribution to the measured value of R_{wire}. The best compromise must be found by careful analysis of the heat flows. Good thermal contact between the system and thermometer always helps. Good thermal contact between the leads and the system, before the leads are allowed to emerge into the surroundings, also helps. A high resistance R_{wire} raises the acceptable value of the electric resistance of the leads and so allows the choice of leads having lower thermal conductance.

The electric current used to measure R_{wire} inevitably raises the temperature of the thermometer and so must be kept as small as possible. In the most precise measurements R_{wire} is measured with two or more different electric currents and the limiting value at zero current is found by extrapolation. That is not usually possible, however, in measurements of temperatures which change with time, as they do for example in calorimetry.

Resistance thermometers have considerable practical importance and we shall have more to say about them in chapter 3. We remind the reader, however, that the definition of temperature given by equation (2) depends on the particular choice of metal, on the precise form chosen for the function $f(R)$, and probably also on the pressure. Other kinds of thermometer will also be discussed in chapter 3. Each kind has its advantages and disadvantages, whether for the precise measurement of *the* temperature of a system, for the precise measurement of small differences of temperature, for the detection of fluctuations of temperature, or for rough measurements of temperature.

If the roles of the small wire and the large bath are interchanged so that the system of interest is the wire rather than the bath, the function of the bath is then to fix the temperature of the wire. We then refer to the bath as a *thermostat*.

§ 1.16 Gas thermometry

Among the almost infinite variety of possible thermometers there is one, called a gas thermometer, which is of special importance. It leads to the definition of a quantity Θ which, as we shall show in chapter 6, is proportional to the thermodynamic

---

† The heat capacity of a thermometer must also be kept low if it is to be used to measure the pre-existing temperature of say a bath of fluid by dipping it into the bath. Unless its heat capacity is negligibly small compared with that of the bath (or unless it happens to have exactly the same temperature as the bath) the thermometer will change the temperature of the bath.

temperature T. The thermodynamic temperature T is of fundamental importance in the formulation of thermodynamics, its definition being such as to make it possible to write the equations of thermodynamics in especially simple forms.

It is found by experiment that the product pV of the pressure p and volume V of an amount of substance n of any real gas can be expressed as a convergent series in ascending powers of the pressure:

$$pV = n\Theta + nBp + nC'p^2 + \cdots ,\qquad (1.16.1)$$

where for any particular gas the coefficients Θ, B, C', $\cdots$, depend only on the empirical temperature θ.† If V is measured as a function of p for a given amount of substance n of a gas maintained in a thermostat at constant empirical temperature θ, and if the measurements extend to low enough pressures to ensure that a plot of pV/n against p has a linear portion at low values of p, then we may extrapolate the plot and so find the limiting value:

$$\lim_{p\to 0} (pV/n) = \Theta , \quad (\theta \text{ constant}),\qquad (1.16.2)$$

of pV/n as p tends to zero at the empirical temperature θ. The results of such experiments on a wide variety of gases show unambiguously that Θ (unlike B, C', $\cdots$) is independent of the nature of the gas. Thus Θ is not only, like B, C', $\cdots$, and like the electric resistance of a wire of given dimensions, a measure of temperature, but is also, unlike B, C', $\cdots$, and unlike the electric resistance of a wire of given dimensions, a measure of the temperature that is *independent of the nature of the working substance*.

Anticipating the conclusion that we shall reach in chapter 6 that Θ is proportional to the thermodynamic temperature T, we write

$$\Theta = RT ,\qquad (1.16.3)$$

where R is a universal constant called the *gas constant*.

Thermodynamic temperature is one of the arbitrarily chosen but internationally agreed base quantities having its own dimension. The SI base unit of thermodynamic temperature is called the kelvin, is denoted by the symbol K, and is defined by the statement:

$$T(H_2O, s+l+g) \stackrel{\text{def}}{=} 273.16 \text{ K} ,\qquad (1.16.4)$$

where $T(H_2O, s+l+g)$ denotes the thermodynamic temperature at the triple point of water, that is to say the unique temperature at which (ice + liquid water + gaseous water) coexist at equilibrium in the absence of air or other impurities. The triple-point temperature of water was chosen because of the relative ease with which it can be reproduced with the highest precision. The value 273.16 K was chosen so as to conform as closely as possible with older definitions of temperature in which there were 100 'degrees' between the ice-temperature, that is to say the temperature of

---

† We have used C' rather than C because we wish to reserve C for a related but different quantity called the third virial coefficient.

equilibrium of (ice + liquid water + air) at standard atmospheric pressure, and the normal boiling temperature of water.†‡

The ratio T_2/T_1, or $T(\theta_2)/T(\theta_1)$, can thus be determined by gas thermometry as the ratio $\Theta(\theta_2)/\Theta(\theta_1)$ of the corresponding values of $\Theta = \lim_{p\to 0}(pV/n)$.

Gas thermometry is not the only way of measuring the ratio T_2/T_1 of two thermodynamic temperatures, but at least from our point of view it is by far the most important way. Gas thermometers have been used at thermodynamic temperatures from about 1.3 K to greater than 1350 K.

Acoustic thermometry, similar to gas thermometry except that the square of the speed of sound in a monatomic gas replaces the product pV, has been used with some experimental advantages to measure the ratio of two thermodynamic temperatures, especially when each is less than about 20 K.

If T_1 is chosen to be $T(H_2O, s+l+g)$ then any thermodynamic temperature T_2 can be expressed as so many kelvins by use of the relation:

$$T_2 = \{T_2/T(H_2O, s+l+g)\} \times 273.16 \text{ K}, \tag{1.16.5}$$

where the ratio $T_2/T(H_2O, s+l+g)$ can be measured, usually by gas thermometry.

The gas constant R has the dimension: $(\text{length})^2(\text{mass})(\text{time})^{-2}(\text{thermodynamic temperature})^{-1}(\text{amount of substance})^{-1}$, and SI unit: $\text{m}^2\,\text{kg}\,\text{s}^{-2}\,\text{K}^{-1}\,\text{mol}^{-1}$ or $\text{J}\,\text{K}^{-1}\,\text{mol}^{-1}$. The value at present recommended§ for R was obtained after consideration of values of $\Theta(T)/T$ for many gases at many temperatures and is

$$R = (8.31441 \pm 0.00026)\,\text{J}\,\text{K}^{-1}\,\text{mol}^{-1}. \tag{1.16.6}$$

Gas thermometry is straightforward in theory but difficult in practice, and is usually undertaken only in national metrological laboratories such as the National Physical Laboratory, Teddington, U.K.; the National Bureau of Standards, Washington, D.C., U.S.A.; and the Physikalisch-Technische Bundesanstalt, Braunschweig, Germany. The reader should turn elsewhere‖ for fuller accounts of gas thermometry.

Obviously any empirical thermometer that gives reproducible enough results can be calibrated by comparison with a gas thermometer and can then be used to measure

---

† The adjective 'normal' here implies 'at standard atmospheric pressure', that is to say at a pressure of 101.325 kPa.

‡ The Celsius temperature θ_C is defined by the relation:

$$\theta_C \overset{\text{def}}{=} T - 273.15 \text{ K}.$$

For Celsius temperatures the kelvin K is often replaced by the degree Celsius °C. Its function is to allow us to say 'the temperature is 23.45 °C' instead of 'the Celsius temperature is 23.45 K' or 'the thermodynamic temperature is 296.60 K'. °C ≡ K. Neither Celsius temperatures nor the degree Celsius will be used in this book.

§ See p. ix.

‖ Stimson, H. F.; Lovejoy, D. R.; Clement, J. R. ETd(I), chapter 2, pp. 19–23.

Temperature: Its Measurement and Control in Science and Industry. American Institute of Physics, Reinhold: New York. 1941.

thermodynamic temperatures. That is what is usually done in practice. We shall discuss practical thermometry in chapter 3.

§ 1.17 Thermodynamic energy

It is a result of experience that the work W needed to bring about a given change of state in an adiabatically enclosed system is independent of the way the work is performed. That is a statement of the law of conservation of energy taken over from electromechanics and applied to adiabatically enclosed thermodynamic systems.

Let W_{21}^{ad} be the work needed to change the state of an adiabatically enclosed system from an initial state defined by the set of independent variables $\mathscr{S}_1$ to a final state defined by the set of independent variables $\mathscr{S}_2$. The relation:

$$W_{21}^{ad} = f(\mathscr{S}_2, \mathscr{S}_1), \tag{1.17.1}$$

then follows from the statement with which we began this section; the work done on an adiabatically enclosed system is a function only of the initial and final states of the system.

Suppose now that we cause the change $\mathscr{S}_1 \to \mathscr{S}_2$ to take place in two stages: $\mathscr{S}_1 \to \mathscr{S}_3$ and then $\mathscr{S}_3 \to \mathscr{S}_2$, and that we measure separately the work W_{31}^{ad} for the first stage and W_{23}^{ad} for the second stage. We then have the two relations:

$$W_{31}^{ad} = f(\mathscr{S}_3, \mathscr{S}_1), \tag{1.17.2}$$

and

$$W_{23}^{ad} = f(\mathscr{S}_2, \mathscr{S}_3), \tag{1.17.3}$$

similar to equation (1). But the work needed to change the state from $\mathscr{S}_1$ to $\mathscr{S}_2$ depends only on $\mathscr{S}_1$ and $\mathscr{S}_2$ and not at all on any intermediate stage $\mathscr{S}_3$. Thus

$$W_{21}^{ad} = W_{31}^{ad} + W_{23}^{ad}, \tag{1.17.4}$$

and by substitution in (4) from (1), (2), and (3):

$$f(\mathscr{S}_2, \mathscr{S}_1) = f(\mathscr{S}_3, \mathscr{S}_1) + f(\mathscr{S}_2, \mathscr{S}_3). \tag{1.17.5}$$

Equation (5) can be satisfied only if

$$\left.\begin{aligned}
f(\mathscr{S}_2, \mathscr{S}_1) &= U(\mathscr{S}_2) - U(\mathscr{S}_1), \\
f(\mathscr{S}_3, \mathscr{S}_1) &= U(\mathscr{S}_3) - U(\mathscr{S}_1), \\
f(\mathscr{S}_2, \mathscr{S}_3) &= U(\mathscr{S}_2) - U(\mathscr{S}_3).
\end{aligned}\right\} \tag{1.17.6}$$

It is easy to see that this is so when we note that in equation (5) any term containing $\mathscr{S}_3$ in $f(\mathscr{S}_3, \mathscr{S}_1)$ must exactly cancel each such term in $f(\mathscr{S}_2, \mathscr{S}_3)$. In particular by using the first of equations (6) in equation (1) we have

$$W_{21}^{ad} = U(\mathscr{S}_2) - U(\mathscr{S}_1). \tag{1.17.7}$$

We call the function U the *energy* or, when we wish to distinguish it from the older electromechanical energy, the *thermodynamic energy*.[†] Equation (7) then becomes

$$W_{21}^{\text{ad}} = U_2 - U_1 = \Delta U, \quad \text{(adiabatic)}. \tag{1.17.8}$$

We have put 'adiabatic' in brackets after equation (8) to remind the reader that not only this equation but all the equations of this section apply only to the work done on an adiabatically enclosed system.

§ 1.18   The first law of thermodynamics

Consider now a system that is *not* enclosed by an adiabatic wall but by a diathermic wall. Then it is a result of experience that the work W_{21} done on the system when the state changes from $\mathscr{S}_1$ to $\mathscr{S}_2$ is not equal to the work W_{21}^{ad} which causes the same change when the system is adiabatically enclosed. In general therefore W_{21} does *not* depend only on the initial and final states and is *not* equal to ΔU. We therefore extend the law of conservation of energy to thermodynamic systems by writing formally

$$\Delta U = W + Q, \tag{1.18.1}$$

where we have dropped the subscript $_{21}$. The quantity Q is called the *heat*. Equation (1) is a statement of the *first law of thermodynamics*.

Heat, like work, is not a property of a system. We talk about work being done on the system by its surroundings and about heat flowing into the system from its surroundings. We might just as well talk about work flowing into the system from its surroundings and about heat being done on the system by its surroundings, however quaint these statements might sound at first. Work is a flow of energy, caused by a difference of pressure or other electromechanical 'force', into the system from its surroundings; heat is a flow of energy, caused by a difference of temperature, into the system from its surroundings.

The energy difference ΔU depends only on the initial and final states of the system. The work W and the heat Q both, however, depend on the way in which the change of state was brought about, that is to say on the *path*. A given change of state involving say an increase of energy ($\Delta U > 0$) can be brought about by a path involving a positive W comparable with ΔU and a small Q of appropriate sign, or by a path involving approximately equal positive values of W and Q, or by a path involving a small W of appropriate sign and a positive Q comparable with ΔU, and so on.

In the special case that the system is adiabatically enclosed equation (1) becomes

$$\Delta U = W, \quad \text{(adiabatic)}, \tag{1.18.2}$$

so that

$$Q = 0, \quad \text{(adiabatic)}, \tag{1.18.3}$$

† Sometimes called *internal energy*: any kinetic energy of the system is not included.

and W in this special case does depend only on the initial and final states in agreement with the conclusions reached in the previous section.

In the special case that the system is diathermically enclosed but isolated from all external agencies that might do work on it, a state of affairs that we shall describe as 'mechanically enclosed', equation (1) becomes

$$\Delta U = Q, \quad \text{(mechanically enclosed)}, \qquad (1.18.4)$$

since

$$W = 0, \quad \text{(mechanically enclosed)}, \qquad (1.18.5)$$

and Q in this special case does depend only on the initial and final states.

In the special case that the system is isolated, that is to say both adiabatically and mechanically enclosed, equation (1) becomes

$$\Delta U = 0, \quad \text{(isolated)}, \qquad (1.18.6)$$

since both

$$W = 0, \quad \text{(isolated)}, \qquad (1.18.7)$$

and

$$Q = 0, \quad \text{(isolated)}. \qquad (1.18.8)$$

We write the equation for an infinitesimal change dU, corresponding to equation (1) for a finite change ΔU, in the form:

$$dU = W + Q. \qquad (1.18.9)$$

We do not write dW or dQ, first because to do so would imply $\int_1^2 dW = W_2 - W_1$ or $\int_1^2 dQ = Q_2 - Q_1$ and no such properties exist as W_1 or W_2 or Q_1 or Q_2 of the state 1 or 2 of a system, and second because W and Q in equation (9) need not be infinitesimal or even small but can be of any magnitude provided that they have appropriate signs and that their *sum* is infinitesimal.

§ 1.19   Measurement of energy difference. Calorimetry. Enthalpy

It is only the *difference* $\Delta U = U_2 - U_1$ between the energies of two states 1 and 2 of a system that can be measured; the energy U of a system is not a measurable quantity and so has no physical significance. We shall now devise an instrument, called a *calorimeter*,[†] for the measurement of an energy difference ΔU.

A calorimeter is merely a vessel equipped for measurement of the work W_{21}^{ad} which would have to be done if the vessel were adiabatically enclosed to bring about the change of state $1 \rightarrow 2$ for which we wish to measure $\Delta U = U_2 - U_1$. Then, since $\Delta U = W_{21}^{ad}$, a measurement of W_{21}^{ad} is a measurement of ΔU.

---

† An instrument for the measurement of a difference in *enthalpy*, a quantity which is closely related to energy and which will be introduced later in this section, is also called a calorimeter.

In modern calorimetry the 'deliberate' work is invariably done electrically (see § 1.9) by passing a measured constant electric current for a measured time through an electric resistor of measured constant resistance buried in the calorimeter; we denote this part of the work by W^{el}. The total work W done on the calorimeter, however, includes (see § 1.9) the 'non-deliberate' work $-\int p_e \, dV$ resulting from any exchange of volume against a non-zero external pressure between the calorimeter and its surroundings, and any other 'additional' work W' done for example in stirring or in the initiation of a chemical reaction in the calorimeter. For calorimetry we may then write equation (1.18.1) in the form:

$$\Delta U = W^{el} - \int p_e \, dV + W' + Q,\qquad(1.19.1)$$

where Q is the unwanted but unavoidable heat leak.

The work term $-\int p_e \, dV$ in equation (1) could of course be measured. In practice, however, it is a nuisance of which we dispose in one of three ways.

If the external pressure p_e is maintained effectively equal to zero, as can often conveniently be done by suspending the calorimeter in an evacuated space, then the integral $-\int p_e \, dV$ vanishes. If the volume of the calorimeter is held constant, by constructing it with rigid walls, then the integral $-\int p_e \, dV$ again vanishes. In either of these special cases equation (1) becomes

$$\Delta U = W^{el} + W' + Q.\qquad(1.19.2)$$

If the calorimeter were maintained in hydrostatic equilibrium with the surrounding atmosphere (as it would be if it were separated from the atmosphere by a movable frictionless piston such as mercury in a U-tube), or if the external pressure p_e were otherwise maintained constant and equal to the pressure p of the contents of the calorimeter, then $-\int_1^2 p_e \, dV = -p(V_2 - V_1)$ and equation (1) becomes

$$\Delta U = U_2 - U_1 = W^{el} - pV_2 + pV_1 + W' + Q,\qquad(1.19.3)$$

or

$$\Delta(U + pV) = (U + pV)_2 - (U + pV)_1 = W^{el} + W' + Q.\qquad(1.19.4)$$

The composite quantity $(U + pV)$ turns up so often that it is convenient to give it a name and a symbol of its own. It is called the *enthalpy* and is denoted by the symbol H. Thus, the equation:

$$\Delta H = H_2 - H_1 = W^{el} + W' + Q,\qquad(1.19.5)$$

corresponds, for a constant-pressure calorimeter, to equation (2), for a constant-volume or zero-external-pressure calorimeter, and differs from it only by the substitution of H for U.

The heat leak Q in equation (1) or (2) or (5) can be made to approach zero by making the calorimeter virtually adiabatic, whether by careful thermal insulation or by means of an 'adiabatic shield'. Thermal insulation can be achieved for example by polishing the outside surface of the calorimeter so as to reduce heat leakage by radiation and by suspending it in an evacuated space so as to reduce heat leakage by thermal conduction and convection. Attempts to achieve thermal insulation are, however, usually frustrated by thermal conduction along the leads which inevitably run from the surroundings to the electric resistor and to the thermometer inside the calorimeter. An 'adiabatic shield' is usually more successful. The calorimeter is designed so that the temperature of its outer surface, which governs the rate of heat exchange with the surroundings, is as nearly uniform as possible. The calorimeter is suspended in an evacuated space surrounded by a shield: a container having as small a mass as possible made of a metal such as copper having a high thermal conductivity and fitted with devices for rapidly changing its temperature, for example an electric (Joule) heating element, and an electric (Peltier) cooling element or a thermal tie to a temperature-controlled bath of a suitable refrigerant. The leads from the calorimeter are thermally tied to the shield. The temperature of the shield is made to follow that of the outer surface of the calorimeter as precisely as possible, and with as little time lag as possible, by means of electronic gear responding quickly to any out-of-balance signal from a rapidly responding difference-thermometer, such as a thermocouple (see § 3.4), straddled between the outer surface of the calorimeter and the adiabatic shield.

When $Q = 0$ so that the calorimeter is adiabatic, equation (2) for a constant-volume or zero-external-pressure calorimeter becomes

$$\Delta U = W^{\text{el}} + W' , \qquad (1.19.6)$$

and equation (5) for a constant-pressure calorimeter becomes

$$\Delta H = W^{\text{el}} + W' . \qquad (1.19.7)$$

§ 1.20   Chemical reaction in an approximately adiabatic calorimeter

Let a chemical reaction proceed in an approximately adiabatic calorimeter from empirical temperature θ_i and extent of reaction ξ_i (see § 2.16) to temperature θ_f and extent of reaction ξ_f. (If $\theta_f > \theta_i$ the reaction is described as exothermic and if $\theta_f < \theta_i$ as endothermic.)

If the volume V of the calorimeter is maintained constant so that no work is done, equation (1.18.1) becomes

$$U(\theta_f, V, \xi_f) - U(\theta_i, V, \xi_i) = 0 + Q_1 \approx 0 . \qquad (1.20.1)$$

Let the temperature be restored from the value θ_f to its initial value θ_i. Then use electric work W^{el} to change it again to θ_f. For this second part of the experiment we then have

$$U(\theta_f, V, \xi_f) - U(\theta_i, V, \xi_f) = W^{\text{el}} + Q_2 \approx W^{\text{el}} . \qquad (1.20.2)$$

Subtraction of (2) from (1) gives

$$\Delta U = U(\theta_i, V, \xi_f) - U(\theta_i, V, \xi_i) = -W^{\text{el}} + (Q_1 - Q_2) \approx -W^{\text{el}} , \qquad (1.20.3)$$

where ΔU is the excess of the energy of the products over that of the reactants at the temperature θ_i and the volume V. The term $(Q_1 - Q_2)$ can be eliminated either by making the calorimeter sufficiently nearly adiabatic so that $Q_1 = Q_2 = 0$, or by arranging the experiment so that Q_1 and Q_2 are equal. Other ways of coping with the heat leaks Q_1 and Q_2 will be described in chapter 4.

Similarly if the pressure of the calorimeter is maintained constant and equal to that of the surroundings we have

$$H(\theta_f, p, \xi_f) - H(\theta_i, p, \xi_i) = 0 + Q_1 \approx 0, \qquad (1.20.4)$$

$$H(\theta_f, p, \xi_f) - H(\theta_i, p, \xi_f) = W^{el} + Q_2 \approx W^{el}, \qquad (1.20.5)$$

so that

$$\Delta H = H(\theta_i, p, \xi_f) - H(\theta_i, p, \xi_i) = -W^{el} + (Q_1 - Q_2) \approx -W^{el}, \qquad (1.20.6)$$

where ΔH is the excess of the enthalpy of the products over that of the reactants at the empirical temperature θ_i and pressure p.

§ 1.21   Chemical reaction in a calorimeter having an isothermal jacket

Let a chemical reaction proceed in a calorimeter having an isothermal† jacket maintained at constant empirical temperature θ_i from temperature θ_i and extent of reaction ξ_i to temperature θ_f and extent of reaction ξ_f, and let us refer to figure 1.4.

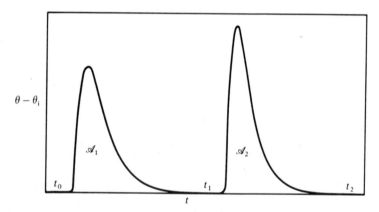

Figure 1.4. Plot of empirical temperature difference $(\theta - \theta_i)$ against time t (a thermogram) for two parts of a calorimetric experiment.

If the volume V of the calorimeter is maintained constant so that no work is done and if the temperature is allowed to return to its initial value θ_i, as in the first part of figure 1.4, then

$$\Delta U = U(\theta_i, V, \xi_f) - U(\theta_i, V, \xi_i) = 0 + Q_1. \qquad (1.21.1)$$

---

† Isothermal means at constant temperature. Such calorimeters are sometimes described as having constant-temperature environments or as 'isoperibol' calorimeters.

If we then supply a measured burst of electric work W^{el} and again allow the temperature to return to its initial values θ_i, as in the second part of figure 1.4, then

$$U(\theta_i, V, \xi_f) - U(\theta_i, V, \xi_f) = 0 = W^{el} + Q_2 . \qquad (1.21.2)$$

If Newton's law of cooling, $dQ/dt = -k(\theta - \theta_i)$, is obeyed with constant k then

$$Q_1 = -k \int_{t_0}^{t_1} (\theta - \theta_i) \, dt = -k\mathcal{A}_1 ,$$

$$Q_2 = -k \int_{t_1}^{t_2} (\theta - \theta_i) \, dt = -k\mathcal{A}_2 , \qquad (1.21.3)$$

where $\mathcal{A}_1$ is the area under the first thermogram, and $\mathcal{A}_2$ that under the second thermogram, in figure 1.4. Equation (2) then gives $k = W^{el}/\mathcal{A}_2$ and equation (1) can then be written in the form:

$$\Delta U = U(\theta_i, V, \xi_f) - U(\theta_i, V, \xi_i) = -W^{el}\mathcal{A}_1/\mathcal{A}_2 , \qquad (1.21.4)$$

so that ΔU can be evaluated from W^{el} and measured values of the areas $\mathcal{A}_1$ and $\mathcal{A}_2$.

Similarly if the pressure of the calorimeter is maintained constant and equal to that of the surroundings we have

$$\Delta H = H(\theta_i, p, \xi_f) - H(\theta_i, p, \xi_i) = -W^{el}\mathcal{A}_1/\mathcal{A}_2 . \qquad (1.21.5)$$

If 'additional' work W' is done during one or both parts of the experiment just described, as for example by N inversions of the calorimeter to cause stirring and to improve the uniformity of θ, a third part of the experiment can be done in which the area $\mathcal{A}_3$ is measured of the thermogram that results from only stirring. The quantity $w' = W'/N$ (assumed constant; that assumption must be tested) can then be eliminated, and equation (5) becomes

$$\Delta H = H(\theta_i, p, \xi_f) - H(\theta_i, p, \xi_i)$$
$$= -W^{el}(\mathcal{A}_1 - \mathcal{A}_3 N_1/N_3)/(\mathcal{A}_2 - \mathcal{A}_3 N_2/N_3) . \qquad (1.21.6)$$

The reaction whose thermogram is shown in the first part of figure 1.4 was exothermic. Had it been endothermic the argument would have been unchanged except that the area $\mathcal{A}_1$ would have been negative and the signs of ΔU or ΔH would have been reversed.

Other ways of coping with heat leaks, particularly when the thermal insulation of the calorimeter is good enough to prevent the temperature from returning to θ_i in a reasonable time but not good enough to allow us to put $Q_1 = Q_2 = 0$, will be described in chapter 4.

§ 1.22 Heat-capacity calorimetry

Consider the difference:

$$\Delta U = U(T_2, V) - U(T_1, V) , \qquad (1.22.1)$$

between the energies of the initial and final states of a system (for example a sample of a solid substance) when its temperature is changed from T_1 to T_2 while its volume V is constrained to be the same in the initial and final states. Let W^{el}(sample + container) be the quantity of work needed to change the temperature of the sample plus its container from T_1 to T_2, and let W^{el}(container) be the corresponding quantity of work for the empty container. Provided that the initial and final states of the container are the same in each experiment, that the volume V of the sample is the same in the initial and final states, that no other work is done (so that $W' = 0$), and that the calorimeter is adiabatic (so that $Q = 0$) or that the heat leak Q is the same whether the calorimeter is full or empty, we have according to equation (1.19.1):

$$\Delta U = U(T_2, V) - U(T_1, V)$$
$$= W^{el}(\text{sample} + \text{container}) - W^{el}(\text{container}) . \qquad (1.22.2)$$

The ΔU of equation (1) or (2) may be written in the alternative forms:

$$\Delta U = U(T_2, V) - U(T_1, V) \equiv \int_{T_1}^{T_2} (\partial U/\partial T)_V \, dT \stackrel{\text{def}}{=} \int_{T_1}^{T_2} C_V \, dT, \quad (1.22.3)$$

where C_V, defined by the relation:

$$C_V \stackrel{\text{def}}{=} (\partial U/\partial T)_V \stackrel{\text{def}}{=} \lim_{T_2 \to T_1} [\{U(T_2, V) - U(T_1, V)\}/(T_2 - T_1)], \quad (1.22.4)$$

is called the *heat capacity at constant volume*. We may obtain values of $C_V(T)$ for the sample, either by measuring ΔU for a given T_1 and a range of values of T_2 and then estimating the slope of a plot of ΔU against T_2, or by measuring ΔU for a small enough $(T_2 - T_1)$ and then evaluating $\Delta U/(T_2 - T_1)$.

The method described above, though sound in principle, is usually hopelessly impractical. Apart from the difficulty of filling the container with a solid or liquid sample, the pressure needed to keep the volume of a solid or liquid constant when the temperature is changed implies a container not only having a volume independent of temperature but also so massive that most by far of the total heat capacity of the (sample + container) would belong to the container rather than to the sample. For example an increase in the temperature of liquid cyclohexane from 300 K to 310 K needs an increase of about 10 MPa in the pressure to prevent any increase in its volume; $(\partial p/\partial T)_V$ is about 1 MPa K^{-1}. For a gas at atmospheric pressure and room temperature $(\partial p/\partial T)_V$ is only about 0.0003 MPa K^{-1} so that a gas is much easier to contain at constant volume than a solid or a liquid. Nevertheless the heat capacity of even the flimsiest container will be much greater than that of the sample of gas. For example the ratio of the heat capacity of a cubical box of side 10 cm made of copper sheet of thickness 0.05 cm to that of a gas contained in it at room temperature and pressure is about 100; the gas will contribute only about 1 per cent to the measured heat capacity of (gas + container).

It is usually much easier in practice to measure the corresponding difference:

$$\Delta H = H(T_2, p) - H(T_1, p)$$

$$= W^{el}(\text{sample} + \text{container}) - W^{el}(\text{container}), \qquad (1.22.5)$$

between the enthalpies of the initial and final states of the sample when its pressure p is constrained to be the same in the initial and final states. The ΔH of equation (5) may be written in the alternative forms:

$$\Delta H = H(T_2, p) - H(T_1, p) \equiv \int_{T_1}^{T_2} (\partial H/\partial T)_p \, dT \overset{\text{def}}{=} \int_{T_1}^{T_2} C_p \, dT, \qquad (1.22.6)$$

where C_p, defined by the relation:

$$C_p \overset{\text{def}}{=} (\partial H/\partial T)_p \overset{\text{def}}{=} \lim_{T_2 \to T_1} [\{H(T_2, p) - H(T_1, p)\}/(T_2 - T_1)], \qquad (1.22.7)$$

is called the *heat capacity at constant pressure*. We may obtain values of $C_p(T)$ for the sample, either by measuring ΔH for a given T_1 and a range of values of T_2 and then estimating the slope of a plot of ΔH against T_2, or by measuring ΔH for a small enough $(T_2 - T_1)$ and then evaluating $\Delta H/(T_2 - T_1)$.

When C_V rather than C_p is needed, for example to test a statistical-mechanical theory of solids, it is usually obtained from an experimental value of C_p by use of the general thermodynamic relation between the heat capacities C_p and C_V that we shall derive in § 6.11.

For gases it is just as difficult, and for the same reason, to measure C_p by the method implied by equation (5) as it was to measure C_V by the method implied by equation (2). The heat capacity at constant pressure of a gas is usually measured by adiabatic flow calorimetry.

§ 1.23   Adiabatic flow calorimeter

Let us consider the apparatus shown diagrammatically in figure 1.5. A fluid flows at constant rate $\dot{n}$ (the amount of substance crossing any plane normal to the direction of flow in unit time) through a thermally insulated tube fitted with a throttle (such as a valve or a porous disk; a throttle is a device for producing a sharp pressure gradient in a flowing fluid), an electric 'worker' on the downstream side of the throttle, a differential thermometer (such as a thermocouple) across the throttle, and a differential manometer across the throttle.

Application of the first law to a sample of fluid containing amount of substance n present initially on the left-hand side at temperature T_1 and pressure p_1 and finally on the right-hand side at temperature T_2 and pressure p_2 gives, since $Q = 0$,

$$U(T_2, p_2) - U(T_1, p_1) = W. \qquad (1.23.1)$$

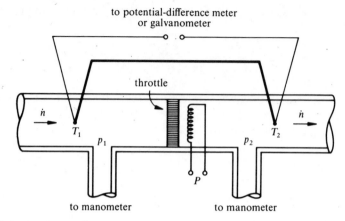

Figure 1.5. An adiabatic flow calorimeter fitted with a throttle and a supply of electric power.

The total work W done on the sample of fluid is given by

$$W = Pn/\dot{n} + p_1 V(T_1, p_1) - p_2 V(T_2, p_2)$$
$$+ nMg(h_2 - h_1) + \tfrac{1}{2}nM(v_2^2 - v_1^2), \tag{1.23.2}$$

where P denotes the constant electric power supplied to the amount of substance n during the time $n/\dot{n}$, M the molar mass of the fluid, g the local acceleration of free fall, $(h_2 - h_1)$ any difference in the heights of the fluid on each side of the throttle, and v_1 and v_2 the velocities of the fluid stream on each side of the throttle.

If $h_2 = h_1$ (as it is in figure 1.5) so that no gravitational work is done, and provided that the velocities v_1 and v_2 of the fluid are low enough or similar enough to make the kinetic-energy term negligible, we obtain

$$W = Pn/\dot{n} + p_1 V(T_1, p_1) - p_2 V(T_2, p_2). \tag{1.23.3}$$

Substitution of (3) in (1) and rearrangement gives

$$H(T_2, p_2) - H(T_1, p_1) = Pn/\dot{n}. \tag{1.23.4}$$

In the special case that no power is supplied to the fluid so that $P = 0$ we describe the experiment as a Joule–Thomson experiment and equation (4) becomes

$$H(T_2, p_2) = H(T_1, p_1). \tag{1.23.5}$$

Adiabatic throttling with $P = 0$ is *isenthalpic*. We measure the quantity:

$$\{T(H, p_2) - T(H, p_1)\}, \tag{1.23.6}$$

and describe

$$(\partial T/\partial p)_H \overset{\text{def}}{=} \mu_{JT} \overset{\text{def}}{=} \lim_{p_2 \to p_1} [\{T(H, p_2) - T(H, p_1)\}/(p_2 - p_1)], \tag{1.23.7}$$

as the *Joule–Thomson coefficient*.

In the special case that the power P is adjusted so that $T_2 = T_1$ we describe the experiment as an isothermal Joule–Thomson experiment and equation (4) becomes

$$H(T, p_2) - H(T, p_1) = Pn/\dot{n} . \qquad (1.23.8)$$

We describe the quantity:

$$(\partial H/\partial p)_T \overset{\text{def}}{=} \phi_{JT} \overset{\text{def}}{=} \lim_{p_2 \to p_1} [\{H(T, p_2) - H(T, p_1)\}/(p_2 - p_1)], \qquad (1.23.9)$$

as the *isothermal Joule–Thomson coefficient.*

In the special case that the throttle is open (or removed) so that $p_2 = p_1$,† we describe the experiment as a heat-capacity experiment and equation (4) becomes

$$H(T_2, p) - H(T_1, p) = Pn/\dot{n} . \qquad (1.23.10)$$

Equation (10) leads by use of equation (1.22.7) to the *heat capacity at constant pressure.* Most heat capacities of gases are measured by this method. The method is also used for liquids.

The three quantities μ_{JT}, ϕ_{JT}, and C_p are not independent. They are related by the '−1 rule' (see appendix I):

$$(\partial T/\partial p)_H (\partial p/\partial H)_T (\partial H/\partial T)_p = -1 , \qquad (1.23.11)$$

so that

$$\mu_{JT} = -\phi_{JT}/C_p . \qquad (1.23.12)$$

If in any of these experiments the apparatus is not truly adiabatic then it can be made effectively so by increasing the flow rate $\dot{n}$ and extrapolating to $1/\dot{n} \to 0$.

Problems for chapter 1

Problem 1.1

The limiting values $\lim_{p \to 0} (pV)$ for a sample of a gas measured at empirical temperatures θ_1 and θ_2 were 45.423 Pa m^3 and 53.394 Pa m^3. Calculate the ratio $T(\theta_2)/T(\theta_1)$ of the corresponding thermodynamic temperatures. If θ_1 is the triple-point temperature of water calculate $T(\theta_2)$.

Problem 1.2

An amount of substance $n = 0.023456$ mol of a gas has a limiting value $\lim_{p \to 0} (pV)$ of 53.272 Pa m^3 at the triple-point temperature of water. Calculate a value for the gas constant R.

---

† Strictly, p_2 can never be equal to p_1 if fluid is flowing down the tube.

Problem 1.3

When 1.29848 g of ethyl acetate initially at 298.150 K was burnt in an excess of oxygen in an adiabatically enclosed bomb calorimeter of constant volume the temperature rose by 2.12807 K. (a) What was the value of ΔU for this part of the experiment? The calorimeter and its contents were cooled to 298.150 K. Then an electric current of 2.13476 A was passed for 75.2613 s through a heating coil of resistance 125.686 Ω in the calorimeter; the temperature rose by 2.78304 K. (b) Calculate the value of ΔU for this part of the experiment. (c) Calculate the value of the excess ΔU of the energy of the products at 298.150 K over that of the reactants at 298.150 K. (d) Calculate the value of the molar energy change ΔU_m for the reaction at 298.150 K.

Problem 1.4

In an isothermal-jacket calorimeter open to the atmosphere 1.0763 g of tetra-methylsilane and 1.7211 g of tetramethylstannane were contained separately at 283.15 K. They were allowed to mix and the thermogram, plotted until the temperature had returned to 283.15 K, had an area $\mathscr{A}_1$. An electric current of 0.068450 A was passed for 14.193 s through a heating coil of resistance 28.132 Ω in the calorimeter, and the thermogram, plotted until the temperature had again returned to 283.15 K, had an area $\mathscr{A}_2$. The ratio $\mathscr{A}_2/\mathscr{A}_1$ was found to be 1.3011. (a) Calculate the enthalpy of mixing the two samples of liquid at 283.15 K. (b) Calculate the molar enthalpy of mixing and the mole fraction x in the mixture $\{(1-x)Si(CH_3)_4 + xSn(CH_3)_4\}$.

Problem 1.5

In an adiabatic constant-pressure calorimeter the work W_2 needed to raise the temperature of (76.1415 g of gadolinium trichloride + container) from 291.73 K to T and the work W_1 needed to raise the temperature of the empty container by the same amount, both at the constant pressure 7.6 kPa, were as follows.

W_2/J	104.42	219.12
W_1/J	37.84	79.41
$T/K - 291.73$	9.872	19.651

Calculate (a) the enthalpy increments $\{H(301.60 \text{ K}, 7.6 \text{ kPa}) - H(291.73 \text{ K}, 7.6 \text{ kPa})\}$ and $\{H(311.38 \text{ K}, 7.6 \text{ kPa}) - H(291.73 \text{ K}, 7.6 \text{ kPa})\}$ for the sample, (b) the heat capacity at constant pressure C_p of the sample at 300.00 K, and (c) the molar heat capacity at constant pressure $C_{p,m}$ of GdCl$_3$ at 300.00 K.

Problem 1.6

When 0.50211 mmol s^{-1} of gaseous H_2O flowed through an adiabatic flow calori-meter fitted with a throttle and electric heater, a power 0.029459 W was needed just to restore the temperature to the inlet temperature of 383.26 K when the pressures were 33.003 kPa on the upstream side and 2.904 kPa on the downstream side of the calorimeter. (a) Calculate the molar enthalpy difference $\{H_m(383.26$ K, 2.904 kPa$) - H_m(383.26$ K, 33.003 kPa$)\}$. (b) Calculate an approximate value for the molar isothermal Joule–Thomson coefficient $(\partial H_m/\partial p)_T$ at 383.26 K.

Chapter 2

Composition, and change of composition, of a phase

§ 2.1 Scope

This chapter is concerned only with the definitions of some quantities which are useful in the study of the dependence of the properties of a mixed phase on its composition, and with relations among those quantities; no thermodynamic principles are involved.

There are two ways in which the composition of a phase might change: by gain or loss of material, or by change of the extent of one or more chemical reactions. Definitions and notation related to the composition of a phase and some important relations for change of composition of a phase are given in §§ 2.2 to 2.14. Definitions and notation for the *extent of reaction* of a chemical reaction are given in §§ 2.15 to 2.17.

§ 2.2 Amount of substance

The *amount of substance* n_B of an entity B in a system is a physical quantity defined by its proportionality to the number N_B of entities B in the system. The entity must be specified, and may be an atom, a molecule, an ion, a radical, an electron, a photon, any other particle, or any *specified* group of such particles. Specified groups need not be confined to those known or thought to exist as independent entities or to groups containing integral numbers of atoms. Thus we may properly speak of an amount of substance of $\frac{1}{2}H_2O$ or of $(H_2 + \frac{1}{2}O_2)$ or of $(H_2 + 0.234O_2)$ or of $(0.075^6Li + 0.925^7Li)$ or of $C(s)$ or of $HgCl(s)$ or of $Fe_{0.91}S(s)$ or of $\frac{1}{5}MnO_4^-(aq)$ or of $(KCl + 123.4H_2O)(l)$. 'The amount of substance of mercury(I) chloride' is, however, ambiguous.

The formula of B must always be given, not only when one speaks of the amount of substance of B, but also whenever one speaks of any derived quantity involving amount of substance, such as concentration, molality, mole fraction, any molar or partial molar quantity, or a rate of reaction.

The proportionality constant L in the relation $N_B = Ln_B$ is called the Avogadro constant (*not* Avogadro 'number'; it is not a number but a number divided by an amount of substance).

The definition of amount of substance, as of any other physical quantity, has nothing to do with any choice of unit, and in particular has nothing to do with the unit of amount of substance called the mole. It is as grammatically incorrect to call n_B the 'number of moles' as it would be to call the mass m_B of B the 'number of grams'.

§ 2.3   The mole

The SI unit of amount of substance is the mole. When one speaks of a mole of B the formula of B must be given, just as it must for any other amount of substance. The definition of the mole is:[†] 'The mole is the amount of substance of a system which contains as many elementary entities as there are atoms in 0.012 kilogram of carbon-12. When the mole is used, the elementary entities must be specified and may be atoms, molecules, ions, electrons, other particles, or specified groups of such particles.' The SI symbol for mole is mol.

   None of the units 'gram-atom', 'gram-molecule', gram-equivalent', 'equivalent', 'gram-ion', 'gram-formula', 'gram-equation', 'faraday', or 'einstein' has any international standing; they are all obsolete 'specialized'[‡] forms of the mole designed to contain some hint about the nature of the entity, and they should all be discarded in favour of the mole. Thus 1 mol of Ar, not 1 'g-atom' of Ar; 1 mol of H_2SO_4, not 1 'g-mol' of H_2SO_4; 1 mol of $\frac{1}{2}H_2SO_4$, not 1 'g-equivalent' or 1 'equiv' of H_2SO_4; 1 mol of SO_4^{2-}, not 1 'g-ion' of SO_4^{2-}; 1 mol of e^-, not 1 'faraday'; 1 mol of γ, not 1 'einstein'. The unit 'normal' and unit-symbol 'N' (newton?) are likewise obsolete 'specialized' forms of the unit $mol\ dm^{-3}$. Thus $c(H_2SO_4) = 0.05\ mol\ dm^{-3}$ or $c(\frac{1}{2}H_2SO_4) = 0.1\ mol\ dm^{-3}$ but not 0.1 'N'.

§ 2.4   Molar quantities

Corresponding to any extensive quantity X of a phase we define an intensive quantity called a molar quantity X_m by the relation:

$$X_m \overset{\text{def}}{=} X / \sum_B n_B .\qquad (2.4.1)$$

For example the molar volume V_m of a phase is the volume V of the phase divided by the total amount of substance of the phase, that is to say divided by the sum $\Sigma_B n_B$ of the amounts of the several substances present in the phase.

   The subscript $_m$ will sometimes be replaced by the chemical symbol for the substance. Thus, X_B denotes the partial molar quantity of the substance B, and X_B^* the molar quantity of the pure substance B (see § 2.7).

   'When there is no risk of ambiguity' the subscript $_m$ for molar is sometimes omitted from the symbol for a molar quantity. It will never be omitted in this book except when it is replaced by the formula of the appropriate substance.

---

   † *Le Système International d'Unités* (*SI*). Bureau International des Poids et Mesures, OFFILIB, 48 rue Gay-Lussac, F 75005, Paris. Revised edition. **1977**. *The International System of Units.* Translation approved by the BIPM, Page, C. H.; Vigoureux, P.: Editors. Her Majesty's Stationery Office: London. **1977**. U.S. Government Printing Office: Washington. **1977**.
   ‡ McGlashan, M. L. *Physicochemical Quantities and Units* (*the Grammar and Spelling of Physical Chemistry*), Second Edition. Royal Institute of Chemistry: London. **1971**.

§ 2.5 Mole fractions

We define the mole fraction x_B of a substance B in a phase by the relation:

$$x_B \overset{\text{def}}{=} n_B / \sum_B n_B . \tag{2.5.1}$$

It follows immediately that

$$\sum_B x_B = 1 . \tag{2.5.2}$$

If there are $\mathscr{C}$ substances present in the phase then, whereas the $\mathscr{C}$ amounts of substance n_B can be varied independently, in view of equation (2) only $(\mathscr{C}-1)$ of the mole fractions x_B can be varied independently. Whereas the n's are extensive quantities, the x's are intensive quantities.

§ 2.6 Choice of independent variables

The most convenient choice of variables for a thermodynamic description of a phase is usually either the two intensive quantities: the thermodynamic temperature T and the pressure p, plus the $\mathscr{C}$ extensive quantities: the amounts of substance n_A, n_B, $n_C, \cdots$; or the $(\mathscr{C}+2)$ intensive quantities: T and p plus the $\mathscr{C}$ partial molar quantities (see § 2.7) X_A, X_B, $X_C, \cdots$. For the first of those choices the $(\mathscr{C}+2)$ variables are all independent; one of them can be regarded as determining the size of the phase. For the second choice we shall find that there is one relation among the $(\mathscr{C}+2)$ variables so that only $(\mathscr{C}+1)$ of them are independent; the description of the phase does not then include its size.

§ 2.7 Partial molar quantities

Corresponding to any extensive quantity X of a phase we define an intensive quantity called the partial molar quantity X_B of the substance B in the phase by the relation:

$$X_B \overset{\text{def}}{=} (\partial X / \partial n_B)_{T,p,n_{A \neq B}} , \tag{2.7.1}$$

where $n_{A \neq B}$ after the partial derivative means that all the n's except n_B are kept constant.† An alternative but clumsy way of writing equation (1) would be

$$X_B = dX / dn_B , \quad (T, p \text{ constant}; n_A \text{ constant for all } A \neq B) . \tag{2.7.2}$$

Whichever way we write X_B, it is the slope at a specified value of n_B of a plot of X against n_B when X has been measured for various values of n_B with T, p, and all the n's except n_B, kept constant at some specified values.

---

† The symbol $\bar{X}_B$ has often been used for our X_B. The bar resolves no ambiguity and so is superfluous.

For any extensive quantity X of a phase for which we may choose as independent variables the set T, p, and the n's, we may write

$$dX = (\partial X/\partial T)_{p,n_B} dT + (\partial X/\partial p)_{T,n_B} dp + \sum_B (\partial X/\partial n_B)_{T,p,n_{A \neq B}} dn_B$$

$$= (\partial X/\partial T)_{p,n_B} dT + (\partial X/\partial p)_{T,n_B} dp + \sum_B X_B dn_B , \qquad (2.7.3)$$

where n_B after the partial derivative means that all the n's are kept constant.

By use of Euler's theorem (see appendix II) it follows immediately from equation (3) that

$$X = \sum_B n_B X_B . \qquad (2.7.4)$$

On division by $\sum_B n_B$ equation (4) becomes

$$X_m = \sum_B x_B X_B . \qquad (2.7.5)$$

The reader should resist any temptation to put 'at constant temperature and pressure' after equation (4) or (5). Given X_B defined by equation (1), equations (4) and (5) are mathematical identities. Nor have we used any thermodynamics. In deriving (4) and (5) we have assumed only (i) that T, p, and the n's form a complete set of independent variables for the phase, and (ii) that T and p are intensive variables while the n's are extensive variables.

For a phase composed of a *pure* substance B any partial molar quantity X_B is identical with the corresponding molar quantity X_m for the phase; in this special case:

$$X_B = X/n_B = X_m , \quad \text{(pure substance B)} . \qquad (2.7.6)$$

We shall denote X_B or X_m for such a pure substance by X_B^* where the superscript * here and throughout this book means 'pure'. Thus for example V_B^* denotes the molar (or for that matter the partial molar) volume of pure substance B.

Differentiation of equation (4) gives

$$dX = \sum_B n_B dX_B + \sum_B X_B dn_B . \qquad (2.7.7)$$

Comparison of equations (3) and (7) leads to the relation:

$$0 = - (\partial X/\partial T)_{p,n_B} dT - (\partial X/\partial p)_{T,n_B} dp + \sum_B n_B dX_B , \qquad (2.7.8)$$

or, at constant temperature and pressure,

$$0 = \sum_B n_B dX_B , \quad (T, p \text{ constant}) , \qquad (2.7.9)$$

so that in view of equation (7)

$$dX = \sum_B X_B dn_B , \quad (T, p \text{ constant}) . \qquad (2.7.10)$$

Division of equation (8) by $\sum_B n_B$ leads to the relation:

$$0 = -(\partial X_m/\partial T)_{p,n_B} \, dT - (\partial X_m/\partial p)_{T,n_B} \, dp + \sum_B x_B \, dX_B ,$$

(2.7.11)

or, at constant temperature and pressure,

$$0 = \sum_B x_B \, dX_B , \quad (T, p \text{ constant}) .$$

(2.7.12)

Differentiation of equation (5) leads in view of equation (11) to

$$dX_m = \sum_B X_B \, dx_B , \quad (T, p \text{ constant}) .$$

(2.7.13)

We shall make much use of equations (8) and (11), and of their special cases (9) and (12) for variations of the composition of a phase at constant temperature and pressure. In particular we shall later find that when $X_m = G_m$ where G_m denotes the molar Gibbs function, equation (11) is called the Gibbs–Duhem equation for the phase and is of fundamental importance in the study of phase equilibria. We shall also make use of equations (5) and (13); in § 2.8 we shall use them to find a method for the measurement of partial molar quantities.

Equations (1) to (13) will amply repay close study. Particular care should be taken to distinguish those equations that are restricted to variations of composition at constant temperature and pressure from those that are not. Experience suggests that care is also needed when, for example, equation (13) is derived from equation (10); one may not simply divide by $\sum_B n_B$ inside the differential operator d, but must use the relations:

$$dX = \left(\sum_B n_B\right) dX_m + X_m \sum_B dn_B ,$$

(2.7.14)

$$dn_B = \left(\sum_B n_B\right) dx_B + x_B \sum_B dn_B .$$

(2.7.15)

§ 2.8   Measurement of partial molar quantities

Equations (2.7.5) and (2.7.13), namely

$$X_m = \sum_B x_B X_B ,$$

(2.8.1)

$$dX_m = \sum_B X_B \, dx_B , \quad (T, p \text{ constant}) ,$$

(2.8.2)

can be used to determine all the partial molar quantities X_B from measurements at given temperature T and pressure p of the molar quantity X_m and of its dependence on the ($\mathscr{C} - 1$) independent mole fractions x_B, where $\mathscr{C}$ is the number of components in the mixture.

The method will be described in detail only for a binary mixture of $\{(1-x)A + xB\}$, where we use the symbol x for x_B so that $x_A = (1-x)$. For this case equations (1) and

(2) become

$$X_m = (1-x)X_A + xX_B, \qquad (2.8.3)$$

$$dX_m = (X_B - X_A)\,dx, \quad (T, p \text{ constant}). \qquad (2.8.4)$$

Solving (3) and (4) for X_A and X_B we obtain

$$X_A = X_m - x(\partial X_m/\partial x)_{T,p}, \qquad (2.8.5)$$

$$X_B = X_m + (1-x)(\partial X_m/\partial x)_{T,p}. \qquad (2.8.6)$$

Thus the partial molar quantities X_A and X_B for a mixture of given mole fraction x can be found from the measured value of the molar quantity X_m and of the slope of a plot of X_m against x.

A simple graphical method for determining X_A and X_B from a plot of X_m against x is illustrated in figure 2.1. The reader should verify for himself that the intercepts at

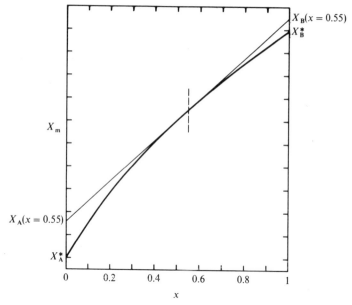

Figure 2.1. Graphical method for the determination of partial molar quantities X_A and X_B for a binary mixture of A and B from measurements of the molar quantity X_m as a function of the mole fraction x of B.

$x = 0$ and at $x = 1$ of the tangent to the curve at a given value of x are the values of the partial molar quantities X_A and X_B respectively for the mixture having that value of x.

There are, however, two disadvantages in the use of the method based on equations (5) and (6). First, when X is a quantity such as enthalpy H or, as we shall find, Gibbs function G or entropy S, the method is not applicable because the quantity itself cannot be measured but only differences of it. Secondly, even when X

is a quantity such as volume V or heat capacity at constant pressure C_p that *can* be measured, the method is an inaccurate one whenever, as is often so, the deviations from linearity of a plot of X_m against x are small compared with the difference $(X_A^* - X_B^*)$ between the end values.

A more generally applicable and more accurate method based on molar quantities of mixing $\Delta_{mix}X_m$ rather than on the molar quantities X_m will now be described.

§ 2.9   Molar quantities of mixing

We define the molar quantity of mixing $\Delta_{mix}X_m$ for a binary mixture by the relation:

$$\Delta_{mix}X_m \stackrel{def}{=} X_m - (1-x)X_A^* - xX_B^*, \tag{2.9.1}$$

where $X_m = X_m(T, p, x)$ is the molar quantity of a binary mixture $\{(1-x)A + xB\}$ at the temperature T and pressure p, and where $X_A^* = X_m(T, p, 0)$ and $X_B^* = X_m(T, p, 1)$ are the molar quantities of the pure substances A and B at the same temperature and pressure.

As an example, the molar volume of mixing $\Delta_{mix}V_m$ can be calculated from measurements of the density ρ of liquid mixtures by use of the relation:

$$\Delta_{mix}V_m = \{(1-x)M_A + xM_B\}/\rho - (1-x)M_A/\rho_A^* - xM_B/\rho_B^*, \tag{2.9.2}$$

where M_A and M_B are the molar masses of A and B, and ρ, ρ_A^*, and ρ_B^* are the densities of a mixture having mole fraction x of B, of pure A, and of pure B, respectively. In this method, however, it is necessary to measure the densities with the highest possible accuracy in order to achieve even a modest accuracy in $\Delta_{mix}V_m$. Direct dilatometric measurement of $\Delta_{mix}V$ gives a much higher ratio of accuracy to effort. A suitable dilatometer is shown in figure 2.2.

In the case of the molar enthalpy of mixing $\Delta_{mix}H_m$ we have no alternative to direct measurement since H_m cannot be measured. Calorimeters suitable for the measurement of $\Delta_{mix}H_m$ will be introduced in § 4.11. Neither molar Gibbs function G_m nor molar entropy S_m can be measured, but we shall find out in chapter 16 how $\Delta_{mix}G_m$ and $\Delta_{mix}S_m$ can be measured.

By use of equation (2.8.3) we can rewrite equation (1) in the form:

$$\Delta_{mix}X_m = (1-x)(X_A - X_A^*) + x(X_B - X_B^*), \tag{2.9.3}$$

and by differentiation of equation (1) and use of equation (2.8.4) we obtain

$$d\Delta_{mix}X_m = \{(X_B - X_B^*) - (X_A - X_A^*)\}\, dx, \quad (T, p\text{ constant}). \tag{2.9.4}$$

Solving (3) and (4) for $(X_A - X_A^*)$ and $(X_B - X_B^*)$ we obtain

$$X_A - X_A^* = \Delta_{mix}X_m - x(\partial\Delta_{mix}X_m/\partial x)_{T,p}, \tag{2.9.5}$$

$$X_B - X_B^* = \Delta_{mix}X_m + (1-x)(\partial\Delta_{mix}X_m/\partial x)_{T,p}. \tag{2.9.6}$$

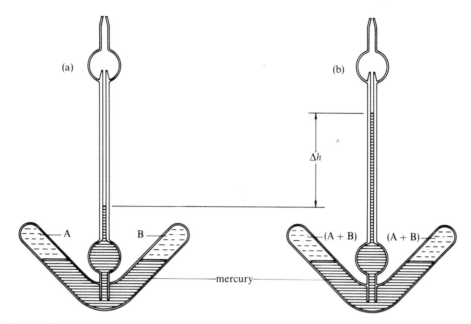

Figure 2.2. A dilatometer for measurements of volume of mixing shown (a), before mixing and (b), after mixing (by rocking to and fro). The dilatometer is filled with the help of a hypodermic syringe. The molar volume of mixing $\Delta_{mix} V_m$ is given (neglecting small terms allowing for the effects of the change of pressure on mixing on the volumes of the liquids and of mercury) by the relation:

$$\Delta_{mix} V_m = \Delta V/(n_A + n_B) = \mathscr{A}\Delta h/(n_A + n_B),$$

where $\mathscr{A}$ is the cross-sectional area of the capillary and n_A and n_B are the amounts of substance of A and of B.

The reader will notice that equations (3) to (6) can be obtained from equations (2.8.3) to (2.8.6) when $\Delta_{mix} X_m$ is substituted for X_m and at the same time $(X_A - X_A^*)$ is substituted for X_A and $(X_B - X_B^*)$ for X_B. With these substitutions the simple graphical method described in the previous section can be used to determine $(X_A - X_A^*)$ and $(X_B - X_B^*)$ as illustrated in figure 2.3.

The method described in this section is usually more convenient than that described in § 2.8 and is moreover applicable for such quantities as enthalpy for which the method of § 2.8 is not applicable. There are some occasions, however, when the method of § 2.8 is applicable and that of the present section is not. These occur when one or more of the X^*'s is unknown. An example is the determination of partial molar volumes for electrolyte solutions for which the volume of the pure liquid solute is inaccessible.

§ 2.10   Mixtures and solutions

The word *mixture* is used to describe a gaseous or liquid or solid phase containing more than one substance, when all the substances are treated in the same way. The

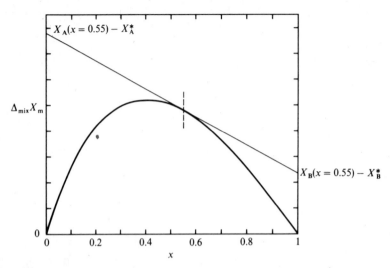

Figure 2.3. Graphical method for the determination of $(X_A - X_A^*)$ and of $(X_B - X_B^*)$ from measurements of the molar quantity of mixing $\Delta_{mix}X_m$ as a function of the mole fraction x of B.

word *solution* is used to describe a liquid or solid phase containing more than one substance, when, for convenience, one (or more) of the substances, called the *solvent*, is treated differently from the other substances, called *solutes*. When, as is often though not necessarily the case, the sum of the mole fractions of the solutes is small compared with unity, the solution is called a *dilute solution*. The superscript $^\infty$ attached to the symbol for a property of a solution will be used to denote the property in the limit of infinite dilution.

§ 2.11   Molality

The composition of a solution (as distinct from a mixture) is usually expressed not in terms of mole fractions but in terms of the *molalities* of the solutes. The molality m_B of a solute B in a solvent A is defined by the relation:

$$m_B \overset{\text{def}}{=} n_B/n_A M_A, \tag{2.11.1}$$

where n_B is the amount of substance of the solute B, and n_A is the amount of substance and M_A the molar mass of the solvent A.

The mole fraction X_B of a solute B is related to its molality m_B by

$$x_B = M_A m_B \Big/ \Big(1 + M_A \sum_B m_B\Big), \tag{2.11.2}$$

or

$$m_B = x_B \Big/ \Big(1 - \sum_B x_B\Big) M_A. \tag{2.11.3}$$

For a solution of a single solute B, equations (2.8.5) and (2.8.6) for the partial molar quantities X_A and X_B can be expressed in terms of the molality rather than the mole fraction in the forms:

$$X_A = X_m - m_B(1 + M_A m_B)(\partial X_m / \partial m_B)_{T,p},\qquad(2.11.4)$$

$$X_B = X_m + \{(1 + M_A m_B)/M_A\}(\partial X_m / \partial m_B)_{T,p}.\qquad(2.11.5)$$

§ 2.12   Apparent molar quantities

The apparent molar quantity $X_{B,\phi}$ of a solute B in a solvent A is defined by the relation:†

$$n_B X_{B,\phi} \stackrel{\text{def}}{=} X - n_A X_A^*,\qquad(2.12.1)$$

which for a solution of a single solute becomes

$$n_B X_{B,\phi} = (n_A + n_B)X_m - n_A X_A^*.\qquad(2.12.2)$$

Expressed in terms of molalities equation (2) becomes

$$m_B M_A X_{B,\phi} = (1 + M_A m_B)X_m - X_A^*.\qquad(2.12.3)$$

Differentiation of equation (2) with respect to n_B at constant T, p, and n_A leads to the equation:

$$X_{B,\phi} + n_B(\partial X_{B,\phi}/\partial n_B)_{T,p,n_A} = (\partial X/\partial n_B)_{T,p,n_A} = X_B,\qquad(2.12.4)$$

or, after substitution from equation (2.11.1) and rearrangement,

$$X_B = X_{B,\phi} + m_B(\partial X_{B,\phi}/\partial m_B)_{T,p}.\qquad(2.12.5)$$

Equation (5) can be used to obtain the partial molar quantity X_B from the sum of the apparent molar quantity and the product of the molality and the slope of a plot of the apparent molar quantity against molality. There is, however, nothing to choose between the use of equation (5) and of equation (2.11.5) for the determination of partial molar quantities. Equation (2.11.5) is useful when the molar quantity X_m has been tabulated or is readily derived as a function of molality; equation (5) is useful when the apparent molar quantity $X_{B,\phi}$ happens to have been tabulated instead of the molar quantity X_m. Apparent molar quantities will not be mentioned again in this book.

§ 2.13   Dilution

The change of a quantity X on dilution of a mixture of amounts of substance n_A of A and n_B of B by an amount of substance Δn_A of A is given by

$$\Delta_{\text{dil}}X = X\{(n_A + \Delta n_A)A + n_B B\} - X\{n_A A + n_B B\} - \Delta n_A X_A^*$$

$$= \Delta_{\text{mix}}X\{(n_A + \Delta n_A)A + n_B B\} - \Delta_{\text{mix}}X\{n_A A + n_B B\}.\qquad(2.13.1)$$

† Other symbols such as $^\phi X$ and ϕ_X have been used for apparent molar quantities.

The quantity obtained by dividing $\Delta_{\text{dil}}X$ by the amount of substance n_B of B is often referred to as the molar change of the quantity on dilution, especially of a solution of the solute B by the solvent A, and is given by

$$\Delta_{\text{dil}}X/n_B = \{(1-x^f)/x^f\}\{X_A(x^f) - X_A^*\} - \{(1-x^i)/x^i\}\{X_A(x^i) - X_A^*\}$$
$$+ \{X_B(x^f) - X_B(x^i)\}$$
$$= \Delta_{\text{mix}}X_m(x^f)/x^f - \Delta_{\text{mix}}X_m(x^i)/x^i, \tag{2.13.2}$$

where x^i and x^f denote the initial and final mole fractions of B.

As $x^f \to 0$ (infinite dilution) equation (2) becomes

$$(\Delta_{\text{dil}}X/n_B)^\infty = -\{(1-x)/x\}\{X_A(x) - X_A^*\} + \{X_B^\infty - X_B(x)\}, \tag{2.13.3}$$

where we have dropped the superscript ^{i} from x.

§ 2.14   Dissolution

The change of a quantity X when an amount of substance n_B of a solid B dissolves in an amount of substance n_A of a liquid A to give a liquid solution is given by

$$\Delta_{\text{soln}}X = X(n_A A + n_B B, l) - n_A X_A^* (l) - n_B X_B^* (s)$$
$$= n_A\{X_A(l, n_B/n_A) - X_A^* (l)\} + n_B\{X_B(l, n_B/n_A) - X_B^* (s)\}. \tag{2.14.1}$$

The quantity obtained by dividing $\Delta_{\text{soln}}X$ by the amount of substance n_B of the solute B is often referred to as the molar change of the quantity on dissolution and is given by

$$\Delta_{\text{soln}}X/n_B = \{(1-x)/x\}\{X_A(l, x) - X_A^* (l)\} + \{X_B(l, x) - X_B^* (s)\}. \tag{2.14.2}$$

As $x \to 0$ (infinite dilution) equation (2) becomes

$$(\Delta_{\text{soln}}X/n_B)^\infty = X_B^\infty (l) - X_B^* (s). \tag{2.14.3}$$

We note that

$$(\Delta_{\text{soln}}X/n_B)^\infty - (\Delta_{\text{dil}}X/n_B)^\infty = \Delta_{\text{soln}}X/n_B. \tag{2.14.4}$$

The formulae of the present section are also used when the solute is a gas.

§ 2.15   Equation for a chemical reaction

We write the equation for any chemical reaction in the general form:

$$0 = \sum_B \nu_B B, \tag{2.15.1}$$

where B denotes the chemical formula of a substance and ν_B the stoichiometric number of the substance B. To ensure that atomic theory is not violated some of the ν_B's must be positive and some negative. We call those substances with positive ν

'products' P and those with negative ν 'reactants' R. Equation (1) can then be written in the more familiar form:

$$\sum_R (-\nu_R)R = \sum_P \nu_P P . \qquad (2.15.2)$$

As an example we write the chemical equation for the reaction between nitrogen, hydrogen, and ammonia in the form:

$$0 = (+2)NH_3 + (-1)N_2 + (-3)H_2$$

$$= 2NH_3 - N_2 - 3H_2 , \qquad (2.15.3)$$

or in the more familiar form:

$$N_2 + 3H_2 = 2NH_3 . \qquad (2.15.4)$$

§ 2.16   Extent of a chemical reaction

For a chemical reaction we define a quantity ξ called the *extent of reaction* by the equation:

$$n_C(\xi) \stackrel{\text{def}}{=} n_C(0) + \nu_C \xi , \qquad (2.16.1)$$

where, for any substance C chosen from the set of substances B reacting according to the chemical equation (2.15.1), ν_C is the stoichiometric number of C, and $n_C(\xi)$ and $n_C(0)$ are the amounts of substance of C present when the extent of reaction is ξ and is 0 respectively. Since $n_C(0)$ is constant it follows that

$$dn_C = \nu_C \, d\xi , \qquad (2.16.2)$$

or for a finite change:

$$\Delta n_C = \nu_C \Delta \xi . \qquad (2.16.3)$$

Thus defined, ξ is independent of which substance C is chosen from the set of substances B. The dimension of ξ is that of amount of substance and its SI unit is the mole.

A measured enthalpy change (for example) for the reaction $0 = \Sigma_B \nu_B B$ depends of course on the change $\Delta \xi$ of the extent of reaction. The quantity $\Delta H / \Delta \xi$ is called the *molar enthalpy change for the reaction* (or more briefly the molar enthalpy of the reaction) and is denoted by ΔH_m. It follows that

$$\Delta H_m = \Delta H / \Delta \xi = \nu_C \Delta H / \Delta n_C . \qquad (2.16.4)$$

§ 2.17   Example

Suppose that as the result of a calorimetric experiment ΔH is found to be -1200 J for the reaction of 25 cm^3 of an acidified solution of oxalate having $c(C_2O_4^{2-}) = 0.16$ mol dm^{-3} (where c denotes concentration) with 20 cm^3 of a solution of permanganate having $c(MnO_4^-) = 0.08 \text{ mol dm}^{-3}$. The equation for the chemical reaction is known to be

$$C_2O_4^{2-}(aq) + \tfrac{2}{5}MnO_4^-(aq) + \tfrac{16}{5}H^+(aq)$$

$$= 2CO_2(g) + \tfrac{2}{5}Mn^{2+}(aq) + \tfrac{8}{5}H_2O(l) . \tag{2.17.1}$$

Since $\Delta n(C_2O_4^{2-}) = -25 \text{ cm}^3 \times 0.16 \text{ mol dm}^{-3} = -0.004 \text{ mol}$, or $\Delta n(MnO_4^-) = -20 \text{ cm}^3 \times 0.08 \text{ mol dm}^{-3} = -0.0016 \text{ mol}$, it follows from equation (2.16.3) that

$$\Delta\xi = \Delta n(C_2O_4^{2-})/\nu(C_2O_4^{2-}) = (-0.004 \text{ mol})/(-1) = 0.004 \text{ mol} ,$$

or

$$\Delta\xi = \Delta n(MnO_4^-)/\nu(MnO_4^-) = (-0.0016 \text{ mol})/(-\tfrac{2}{5}) = 0.004 \text{ mol} ,$$

so that $\Delta\xi$ is, as it should be, independent of which of the reacting substances we focus our attention on. We then obtain from equation (2.16.4), for the reaction specified in equation (1):

$$\Delta H_m = \Delta H/\Delta\xi = (-1200 \text{ J})/(0.004 \text{ mol}) = -300 \text{ kJ mol}^{-1} .$$

Had we chosen to specify the chemical equation in the form:

$$5C_2O_4^{2-}(aq) + 2MnO_4^-(aq) + 16H^+(aq) = 10CO_2(g) + 2Mn^{2+}(aq) + 8H_2O(l) ,$$

then we should of course have found $\Delta H_m = -300 \times 5 \text{ kJ mol}^{-1} = -1500 \text{ kJ mol}^{-1}$.

Thus 'mol' in such a context is the SI unit of the physical quantity 'extent of the specified reaction'; the 'mol^{-1}' should never be omitted.

Problems for chapter 2

Problem 2.1

You are given the following measured values of the density ρ of liquid mixtures $\{(1-x)H_2O + xCH_3OH\}$ of water + methanol at 273.15 K and atmospheric pressure.

x	$\dfrac{\rho}{\text{g cm}^{-3}}$	$\dfrac{V_m}{\text{cm}^3 \text{ mol}^{-1}}$	$\dfrac{\Delta_{mix} V_m}{\text{cm}^3 \text{ mol}^{-1}}$	x	$\dfrac{\rho}{\text{g cm}^{-3}}$	$\dfrac{V_m}{\text{cm}^3 \text{ mol}^{-1}}$	$\dfrac{\Delta_{mix} V_m}{\text{cm}^3 \text{ mol}^{-1}}$
0	0.9971	18.068	0	0.6127	0.8579		
0.1144	0.9665	20.302	−0.361	0.6924	0.8421	32.928	−0.835
0.1974	0.9479	21.926	−0.616	0.7845	0.8246	35.195	−0.656
0.2487	0.9366	22.960	−0.744	0.8923	0.8051	37.924	−0.370
0.3435	0.9153			1	0.7866	40.735	0
0.4945	0.8824	28.277	−0.999				

Values of the molar volume V_m and the molar change of volume on mixing $\Delta_{mix} V_m$ have been calculated for all except two of the measurements. (a) Calculate the missing values. (b) Plot V_m against x and determine the partial molar volumes V_A

and V_B at $x = 0.4$, where $A = H_2O$ and $B = CH_3OH$. (c) Plot $\Delta_{mix}V_m$ against x and determine $(V_A - V_B^*)$ and $(V_B - V_B^*)$, and thence V_A and V_B, at $x = 0.4$. (d) Compare the methods and results of (b) and (c).

Problem 2.2

Measurements of the molar enthalpy change on mixing $\Delta_{mix}H_m$ of liquid mixtures $\{(1-x)C_6H_{12} \text{ (cyclohexane)} + xC_6H_{14} \text{ (hexane)}\}$ at 298.15 K and atmospheric pressure can be fitted within experimental accuracy by the formula:

$$\Delta_{mix}H_m/J \text{ mol}^{-1} = x(1-x)\{864.67 + 249.51(1-2x) + 99.06(1-2x)^2$$
$$+ 33.22(1-2x)^3\}.$$

Calculate $\Delta_{mix}H_m$, $(H_A - H_A^*)$, and $(H_B - H_B^*)$, where $A = C_6H_{12}$ and $B = C_6H_{14}$, at $x = 0, 0.25, 0.5, 0.75$, and 1. Plot each of these three quantities against x on the same graph.

Problem 2.3

Prove that the equations:

$$V_A = V_m - x(\partial V_m/\partial x)_{T,p}, \quad \text{and} \quad V_B = V_m + (1-x)(\partial V_m/\partial x)_{T,p},$$

can be expressed in the forms:

$$V_A = (M_A/\rho) + \{mM_A(1 + mM_B)/\rho^2\}(\partial\rho/\partial m)_{T,p},$$

and

$$V_B = (M_B/\rho) - \{(1 + mM_B)/\rho^2\}(\partial\rho/\partial m)_{T,p},$$

where m denotes the molality of B in a solution of B in the solvent A, and ρ denotes the density of the solution.

At 298.15 K the density ρ of solutions of KCl (B) in H_2O (A) is given for $m \leqslant 1.5$ mol kg^{-1} by the formula:

$$\rho/g\,cm^{-3} = 0.9971 + 0.0472(m/m^\ominus) - 0.0031(m/m^\ominus)^2 + 0.0003(m/m^\ominus)^3,$$

where $m^\ominus = 1$ mol kg^{-1}. Calculate the partial molar volumes V_A and V_B at $m = 0$, $0.1, 0.5$, and 1.0 mol kg^{-1}. Plot $10^2(V_A - V_A^*)$ and $(V_B - V_B^\infty)$ against m on the same graph.

Chapter 3

Practical thermometry

§ 3.1 Liquid-in-glass thermometers

Mercury-in-glass thermometers are used most commonly for rough measurements in the range 235 K (the freezing temperature of mercury) to about 600 K. Their reproducibility is limited by the dependence of the volumetric properties of glass on its recent thermal history. Beckmann thermometers have a scale covering only 5 K or sometimes only 1 K but can be set by use of a mercury reservoir so that any chosen temperature gives a reading on the scale. They are especially useful for measurements of small differences of temperature or for the detection of fluctuations of temperature for example in a thermostat. Ethanol-in-glass or isopentane-in-glass thermometers have been used down to the freezing temperature of the working substance (160 K for ethanol and 113 K for isopentane).

§ 3.2 Platinum resistance thermometers

Platinum resistance thermometers are widely used as secondary standards† for the measurement of temperatures between about 15 K and about 900 K. Provided that the platinum wire is highly pure and has been well annealed, and that the thermometer is so constructed as to ensure that the wire is as nearly strain-free as possible, platinum resistance thermometers hold their calibration for long periods even if they are subjected to large temperature cycles, and can be used to measure temperatures to within 0.001 K or better. They are commonly made so as to have a resistance of about 25 Ω at room temperature, in which case the resistance varies from about 0.05 Ω at 15 K to about 80 Ω at 900 K. The sensitivity $R^{-1}\,dR/dT$ varies from about +0.15 K^{-1} at 15 K, through about +0.004 K^{-1} at 273 K, to about +0.001 K^{-1} at 900 K. The resistance becomes uncomfortably low at the lowest temperatures (and the sensitivity falls dramatically from its maximum value at about 17 K as the temperature is lowered from 17 K towards 0). The sensitivity becomes rather low at high temperatures.

In high-precision platinum resistance thermometry the resistance of the thermometer was until recently obtained by direct-current comparison with a Smith or Mueller bridge, the balance point being detected with a galvanometer. During the last decade or so alternating-current measurements of the ratio of the resistance of

† Platinum resistance thermometers are primary standards for the measurement of 'International Practical Temperatures', which have been adjusted so as to be as close as possible to thermodynamic temperatures.

the thermometer to that of a standard resistor by means of a transformer-ratio bridge have been used with some advantages, especially those of relative freedom from the 'parasitic' effects that are so difficult to eliminate in precise d.c. work, and of greater sensitivity leading to a reduction of the electric current that it is necessary to pass through the thermometer.†

Platinum resistance thermometers are also used under less exacting conditions for less precise measurements of temperature and as sensors for temperature controllers. Other metals such as nickel and copper have also been used to make resistance thermometers; they are cheaper and, at temperatures around room temperature, more sensitive than platinum resistance thermometers, but they are less reproducible after large temperature cycles and so are unsuitable for use as secondary standards.

§ 3.3   Semi-conductor thermometers

Thermistors are resistance thermometers made from semi-conducting mixtures of oxides of such metals as copper, manganese, and nickel. The variation of electric resistance R with temperature T is given approximately for thermistors by the formula: $R(T) = A \exp(B/T)$, where A and B are constants and typically $B \approx 3500$ K. Thermistors have the advantage of smallness and high sensitivity (compare $R^{-1} dR/dT = -0.045$ K^{-1} for a typical thermistor with the value $+0.004$ K^{-1} for platinum at 273 K; note also the negative dR/dT characteristic of semi-conductors). They can be obtained with any of a wide range of room-temperature resistances; those of high-resistance have the advantage that leads of relatively high electric resistance, and so of relatively high thermal resistance, can be used. The reproducibility of thermistors after large temperature cycles is not yet good enough to allow their use as secondary standards. They are nevertheless most useful as sensors for temperature controllers and for the measurement of small changes of empirical temperature, especially in calorimetry where only $\Delta\theta$ and not ΔT need be measured (see chapter 4).

Other semi-conductors have also been used as resistance thermometers. Germanium resistance thermometers made from single crystals of germanium doped with arsenic or gallium are especially useful at temperatures between about 1 K, where they have high sensitivity, and about 35 K. They can now be made reproducible enough for use as secondary standards, extending downwards (and conveniently overlapping) the useful range of platinum. Carbon resistance thermometers have been used with similar success at temperatures down to less than 1 K.

§ 3.4   Thermocouples

A thermocouple is shown diagrammatically in figure 3.1. A length of wire of a metal A (say 'constantan' – an alloy of copper and nickel with traces of manganese) is

---

† Martin, J. F. *SPRCT(I)*, chapter 4, p. 136.

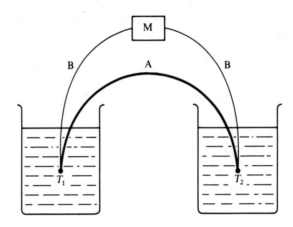

Figure 3.1. A thermocouple of metals A and B.

joined at either end (by soldering, or better by spot-welding) to a length of wire of a different metal B (say copper). The wire B is cut and the ends are attached to an electric-potential-difference meter M (such as a potentiometer + galvanometer, or in modern work an electronic 'digital voltmeter' of high input impedance). When one junction is maintained at a temperature T_1 and the other at a temperature T_2, an electromotive force E is set up which ideally depends only on T_1 and T_2. If T_1 is the constant temperature of a reference bath (such as ice + water + water vapour in a 'triple-point cell' at $T_1 \stackrel{\text{def}}{=} 273.16$ K, or ice + water saturated with air at atmospheric pressure so that $T_1 = 273.150$ K) then $E = f(T_2)$. If the thermocouple has been calibrated (and if it holds its calibration after temperature cycles) then T_2 can be calculated from the measured value of E. If N thermocouples are connected in series then $E(N) = NE(1) = Nf(T_2)$; such a multiple thermocouple is called a thermopile. The inevitably higher thermal conductance of a thermopile is a disadvantage that often outweighs the advantage of higher sensitivity.

Thermocouples of one kind or another have been used at temperatures from about 1 K to about 2600 K. For the most part they cannot be made highly reproducible, though their reproducibility can be greatly improved by careful annealing especially of the region near a junction after enclosing it in a rigid tube so as to prevent flexing. Platinum-to-(platinum + mass fraction 0.1 of rhodium) thermocouples can, however, be made highly reproducible and are used as secondary standards at temperatures from about 900 K to about 1340 K. With this exception, most thermocouples are used to measure temperature differences, especially small temperature differences. They have the advantages of smallness, of high sensitivity $E/\Delta T$, and of needing no electric current through their junctions. They are especially useful as sensors for temperature controllers and for the measurement of small changes in temperature from a base temperature in calorimetry (see §§ 1.19 to 1.23 and chapter 4), or in such measurements as the lowering of the freezing temperature of a dilute solution (see § 18.6).

The platinum-to-(platinum + rhodium) thermocouple has a relatively low sensitivity $E/\Delta T$ of about 10 μV K^{-1}; it is nevertheless widely used at high temperatures because of its stability. For the copper-to-constantan thermocouple $E/\Delta T$ varies from about 15 μV K^{-1} at 75 K through about 40 μV K^{-1} at room temperature to about 60 μV K^{-1} at 625 K; it is widely used in physicochemical experiments in that temperature range. At temperatures greater than 625 K, at which copper oxidizes rapidly in air, or when the high thermal conductivity of copper is a disadvantage, thermocouples of the 'chromel'-to-'alumel' type are often used; their sensitivity is similar to that of copper-to-constantan but they can be used up to about 1640 K. The chromel-to-constantan thermocouple combines the advantages of slightly higher sensitivity and much lower thermal conductivity than copper-to-constantan and can be used up to about 1775 K. The iron-to-constantan thermocouple with $E/\Delta T$ varying from about 26 μV K^{-1} at 85 K to about 63 μV K^{-1} at 1075 K is the thermocouple most commonly used in industrial applications.

§ 3.5 Quartz thermometers

When the resonant frequency of a quartz crystal is used as a frequency standard, the axis along which the crystal is cut is chosen so at to make the temperature dependence of the resonant frequency as small as possible and the crystal is housed in an efficient thermostat. In the quartz thermometer, that application is turned inside out: the crystal is cut so as to maximize the temperature dependence of the resonant frequency and an oscillator is used to measure the resonant frequency of the crystal, which is a nearly linear function of temperature. The instrument† can be used either with one probe to measure a temperature in the range 195 K to 525 K, or even more usefully with two probes to measure a temperature difference. The instrument can be calibrated to within +0.02 K as an absolute thermometer and so is not yet quite suitable as a secondary standard. As a difference thermometer, however, its short-term stability is better than 0.1 mK and its resolution (digital display) is 0.1 mK.

§ 3.6 Vapour-pressure thermometers

Vapour-pressure thermometers depend on the fact that the equilibrium ('vapour') pressure p^{l+g} of a system containing two phases, liquid and gas, of a pure substance depends only, and depends strongly, on the temperature T. For example $dp^{l+g}/dT \approx$ 2400 Pa K^{-1} for n-pentane at 298 K. Published high-precision measurements of $p^{l+g}(T)$, usually emanating from a standardizing laboratory, have sometimes been used to measure $T(p^{l+g})$, especially by those without access to the relatively expensive instruments needed to measure temperatures directly with high precision. A vapour-pressure cell connected to a mercury-filled manometer is rather a clumsy thermometer for general use; modern pressure transducers, however, make it possible for the vapour-pressure cell to be confined to a relatively small space, the

† Hewlett-Packard, 3200 Hillview Avenue, Palo Alto, California 94304, U.S.A.: Model 2801A.

outside of the transducer being connected through tubing filled with a 'permanent' gas to an external manometer. We shall find in § 3.8 that ^{4}He and ^{3}He vapour-pressure thermometers are used as secondary standards at temperatures between about 0.2 K and 5.2 K.

§ 3.7 Other thermometers

Magnetic measurements on paramagnetic salts are used to measure the ratio of two thermodynamic temperatures each less than about 5 K. At the other extreme, optical pyrometry (see also the following section) is used to measure the ratio of two temperatures each greater than about 1300 K.

For further details about practical thermometry the reader should consult specialized monographs.[†]

§ 3.8 The International Practical Temperature Scale

The practical difficulties associated with the direct realization of thermodynamic temperatures by gas thermometry led to the adoption in 1927 of an International Temperature Scale designed to lead to values as nearly as possible equal to the corresponding values of thermodynamic temperature. The Scale was revised in 1948 and again, extensively, in 1968. The International Practical Temperature Scale of 1968 (IPTS-68) has been so chosen that the temperatures T_{68} measured on it are identical within the bounds of the present accuracy of measurement with the corresponding values of the thermodynamic temperature T.

The IPTS-68 is based on values of T_{68} assigned to the temperatures of ten examples ('fixed points'), chosen for their reproducibility, of an equilibrium between two or three phases of a pure substance, together with the value 273.16 K assigned by definition (of the kelvin) to the triple-point temperature of water, and on standard instruments calibrated at those temperatures. The equilibria, the assigned values of T_{68}, and the corresponding values of $(T - T_{68})$ are given in table 3.1.

At temperatures between any two 'fixed points' the IPTS-68 is defined by specified interpolation formulae which establish the relation between T_{68} and the reading of a thermometer of specified kind and quality. The defining formulae contain adjustable parameters, the values of which are determined from the results of measurements made with the specified thermometers at 'fixed points'.

At temperatures between 273.15 K and 903.89 K (the normal melting temperature of antimony), the specified thermometer is a platinum resistance thermometer of specified quality, and the specified formula contains three adjustable parameters the values of which are determined from values of the resistance of the thermometer measured at the triple-point temperature of water, the normal boiling temperature of water (or the normal melting temperature of tin; see footnote c to table 3.1), and

† Sturtevant, J. M. *Techniques of Chemistry*, Volume 1, *Physical Methods of Chemistry*, Weissberger, A.; Rossiter, B. W.: Editors. Wiley-Interscience: New York. **1971**, Part V, chapter I, pp. 1–22.
Stimson, H. F.; Lovejoy, D. R.; Clement, J. R. *ETd(I)*, chapter 2, pp. 15–57.

Table 3.1. Defined temperatures T_{68} of the International Practical Temperature Scale of 1968 (IPTS-68).

Equilibrium	T_{68}/K	$(T - T_{68})/\mathrm{K}$
$^1H_2(s) + {}^1H_2(l) + {}^1H_2(g)$ [a]	13.81	0.00 ± 0.01
$^1H_2(l, 33.3306\ \mathrm{kPa}) + {}^1H_2(g, 33.3306\ \mathrm{kPa})$ [a]	17.042	0.00 ± 0.01
$^1H_2(l, 101.325\ \mathrm{kPa}) + {}^1H_2(g, 101.325\ \mathrm{kPa})$ [a]	20.28	0.00 ± 0.01
$Ne(l, 101.325\ \mathrm{kPa}) + Ne(g, 101.325\ \mathrm{kPa})$	27.102	0.00 ± 0.01
$O_2(s) + O_2(l) + O_2(g)$	54.361	0.00 ± 0.01
$O_2(l, 101.325\ \mathrm{kPa}) + O_2(g, 101.325\ \mathrm{kPa})$	90.188	0.00 ± 0.01
$H_2O(s) + H_2O(l) + H_2O(g)$ [b]	273.16	0 (exactly)
$H_2O(l, 101.325\ \mathrm{kPa}) + H_2O(g, 101.325\ \mathrm{kPa})$ [b,c]	373.15	0.000 ± 0.005
$Sn(s, 101.325\ \mathrm{kPa}) + Sn(l, 101.325\ \mathrm{kPa})$ [c]	505.1181	0.000 ± 0.015
$Zn(s, 101.325\ \mathrm{kPa}) + Zn(l, 101.325\ \mathrm{kPa})$	692.73	0.00 ± 0.03
$Ag(s, 101.325\ \mathrm{kPa}) + Ag(l, 101.325\ \mathrm{kPa})$	1235.08	0.0 ± 0.2
$Au(s, 101.325\ \mathrm{kPa}) + Au(l, 101.325\ \mathrm{kPa})$	1337.58	0.0 ± 0.2

[a] For equilibrium-hydrogen consisting at these temperatures of nearly pure para-hydrogen.
[b] The water used should have the nuclidic composition of ocean water.
[c] These may be used as alternatives.

the normal melting temperature of zinc. At temperatures between 13.81 K and 273.15 K the specified thermometer is the same high-quality platinum resistance thermometer, but values of T_{68} are obtained from a tabulated reference function $W_{\mathrm{CCT-68}}(T_{68})$ used with the appropriate specified deviation formula with parameters determined by measurements at the appropriate fixed points. At temperatures between 903.89 K and 1337.58 K (the normal melting temperature of gold) the standard thermometer is a platinum-to-(platinum + mass fraction 0.1 of rhodium) thermocouple of specified quality, and the specified formula contains three adjustable parameters the values of which are determined from values of the e.m.f. of the thermocouple at the normal melting temperatures of antimony,[†] silver, and gold. At temperatures above 1337.58 K the standard thermometer is an optical pyrometer calibrated at the normal melting temperature of gold and used in conjunction with Planck's radiation formula, derivable from quantal statistical thermodynamics, with a specified value for the second radiation constant: $c_2 = hc/k = 0.014388$ m K.

The IPTS-68 is not defined at temperatures below 13.81 K. The '1958 ^{4}He Scale' recommended by the International Committee of Weights and Measures is defined by a table of vapour pressures of ^{4}He at temperatures up to 5.2 K. The '1962 ^{3}He scale' recommended by the International Committee of Weights and Measures is defined by a table of vapour pressures of ^{3}He at temperatures up to 3.3 K.

Further details of the IPTS-68 can be found in official publications.[‡]

[†] This secondary fixed point (903.89 K) is used to pass from the platinum resistance thermometer to the thermocouple.

[‡] Échelle Internationale Pratique de Température de 1968. *Comptes Rendus de la 13me Conférence Générale des Poids et Mesures* **1967–68**, Annex 2. *Metrologia*, **1969**, 5, 35.

The International Practical Temperature Scale of 1968 (English version of the Official French Text), HMSO: London. **1969**.

Rossini, F. D. *J. Chem. Thermodynamics* **1970**, 2, 447. *See also* Douglas, T. B. *J. Res. Natl. Bur. Std.* **1969**, 73A, 451.

Chapter 4

Practical calorimetry

§ 4.1 Methods of coping with heat leaks

When it is inconvenient either to construct an adiabatic calorimeter (§ 1.20), or for an isothermal-jacket calorimeter to follow the thermogram until the temperature has returned to that of the surroundings (§ 1.21), methods must be devised for coping with the heat leaks Q. All such methods depend in practice on the assumption that the calorimeter follows Newton's law of cooling:

$$Q = -k \int (T - T_e) \, dt, \qquad (4.1.1)$$

where t denotes time, T the temperature of the outer surface of the calorimeter and T_e that of its thermal environment, and k a constant. The temperature T has been supposed uniform over the outer surface of the calorimeter; ensuring that this is so is a most important consideration in the design of a calorimeter. Newton's law of cooling can easily and should always be tested for any isothermal-jacket calorimeter by comparison of the areas of thermograms resulting from experiments in which different measured quantities of work are done on the calorimeter at different rates.

Once the constant k has been determined for a calorimeter, the heat leak Q can be obtained from a plot of temperature against time for any experiment made with it. Such a thermogram is nowadays often recorded automatically on a chart recorder or in digital form on tape. The quantity plotted as ordinate on a thermogram is often the e.m.f. of a thermocouple or the electric resistance of a resistance thermometer rather than the thermodynamic temperature itself. Provided that the quantity plotted is an accurately linear function of thermodynamic temperature, it is unnecessary to convert the ordinate to thermodynamic temperature since only ratios of the areas under thermograms are needed.

In some kinds of calorimeter continuous stirring is used throughout the whole experiment, or a constant electric power is dissipated by a resistance thermometer throughout the whole experiment, and consequently the temperature does not return to T_e, but to a higher value T_c called the convergence temperature. For example a 'bomb calorimeter' most commonly consists of a 'bomb' immersed in a can of stirred water, the whole being separated by an air-filled or evacuated space from its thermal environment. The function of the stirred water is to ensure that the temperature of the outside surface of the calorimeter is always uniform; if the bomb itself is used as the calorimeter the outside surface is likely to develop 'hot spots' during a chemical reaction such as a fast combustion.† For such a calorimeter with an 'additional' constant power input dW'/dt from continuous stirring or from any other source, we have using equation (1) the relation:

$$d(W' + Q)/dt = dW'/dt - k(T - T_e).$$ (4.1.2)

When such a calorimeter has reached a steady state $d(W' + Q)/dt = 0$, so that

$$dW'/dt = k(T_c - T_e),$$ (4.1.3)

where T_c is the steady-state or convergence temperature. Substitution of (3) into (2) yields

$$d(W' + Q)/dt = -k(T - T_c),$$ (4.1.4)

which on integration gives

$$W' + Q = -k \int (T - T_c) \, dt.$$ (4.1.5)

For such a calorimeter, therefore, equation (1) remains valid when Q is replaced by $(W' + Q)$ and T_e is replaced by T_c, the areas $\mathscr{A}$ now being those obtained from thermograms with $(T - T_c)$ as ordinate in place of $(T - T_e)$, and ΔH now being determined at the temperature T_c rather than at T_e.

It is hard to understand why the method implied in § 1.21 has not been more widely used. Whereas for any realistic time-scale that method implies a relatively large value of k, most thermochemists have preferred to put a good deal of effort into making k as small as possible, usually by evacuating a gap between the calorimeter and its constant-temperature environment and by silvering both walls of the gap, and then to correct for heat leaks as will be described below without waiting for the temperature of the calorimeter to return to that of the steady state.

The method implied in § 1.21 has nevertheless been used, especially in 'heat-flow' or 'conduction' calorimeters and especially for the calorimetry of slow processes. In a typical heat-flow calorimeter a multi-junction thermocouple, with one set of junctions distributed over the surface of the calorimeter vessel and the other set connected to a massive copper block, acts both as a heat-conductor and as a sensitive differential thermometer. Futher details can be found in the literature.‡

---

† Nevertheless 'aneroid' bomb calorimeters consisting of only the bomb itself, separated from its isothermal environment by an evacuated space, have been successfully used. See for example Meetham, A. R.; Nicholls, J. A. *Proc. Roy. Soc. A* **1960**, 256, 384; Adams, G. P.; Carson, A. S.; Laye, P. G. *Trans. Faraday Soc.* **1969**, 65, 113.

‡ Sturtevant, J. M. *Techniques of Chemistry*, Volume 1, *Physical Methods of Chemistry*, Weissberger, A.; Rossiter, B. W.: Editors. Wiley-Interscience: New York. **1971**, Part V, chapter VII, pp. 349–425.

Calvet, E.; Prat, H. *Récents Progrès in Microcalorimétrie*. Dunod: Paris. **1958**; *Recent Progress in Microcalorimetry*, Skinner, H. A.: Translator. Pergamon: London. **1963**.

Calvet, E. *ETc*(*I*), chapter 12, pp. 237–286; *ETc*(*II*), chapter 17, pp. 385–410.

Skinner, H. A. *Biochemical Microcalorimetry*, Brown, H. D.: Editor. Academic Press: New York. **1969**, p. 7.

§ 4.2   Methods of coping with heat leaks in 'small k' calorimetry

In 'small k' calorimetry the thermogram, analogous to that shown in figure 1.5 for 'large k' calorimetry, is like that shown in figure 4.1. Point 1 relates to a time t_1

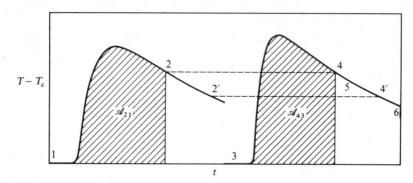

Figure 4.1. Thermogram for the two stages of an experiment in an isothermal-jacket calorimeter with a relatively low value of k.

before the initiation of a chemical reaction, and point 3 to a time t_3 before the switching on of electric power. Point 2 relates to a time t_2 appreciably after the reaction has stopped, and point 4 to a time t_4 at which the temperature T_4 is identical to the temperature T_2 and appreciably after the electric power has been switched off; at the times t_2 and t_4 the temperature is and has been drifting towards the steady-state temperature T_c in response only to the temperature T_e of the environment. The experiment can always be arranged so that the pair of points 2 and 4 can be chosen so as to satisfy these criteria. If there is any doubt about the completeness of the effect of the reaction or of the effect of passage of electric current on the temperature of the calorimeter, a pair of later points 2′ and 4′ can be chosen; the same result should be obtained for $\Delta H(T)$. We have, for the first stage:

$$H_2 - H_1 = H(T_2, p, \xi_f) - H(T_c, p, \xi_i) = -k\mathscr{A}_{21} , \qquad (4.2.1)$$

and for the second stage:

$$H_4 - H_3 = H(T_4, p, \xi_f) - H(T_c, p, \xi_f) = W^{el}_{43} - k\mathscr{A}_{43} , \qquad (4.2.2)$$

so that in view of the identity of T_2 and T_4

$$\Delta H(T_c, p) = H(T_c, p, \xi_f) - H(T_c, p, \xi_i) = -W^{el}_{43} + k(\mathscr{A}_{43} - \mathscr{A}_{21}) . \qquad (4.2.3)$$

The required enthalpy change $\Delta H(T_c, p)$ can be evaluated if the value of k has been found for the calorimeter, for example by allowing the temperature T in the second stage to drift right back to T_c and using the relation:

$$W^{el}_{\infty 3} = k\mathscr{A}_{\infty 3} , \qquad (4.2.4)$$

or by using the relation:

$$k = C(t_6 - t_5)^{-1} \ln\{(T_5 - T_c)/(T_6 - T_c)\}, \qquad (4.2.5)$$

where the subscripts 5 and 6 relate to points such as those marked in figure 4.1, and C is the heat capacity of (calorimeter + contents) obtained from separate measurements or calculated as the sum of the heat capacities of the component parts. The smaller the value of k for the calorimeter, and the smaller the difference between the areas $\mathscr{A}_{21}$ and $\mathscr{A}_{43}$ for the experiment, the less accurately need k be known. No difficulties arise if each stage of the experiment is begun at a temperature other than T_c, provided that the times t_1 and t_3 are chosen so as to be earlier than the beginning of the reaction and the beginning of the electrical working respectively, so that $T_1 = T_3$, and provided that it is recognised that ΔH is then measured at the temperature $T_1 = T_3$ and not at T_c. The method just described for an exothermic reaction can also be used for an endothermic reaction if a measured quantity W_{21}^{el} of electrical work, sufficient to ensure that $T_2 > T_c$, is done during the first stage.

The method used above to analyse the results of experiments on 'small k' isothermal-jacket calorimeters is not, however, widely used. Most thermochemists prefer to estimate from the results at each stage what the temperature change would have been if k had been zero, that is to say if the calorimeter had been adiabatic. For the experiment described in figure 4.1 we then have

$$\{T_2(\text{corr.}) - T_1\} = \{T_2(\text{obs.}) - T_1\} + (k/C) \int_{t_1}^{t_2} (T - T_c)\, dt, \qquad (4.2.6)$$

for the first stage, and

$$\{T_4(\text{corr.}) - T_3\} = \{T_4(\text{obs.}) - T_3\} + (k/C) \int_{t_3}^{t_4} (T - T_c)\, dt, \qquad (4.2.7)$$

for the second stage. In equations (6) and (7) we have assumed that the heat capacity C of (calorimeter + contents) is constant over the relevant temperature range and independent of the extent of reaction. We have also lost the means precisely to determine the temperature at which ΔH was measured. The best we can do is to write

$$\Delta H\{T \approx (T_1 + T_2)/2, p\} = -W_{43}^{el}\{T_2(\text{corr.}) - T_1\}/\{T_4(\text{corr.}) - T_3\}. \qquad (4.2.8)$$

Further details of methods of coping with heat leaks, often based on the approximation that the heat capacity C is constant, can be found in the literature.[†] Here we shall describe only Dickinson's method.

[†] Challoner, A. R.; Gundry, H. A.; Meetham, A. R. *Phil. Trans. Roy. Soc. London A* **1955**, 247, 553.
Coops, J.; Jessup, R. S.; van Nes, K. *ETc(I)*, chapter 3.
Good, W. D.; Scott, D. W.; Waddington, G. *J. Phys. Chem.* **1956**, 60, 1080.
Stout, J. W. *ETd(I)*, chapter 6, pp. 232–251.
Oetting, F. L. *J. Chem. Thermodynamics* **1970**, 2, 727.
Gunn, S. R. *J. Chem. Thermodynamics* **1971**, 3, 19.

§ 4.3   Dickinson's method of coping with heat leaks

This method[†] of calculating $(T_2 - T_1)$(corr.) from $(T_2 - T_1)$(obs.) also depends on the assumption that

$$(T_2 - T_1)(\text{corr.}) = (T_2 - T_1)(\text{obs.}) + (k/C) \int_{t_1}^{t_2} (T - T_\text{e}) \, dt, \qquad (4.3.1)$$

where T_1 and T_2 are the temperatures at times t_1 and t_2 in the 'fore' and 'after' parts of a calorimetric experiment when the temperature is drifting towards the steady-state temperature T_c in response only to the temperature T_e of the environment. A time t_D between t_1 and t_2 is chosen so that

$$\int_{t_1}^{t_\text{D}} (T - T_1) \, dt + \int_{t_\text{D}}^{t_2} (T - T_2) \, dt = 0 . \qquad (4.3.2)$$

It is then easily proved that the correction term in equation (1) can be written in the form:

$$(k/C) \int_{t_1}^{t_2} (T - T_\text{c}) \, dt = -\{ (dT/dt)_{t_1}(t_\text{D} - t_1) + (dT/dt)_{t_2}(t_2 - t_\text{D}) \} . \qquad (4.3.3)$$

According to equation (2) the time t_D is chosen so that the horizontally and vertically shaded areas in figure 4.2(a) are equal. The correction term is then evaluated as the

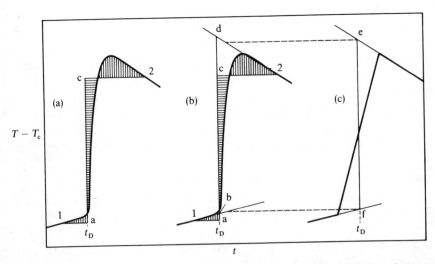

Figure 4.2. Dickinson's method of correcting for heat leaks in an isothermal-jacket calorimeter.

[†] Dickinson, H. C. *Bull. Natl. Bur. Stand. (U.S.)* **1944**, 11, 189.

right-hand side of (3) from values of the slopes $(dT/dt)_{t_1}$ and $(dT/dt)_{t_2}$ of the thermogram at times t_1 and t_2. Dickinson's method, summarized by equations (2) and (3), is rather tedious but is widely used. Its main advantage over the method summarized by equation (4.2.3) seems to be the avoidance of any need to evaluate k for the calorimeter.

Dickinson's method takes a somewhat simpler form if the 'fore' and 'after' parts of the thermogram can plausibly be replaced by straight lines as in figure 4.2(b). The right-hand side of equation (3) may then be written in the form:

$$-(T_b - T_a) + (T_d - T_c) = -(T_b - T_1) + (T_d - T_2)$$
$$= (T_d - T_b) - (T_2 - T_1), \qquad (4.3.4)$$

so that

$$(T_2 - T_1)(\text{corr.}) = (T_d - T_b). \qquad (4.3.5)$$

The corrected temperature change is thus obtained from the intercepts at $t = t_D$ of the linearly extrapolated 'fore' and 'after' parts of the thermogram, the time t_D being chosen according to Dickinson's criterion expressed by equation (2).

Dickinson's method takes its simplest form for the highly idealized thermogram shown in figure 4.2(c) in which the 'middle' part of the thermogram is also replaced by a straight line. The Dickinson time t_D is then equal to the 'mid-time' $(t_1 + t_2)/2$ and no tedious matching of areas is needed. For more realistic thermograms, however, the 'mid-time' method is often a poor approximation, as can be seen by study of figure 4.2(b) where it would lead to an error $\{(T_d - T_b) - (T_e - T_f)\}$ comparable with the difference $\{(T_d - T_b) - (T_2 - T_1)\}$ that it was being used to estimate.

§ 4.4   Bunsen calorimetry

Sometimes a calorimeter includes an outer annular space filled with a mixture of the liquid and solid phases (or more rarely liquid + gas) of a pure substance, carefully maintained at equilibrium, and arranged so that any volume change displaces mercury in a capillary tube. If the calorimeter is separated by an evacuated space from another annular space containing the same equilibrium mixture, or from any other thermostat at the same temperature, then $Q = 0$. The two parts of a calorimetric experiment are then described by equations (1.20.1) to (1.20.3) or (1.20.4) to (1.20.6) with the approximation signs replaced by equality signs and the empirical temperatures replaced by the volumes of the two-phase mixture. In Bunsen's calorimeter (ice + water) at 273.15 K was used. In other calorimeters of this kind use has been made of (solid + liquid) diphenyl ether (300.02 K) or other substances with suitable melting temperatures. Care must be taken to ensure equilibrium between the solid and liquid phases, usually by freezing a solid sheath over the calorimeter vessel and then partly melting it from the inside so as to produce a free mantle of the solid separated from the calorimeter vessel by a thin layer of the liquid phase. Any calorimeter of this kind can of course be used at only one temperature.

§ 4.5 Measurements of heat capacities

For solids (and for dense fluids of low volatility) measurements of C_p are commonly made from about room temperature down to temperatures as low as 4 K, or from about room temperature up to 1000 K or more, in calorimeters with adiabatic shields. More specialized designs are used from about 20 K down to temperatures as low as 0.05 K. A typical adiabatic heat-capacity calorimeter is shown in figure 4.3. Calorimeters with isothermal shields, in which corrections for heat leaks must be made as outlined in §§ 4.1 to 4.3, are usually used from room temperature down to about 4 K. At temperatures of 800 K and upwards even to about 3000 K 'drop calorimeters' are usually used.

In a drop calorimeter the sample is suspended in a vertically mounted tube furnace which determines its original temperature T, and is then dropped into a calorimeter at a (usually lower) temperature T_i. If $W^{el} = W' = Q = 0$ then the final temperature T_f of the calorimeter is determined by the relation:

$$\{H(T_f) - H(T_i)\}_{\text{calorimeter}} = \{H(T) - H(T_f)\}_{\text{sample}} . \tag{4.5.1}$$

If electrical work W^{el} is then needed to raise the temperature of the (calorimeter + sample) from T_i to T_f we have

$$\{H(T_f) - H(T_i)\}_{\text{calorimeter}} + \{H(T_f) - H(T_i)\}_{\text{sample}} = W^{el} . \tag{4.5.2}$$

Substitution from (1) into (2) then gives

$$\{H(T) - H(T_i)\}_{\text{sample}} = W^{el} . \tag{4.5.3}$$

Any necessary corrections for heat leaks or for 'additional' work are made according to the methods described in §§ 4.1 to 4.3.

There is an unavoidable space saturated with vapour in the container of a solid (or often of a fluid) sample.† The measured heat capacity is thus C_{sat}, the heat capacity of the (solid + gas) at the equilibrium vapour pressure $p^{s+g}(T)$. The correction is given by

$$C_p - C_{\text{sat}} = -n^g\{C_{p,m}(g) - C_{p,m}(s)\} - n[V_m(s) - T\{\partial V_m(s)/\partial T\}_p]\Delta_s^g H_m / T\Delta_s^g V_m$$
$$- n^g\{1 - T(\partial \ln \Delta_s^g V_m/\partial T)_p\}\Delta_s^g H_m / T - \Delta_s^g H_m(\partial n^g/\partial T)_{\text{sat}} , \tag{4.5.4}$$

where n denotes the total amount of substance of the sample and n^g the amount of substance present in the gas. We shall learn how to derive equation (4) in § 9.7. In the meantime we note that for $n^g \ll n$ all the correction terms can usually be estimated except the last which demands a knowledge of the volume of the calorimeter and its temperature coefficient.

For fluids having appreciable vapour pressures any vapour space must be avoided. The degassed fluid can be introduced into the container with no vapour space by

---

† Helium at about 5 to 10 kPa is usually added to the container to raise the thermal conductance. In most cases it does not interact with the sample and so may be regarded as part of the container. No helium is necessary if the vapour pressure of the sample is greater than about 10 kPa, but then the corrections under discussion must be made.

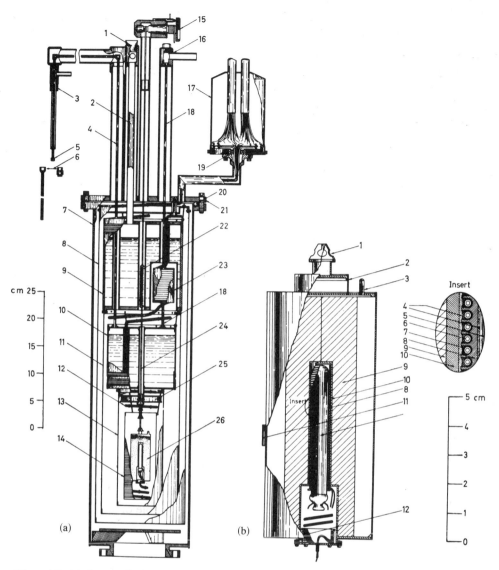

Figure 4.3. (a), Cryostat for low-temperature adiabatic calorimetry, and (b), cross-sectional diagram of a calorimeter for solid samples. (a): 1, Liquid-nitrogen inlet and outlet connector; 2, liquid-nitrogen filling tube; 3, sleeve fitting to liquid-helium transport Dewar; 4, liquid-helium transfer tube; 5, screw fitting at the inlet of the liquid-helium transfer tube; 6, liquid-helium transfer-tube extender and cap; 7, brass vacuum jacket; 8, outer 'floating' radiation shield; 9, liquid-nitrogen tank; 10, liquid-helium tank; 11, nitrogen radiation shield; 12, bundle of lead wire; 13, helium radiation shield; 14, adiabatic shield; 15, windlass; 16, helium-exit connector; 17, copper shield for terminal block; 18, helium-exit tube; 19, vacuum seal and terminal plate for leads; 20, O-ring gasket; 21, cover plate; 22, coil spring; 23, 'economizer' (exchanger for effluent helium vapour); 24, supporting braided silk line; 25, 'floating' ring; 26, calorimeter assembly. (b): 1, Thermal-contact cone; 2, monel cupola; 3, solder-capped monel tube for helium seal-off; 4, grease for thermal contact; 5, capsule-type platinum resistance thermometer; 6, fibre-glass insulated 0.08 mm Advance wire; 7, 'Formvar' enamel; 8, gold-plated copper heater core; 9, copper vanes; 10, gold-plated copper heater sleeve; 11, differential-thermocouple sleeve; 12, spool for thermal equilibration of the leads with the calorimeter.

(Reproduced with permission from Westrum, E. F., Jr.; Furukawa, G. T.; McCullough, J. P. *ETd(I)*, chapter 5, pp. 138 and 153.)

displacement of mercury and can be kept there in the absence of vapour by applying sufficient pressure to the mercury. If a container is completely filled with fluid then (supposing the volume of the container to have a negligibly small temperature coefficient) C_V would be measured; this is feasible only for fluids of high compressibility. For gases at pressures near atmospheric or below, the heat capacity of the sample is too small a fraction of that of the (sample + container). The only practical choice is then the measurement of C_p by flow calorimetry (§ 1.23), which is convenient and much used for fluids in general.

Further details of heat-capacity measurements can be found in the literature.†

§ 4.6 Measurements of enthalpies of transition

When the temperature of a pure substance is raised through a phase transition, whether between two crystalline forms, between solid and liquid, or between liquid and gas, the enthalpy is increased while the temperature remains constant so long as the two phases coexist. The temperature can rise again only when the whole of the low-temperature form has been converted into the high-temperature form. The enthalpy change at the transition temperature is called the enthalpy of transition (or of melting or of evaporation) and is denoted by $\Delta_\alpha^\beta H$ (or $\Delta_s^l H$ or $\Delta_l^g H$).

Enthalpies of transition between two crystalline forms and enthalpies of melting are often measured in the same experiments in the same apparatus as that discussed in the previous section in connexion with heat capacities. Indeed if those experiments are correctly thought of as measurements of enthalpy increments with temperature, an enthalpy of transition appears simply as a vertical step during the measurements. Occasionally special calorimeters are designed for measurements of enthalpies of melting. Because of the great change of volume, however, special calorimeters are usually needed for measurements of enthalpies of evaporation. These are sometimes combined with flow calorimeters for the measurement of the heat capacities at constant pressure of vapours; an example of such a calorimeter is shown in figure 4.4.‡

As an example of the kind of results obtained from measurements made in calorimeters like those described in §§ 4.5 and 4.6 we have plotted in figure 4.5 the enthalpy as a function of temperature at constant pressure for CO, and certain

† Westrum, E. F., Jr.; Furukawa, G. T.; McCullough, J. P. *ETd(I)*, chapter 5, pp. 133–214.
Stout, J. W. *ETd(I)*, chapter 6, pp. 215–261.
Hill, R. W.; Martin, D. L.; Osborne, D. W. *ETd(I)*, chapter 7, pp. 263–292.
Douglas, T. B.; King, E. G. *ETd(I)*, chapter 8, 293–331.
West, E. D.; Westrum, E. F., Jr. *ETd(I)*, chapter 9, pp. 333–367.
McCullough, J. P.; Waddington, G. *ETd(I)*, chapter 10, pp. 369–394.
Cruickshank, A. J. B.; Ackermann, Th.; Giguère, P. A. *ETd(I)*, chapter 12, pp. 421–535.
Martin, J. F. *SPRCT(I)*, chapter 4, pp. 133–161.
‡ McCullough, J. P.; Waddington, G. *ETd(I)*, chapter 10, pp. 369–394.
Ginnings, D. C.; Stimson, H. F. *ETd(I)*, chapter 11, pp. 395–420.
Counsell, J. F. *SPRCT(I)*, chapter 6, pp. 204–217.

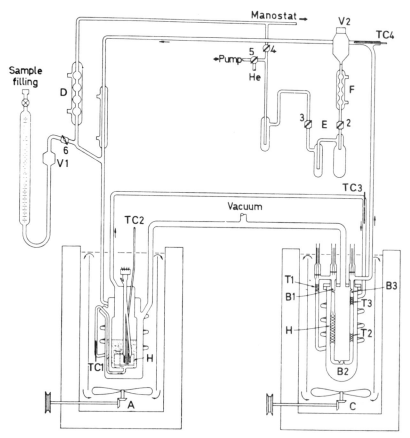

Figure 4.4. A calorimeter for the measurement of enthalpies of evaporation (left-hand part) and of the heat capacities at constant pressure of vapours (right-hand part). A. Thermostat for calorimeter to measure $\Delta_l^g H$; C, thermostat for flow calorimeter to measure $C_p(g)$; D, F, condensers; E, sample receiver; 2, 3, sample-receiver valves; 4, 5, 6, valves; V1, V2, solenoid-operated valves; T1, T2, T3, platinum resistance thermometers (not shown: a thermometer in each of the calorimeter vessels); H, heaters; TC1, TC2, TC3, TC4, thermocouples; B1, B2, B3, indentations to baffle the flow of vapour.

(Reproduced with permission from McCullough, J. P.; Waddington, G. *ETd(I)*, chapter 10, p. 373.)

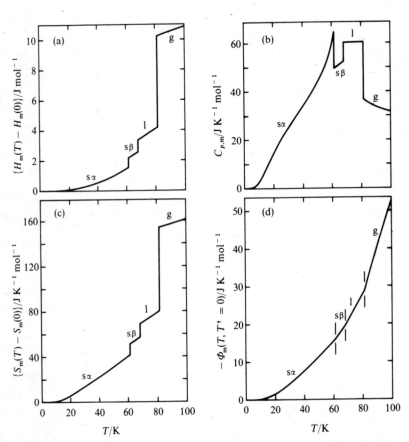

Figure 4.5. The molar enthalpy of CO from $T = 0$ to $T = 100$ K at a constant pressure of 101.325 kPa, and certain derived quantities, plotted against temperature T. (a), Molar-enthalpy increments $\{H_m(T) - H_m(0)\}$; (b), molar heat capacity at constant pressure $C_{p,m} = (\partial H_m/\partial T)_p$; (c), the integral $\int_{H_m(0)}^{H_m(T)} T^{-1} \, dH_m = \int_0^T (C_{p,m}/T) \, dT$, which we shall later learn to recognize as the molar entropy increment $\{S_m(T) - S_m(0)\}$; and (d), the molar Giaque function $\Phi_m(T, T^\dagger = 0) \overset{\text{def}}{=} -\{H_m(T) - H_m(T^\dagger = 0)\}/T + \{S_m(T) - S_m(0)\}$. The experimental gap from $T = 0$ to the lowest temperature $T = 14.36$ K at which measurements have been made was filled by Debye's formula $C_{p,m} = AT^3$ for crystalline solids at very low temperatures.

derived quantities. Of those, the heat capacity has already been dealt with, the entropy will be discussed in chapter 6, and the Giauque function in chapter 11.†

§ 4.7   Measurements of energies or enthalpies of chemical reaction

A value of the energy change $\Delta_r U$ for a chemical reaction can easily be converted to the corresponding value of the enthalpy change $\Delta_r H$ for the chemical reaction, and *vice versa*, by evaluation of the change $\Delta_r(pV)$.‡ Only the more convenient of $\Delta_r U$ and $\Delta_r H$ needs to be measured. Happily it is not necessary to measure one of these quantities for every chemical reaction; it is sufficient to measure one of them for every *independent* chemical reaction.§

    The quantity usually tabulated is the standard molar enthalpy (change) of *formation* $\Delta_f H_m^{\ominus}$ of each substance at 298.15 K. Values of $\Delta_f H_m^{\ominus}(T)$ at temperatures T other than 298.15 K may be found from tabulated values of the standard molar enthalpy increment $\{H_m^{\ominus}(T) - H_m^{\ominus}(298.15\ K)\}$ measured as described in § 4.5 for each participating substance. The conventional meaning of the word 'standard' and its symbol $^{\ominus}$ will be introduced in chapter 11 and explained for gases in chapter 12, for liquids and solids in chapter 16, and for solutes and for the solvent in solutions in chapter 18; in the meantime we note first that standard thermodynamic functions will be defined so that each depends only on temperature, and second that the 'Washburn' correction (see § 12.10) of $\Delta_r U$ or $\Delta_r H$ to $\Delta_r H^{\ominus}$, though important, is usually small.‖ The reaction described as 'formation' is by convention that in which the substance is formed from its elements each in the form most stable¶ at the

---

† The molar Giauque function is often written as "$-\{G_m(T) - H_m(T^{\dagger})\}/T$" which would be identical with

$$\Phi_m(T, T^{\dagger}) \overset{\text{def}}{=} -\{H_m(T) - H_m(T^{\dagger})\}/T + \{S_m(T) - S_m(0)\} = -\{G_m(T) - H_m(T^{\dagger})\}/T - S_m(0),$$

only if $S_m(0)$ were zero. We shall find, as we did for enthalpies, that it is only differences of entropy that can be measured. We thus cannot know the value of $S_m(0)$. It is true, however, that the algebraic sum $\Sigma_B \nu_B S_B(0)$ for a chemical reaction $0 = \Sigma_B \nu_B B$ can be measured, and that it is nearly always equal to zero. It is believed that it would always be zero if the solids on which the measurements were made were always perfectly ordered crystals as $T \to 0$; that is called *Nernst's heat theorem*. If Nernst's heat theorem were always obeyed then it would not matter if $S_m(0)$ were omitted from the right-hand side of the displayed equation, though this would not imply that $S_m(0) = 0$. It does, however, matter for any reaction involving CO. It also matters for reactions involving some other molecules like NNO, NO, H_2O, and $Na_2SO_4 \cdot 10H_2O$. This subject will be further discussed in chapter 15.

‡ For reactions not involving gases $\Delta_r(pV) = \Sigma_B (n_B^f p^f V_B^f - n_B^i p^i V_B^i)$, where n_B denotes the amount of substance of B, p the pressure, and V_B the partial molar volume of B, is always small and often negligible; for reactions involving gases $\Delta_r(pV) \approx \xi RT \Sigma_{B(g)} \nu_{B(g)}$ where ξ denotes extent of reaction and where the stoichiometric numbers $\nu_{B(g)}$ are counted in the summation only for gaseous substances. Extent of reaction was defined in § 2.16.

§ For example, only two of the three reactions $C + \frac{1}{2}O_2 = CO$, $C + O_2 = CO_2$, and $CO + \frac{1}{2}O_2 = CO_2$ are independent; the third is equal to the second minus the first. The saving is immense. Compare one experiment for each substance with one experiment for each reaction.

‖ The corresponding differences between $\Delta_r G$ and $\Delta_r G^{\ominus}$ and between $\Delta_r S$ and $\Delta_r S^{\ominus}$ are *not* usually small.

¶ Exceptionally, the state chosen for phosphorus is the white form, which has been better characterized than the more stable red form.

temperature in question and atmospheric pressure. For example $\Delta_f H_m^\ominus$ ($C_6H_5CO_2H$, s, 298.15 K) is shorthand for the standard molar enthalpy change at 298.15 K for the reaction:

$$7C(graphite) + 3H_2(g) + O_2(g) = C_6H_5CO_2H(s) \,. \tag{4.7.1}$$

Values of $\Delta_f H_m^\ominus$ are obtained from values of $\Delta_r H_m^\ominus$ for reactions that can be carried out in calorimeters. For example $\Delta_f H_m^\ominus$($C_6H_5CO_2H$, s, 298.15 K) cannot be measured directly, but can be found algebraically from values of $\Delta_r H_m^\ominus$(298.15 K) for the reactions:

$$C_6H_5CO_2H(s) + \tfrac{15}{2}O_2(g) = 7CO_2(g) + 3H_2O(l) \,, \tag{4.7.2}$$

$$C(graphite) + O_2(g) = CO_2(g) \,, \tag{4.7.3}$$

and

$$H_2(g) + \tfrac{1}{2}O_2(g) = H_2O(l) \,, \tag{4.7.4}$$

each of which can be carried out completely and unambiguously in a calorimeter. The algebraic manipulation is said to be according to Hess's law, though a separate name is hardly needed for such an obvious application of the first law. It is nevertheless worth remarking that Hess's law applies rigorously and generally only to standard enthalpy changes; the enthalpy changes actually measured always depend, at least to a small extent, on composition and on pressure, as well as on temperature.

Any chemical reaction studied calorimetrically must proceed according to a known chemical equation, and the change of the extent of reaction (see § 2.16) must be accurately known or measured, by ensuring the purity of the starting materials and usually by chemical analysis of the contents of the calorimeter after the reaction. The accuracy with which enthalpies of reaction are determined often depends more on factors like those than on calorimetric accuracy. Much chemical ingenuity has been put into finding reactions that will, or can be made to, go cleanly in a calorimeter, and to the devising of conditions that will ensure that the enthalpies of phase transition, of dissolution, of mixing, and of compression, needed to make the Washburn correction from ΔU or ΔH to $\Delta H^\ominus$ are sufficiently accessible.

§ 4.8 Measurements of energies or enthalpies of combustion

The most nearly general kind of reaction that can usefully be carried out in a calorimeter is combustion in oxygen, as in reactions (4.7.2) to (4.7.4). The energy of combustion $\Delta_c U$ or enthalpy of combustion $\Delta_c H$ can be corrected to the standard molar enthalpy of combustion $\Delta_c H_m^\ominus$ (see § 12.10). Values of $\Delta_c H_m^\ominus$ must be obtained very accurately because the required values of $\Delta_f H_m^\ominus$ are obtained from usually small differences between large values of $\Delta_c H_m^\ominus$. For example: $\Delta_c H_m^\ominus$(C_6H_6, l, 298.15 K) = -3267 kJ mol^{-1} while $\Delta_f H_m^\ominus$(C_6H_6, l, 298.15 K) = $(3267 - 6 \times 393.5 - 3 \times 285.8)$ kJ mol^{-1} = $+49$ kJ mol^{-1}.

When the combustion is carried out in a bomb calorimeter it leads, since the volume is constant, to a value of the energy change $\Delta_c U$. The requirement of high accuracy can be met in spite of the inevitably high total heat capacity of a bomb calorimeter that must be strong enough to withstand pressures of 10 MPa or so. For example the combustion of 1 g of benzoic acid in a bomb calorimeter of total heat capacity $16000\ \text{J K}^{-1}$ (equivalent to about 10 kg of steel + 2.7 kg of water) gives a temperature rise of 1.7 K. A typical bomb is shown in figure 4.6. The bomb is either

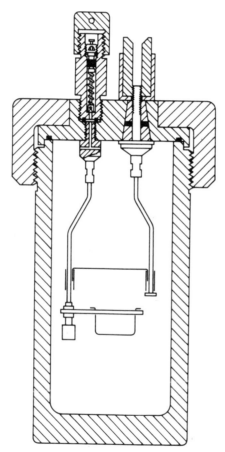

Figure 4.6. A typical combustion bomb.
(Reproduced with permission from A. Gallenkamp & Co. Ltd.)

immersed in a can of stirred water, as shown in modern form in figure 4.7, or is suspended in an evacuated jacket (when it is called 'aneroid'; see the footnote on p. 49). A small volume of water is usually added to the bomb to ensure that any solution is dilute enough to facilitate the Washburn correction. Such a 'static' bomb calorimeter can be used for the combustion of substances containing only C, H, and

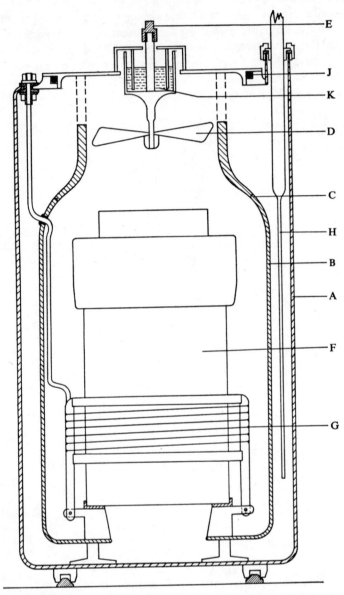

Figure 4.7. A complete bomb calorimeter. A, Rhodium-plated copper can; B, cylindrical stainless-steel shield (a poor thermal conductor), constructed in two parts; C, water-tight seal between the two parts of B; D, stirrer; E, bush; F, bomb; G, electrical heater; H, platinum resistance thermometer; J, neoprene O-ring seal; K, oil-seal.

(Reproduced with permission from Gundry, H. A.; Harrop, D.; Head, A. J.; Lewis, J. B. *J. Chem. Thermodynamics* **1969**, 1, 321.)

O (giving as products CO_2 and H_2O, usually in two phases: a gaseous mixture of O_2, CO_2, and H_2O, and a liquid phase of H_2O saturated with O_2 and CO_2; the absence of CO and of C (soot) should be verified or if present allowed for), or of substances containing only C, H, O, and N (in which most of the N appears in the products as N_2 but some as aqueous HNO_3 and sometimes HNO_2; the amounts of HNO_3 and of any HNO_2 must be estimated chemically), or of metals (burnt without added water) that form only one oxide, or of some carbides, nitrides, lower oxides, and a few organometallic substances such as tin alkyls. Encapsulation of the sample, in a glass ampoule or in a plastic bag, is usually necessary to prevent evaporation or adsorption of water or premature reaction with oxygen. An excess of purified oxygen is used, commonly at an initial pressure of about 3 MPa. The sample is fired by passage of an electric current through a platinum wire suspended close to the sample, often with the help of a cotton or polyethene 'fuse'. The work needed to ignite the sample and the energy of combustion of any fuse must of course be known.

It is much less expensive and, except in the most expert hands, more accurate to forego electrical calibration and to make bomb-calorimetric measurements relative to others made on benzoic acid in the same calorimeter. Benzoic acid can be obtained certified for this purpose. The energy of combustion of this benzoic acid is very accurately known; three determinations in quite differently designed calorimeters gave for the specific energy of combustion $\Delta_c u$ under 'standard bomb conditions' the values:[†]

	$-\Delta_c u/\text{J g}^{-1}$
Churney and Armstrong	26434.4 ± 3.3
Mosselmann and Dekker	26432.7 ± 1.6
Gundry, Harrop, Head, and Lewis	26434.4 ± 1.8

For sufficiently volatile substances, including gases, flame calorimetry can be used instead of bomb calorimetry. In a flame calorimeter streams of the gaseous sample (if necessary mixed with argon and pre-mixed with oxygen) and oxygen are burnt at a jet in a combustion vessel like that shown in figure 4.8. The combustion vessel is immersed in a can of stirred water, the whole constituting the calorimeter, much as for a static bomb calorimeter. The pressure is usually close to atmospheric (which makes the Washburn corrections easier) and constant so that the quantity measured is $\Delta_c H$ rather than $\Delta_c U$. The gaseous products are passed through a spiral so as to bring them to thermal equilibrium with the calorimeter, and are then removed for analysis and determination of the extent of reaction.

For other substances a moving-bomb calorimeter, such as that shown in figure 4.9, has great advantages. By continuous rotation end-over-end and at the same time about the cylindrical axis of the bomb, and usually by the addition of rather more water to the bomb, the walls of the bomb are thoroughly washed by the solution of combustion products, which can then be analysed after the experiment with

[†] Churney, K. L.; Armstrong, G. T. *J. Res. Nat. Bur. Stand. (U.S.) A* **1968**, 72, 453.
Mosselman, C.; Dekker, H. *Rec. Trav. Chim. Pays Bas* **1969**, 88, 161.
Gundry, H. A.; Harrop, D.; Head, A. J.; Lewis, G. B. *J. Chem. Thermodynamics* **1969**, 1, 321.

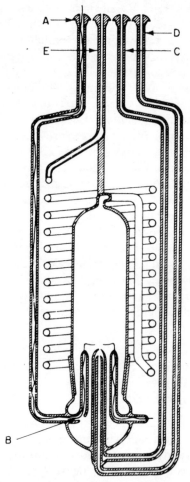

Figure 4.8. Combustion vessel for flame calorimetry. The main supply of oxygen enters at A and is directed through orifice B to the bottom of the combustion chamber. Pre-mixing oxygen enters at C, and the substance, carried if necessary in a stream of argon, enters at D. The mixture is ignited by a spark across the 4 mm gap between two platinum wires placed about 1 mm above the jet. The exit gases leave the chamber at its top and pass through a heat-exchange spiral before leaving the vessel at E.

(Reproduced with permission from Pilcher, G.; Skinner, H. A.; Pell, A. S.; Pope, A. E. *Trans. Faraday Soc.* **1963**, 59, 316.)

confidence that it is uniform and in equilibrium with the gaseous mixture in the bomb. Substances containing S, Cl or Br, I, F, B or Si, and P can then be studied, the products being respectively aqueous H_2SO_4 (in the presence of a small amount of N_2 or of $CO(NH_2)_2$ as auxiliary substance), aqueous HCl or HBr (in the presence of a reducing agent such as As_2O_3), I_2 (also possible in static-bomb calorimetry), aqueous HF (if the atomic ratio $n(F)/n(H) < 1$; for highly fluorinated substances CF_4 is also produced; such substances must be burnt with an auxiliary substance such as a

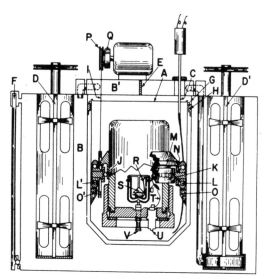

Figure 4.9. A moving-bomb calorimeter. A, Calorimeter; B,B', constant-temperature jacket; C, thermometer; D,D', stirrers; E, partition to direct the circulation of water in the jacket lid B'; F, escape for excess water; G, three pegs resting in grooves in plastic blocks H; I, rotatory-mechanism drive wire; J, ball bearing mounted in yoke K which can rotate in ball bearings L,L', resting in grooves in copper blocks O,O'; M, N, gears causing the bomb to rotate about its cylindrical axis; P, pulley; Q, clutch plate.
(Reproduced with permission from Waddington, G.; Sunner, S.; Hubbard, W. N. *ETc(I)*, chapter 7, p. 156.)

hydrocarbon of high molar mass), aqueous (HBF_4 and HF) or aqueous (H_2SiF_6 and HF) (in the presence of a fluorine-containing substance such as polyvinylidene fluoride together with aqueous HF), and aqueous H_3PO_4. Various chemical tricks are similarly used to convert the products of combustion of organometallic substances to thermochemically known metal-containing solutions.

Conventional bomb calorimeters demand samples of mass 1 to 2 g, much too large to allow the determination of enthalpies of combustion of many of the most interesting products of the synthetic organic chemist. Miniature bomb calorimeters suitable for burning as little as 10 mg of material have recently given accuracies comparable with those obtainable with conventional bombs.[†]

Combustion in fluorine, instead of oxygen, has opened many pathways to accurate enthalpies of formation, though it can be regarded as safe only in the most highly specialized laboratories. A conventional bomb is usually used, though it is made of materials (like nickel) inert towards fluorine and is divided by a breakable partition into two compartments initially to separate the fluorine and the sample; most substances react spontaneously with fluorine to give CF_4, HF, O_2, N_2, BF_3, SiF_4, PF_5, SF_6, or for metals the highest fluoride. Fluorine has also been used in place of oxygen in flame calorimetry.

[†] See for example Månsson, M. *J. Chem. Thermodynamics* **1973**, 5, 721.

Combustions in Cl_2, NF_3, H_2, and NH_3 also have important though more limited applications.

Full accounts of combustion calorimetry can be found in the literature.† We conclude this section with two quotations, the first by Head (from p. 101 of the first reference below): "Indeed, elaborate calorimetric methods are superfluous unless the criteria of purity of sample and completeness of combustion have been met.", and the second by Good and Scott (from p. 71 of the third reference below): "Control of the combustion reaction is more of an art than a science. The more tricks the thermochemist has up his sleeve, and the more he has the ingenuity to devise, the better he is equipped to deal with a fickle and perverse Nature."

§ 4.9   Measurements of enthalpies of reactions other than combustion

For reactions other than combustion the enthalpy changes $\Delta_r H$ are commonly measured in calorimeters at constant (often atmospheric) pressure surrounded as far as possible by a highly evacuated space, the whole being immersed in a precise thermostat. The purities of samples, the avoidance of premature reaction, the avoidance of side reactions (including those with the material of the calorimeter vessel and with air remaining in the calorimeter), the determination of the extent of reaction, and the effects of evaporation consequent on breaking an ampoule or on reaction, are often more important than purely calorimetric technique. The wide variety of reactions, having molar enthalpies of reaction of either sign and varying from a few J mol^{-1} or less to many kJ mol^{-1}, leads to a wide variety of calorimeters.

The result of a detailed study of the best design for a generally useful and accurate 'reaction' calorimeter is shown in figure 4.10, the calorimeter having been designed for reasonably fast reactions (complete in less than 1 h) with a measured $|\Delta H|$ of about 200 J or less, to hold about 100 cm^3, and to give a reproducibility of 0.1 per cent or 0.02 J whichever is greater. A thin-walled (0.5 mm) open can machined from one piece of brass was silver-soldered to a brass lid. The bottom had a central rod to support an ampoule-breaking pin. The lid was provided with two thin-walled tubes, one for a thermistor and the other for an electric heater, and had a central wide opening with two thin brass rings 0.7 mm apart on its outside. The interior of the calorimeter vessel was chromium plated and then coated with polytetrafluoroethene (PTFE). The vessel hung from the lid of the surrounding brass jacket by a thin-walled (0.7 mm) glass tube which fitted into the groove between the two brass rings on the lid of the vessel and into two matching brass rings on the bottom of the jacket lid, and was sealed into place with epoxy resin. The jacket could then be evacuated to high vacuum. A snugly fitting PTFE-coated brass gasket was introduced round the stirrer shaft so as further to isolate the contents of the calorimeter from the jacket. The combined ampoule holder and stirrer was made from PTFE-coated stainless steel and

---

† Head, A. J. *SPRCT*(*I*), chapter 3, pp. 100–118.
ETc(*I*), chapters 4 to 10.
ETc(*II*), chapters 1 to 7.
ECTd(*I*). (The whole of this volume is devoted to combustion calorimetry.)

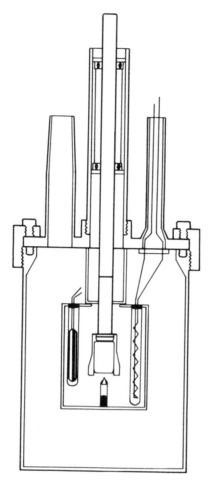

Figure 4.10. A carefully designed 'reaction' calorimeter.
(Reproduced with permission from Sunner, S.; Wadsö, I. *Acta Chem. Scand.* **1959**, 13, 97.)

fastened to a PTFE rod, the upper end of which could be screwed to the steel spindle of the stirrer motor. The whole stirrer was carefully centred, and could be unscrewed from the lid of the jacket. The electrical connexions were passed through the jacket by metal-to-glass seals soldered to its lid. The resulting calorimeter had great advantages over conventionally modified Dewar flasks and the like, mainly in the short time (2 min) needed for its thermal equilibration (a Dewar flask, though specially designed, was found still to need 60 min) and in its much better reproducibility of $|\Delta H|$.

In the calorimetric study of hydrogenations, and of other reactions involving gases, a catalyst must often be present, and both any incoming and outgoing gases must be brought thoroughly to thermal equilibrium with the calorimeter by passage through heat-exchange coils within the calorimeter. Outgoing gas will be more or less saturated with the vapour of any volatile liquid from the calorimeter; a correction must be made for the enthalpy of evaporation. Better accuracy can usually be obtained by containing the gas throughout in a closed 'bomb-type' calorimeter.

For slow reactions the calorimeter can be surrounded by an adiabatic shield (see p. 19), or use can be made of a Bunsen isothermal phase-change calorimeter (see § 4.4), or of a 'heat-flow' or 'conduction' calorimeter (see p. 49).

Accounts can be found† of all these kinds of 'reaction' calorimeter, and of others such as flow calorimeters used for the measurement of enthalpies of reaction between fluids, microcalorimeters for the measurement of values of $|\Delta H|$ less than about 1 J, calorimeters for the measurement of enthalpies of adsorption of gases on solids or of liquids on solids (wetting), 'hot-zone' and 'explosion' calorimeters, calorimeters specially designed for the study of reactions between metals or refractory materials or fused salts at high temperatures, and commercially available 'differential scanning calorimeters' useful especially for preliminary studies of phase transitions in solids or of thermal decompositions.

§ 4.10 Measurements of enthalpies of dissolution and of dilution

Since dissolution of a solid in a liquid is merely a particular example of a chemical reaction not much more need be said, except to point to two problems that cause greater (or more obvious) errors than one meets in the calorimetry of reactions in general.

The first problem arises from the 'cold spots' (or 'hot spots' for a substance that dissolves exothermically) that come from the difficulty of stirring sufficiently vigorously two phases of widely different densities. A closed thin-walled 'bomb-type' calorimeter of the kind shown in figure 4.11 can be rocked backwards and forwards through angles of π or more and so has the advantage of securing that uniformity of the temperature of the outside surface of the calorimeter without which corrections for heat leaks cannot be made accurately. Indeed, the advantages of the rocking calorimeter are by no means confined to the measurement of enthalpies of dissolution.

The second problem arises from the condensation of solvent from the vapour in response to the dissolution of a solid in a liquid in the presence of a vapour space (or of evaporation on opening an ampoule). When an involatile solute is dissolved in a volatile solvent in the presence of a vapour space of volume V^g the error due to condensation of solvent is given approximately by

$$\delta \Delta H \approx -(2xp^* V^g / RT)\Delta_l^g H_m^* , \qquad (4.10.1)$$

where x denotes the mole fraction of the solute after dissolution, p^* the vapour pressure of the pure solvent, and $\Delta_l^g H_m^*$ the molar enthalpy of evaporation of the

---

† Head, A. J. *SPRCT(I)*, chapter 3, pp. 118–132.
ETc(I), chapters 11 to 13.
ETc(II), chapters 8 to 13 and 16 to 19.

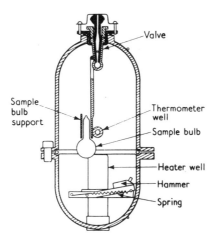

Figure 4.11. A rocking closed calorimeter suitable for measurements of enthalpies of dissolution, and of other enthalpies of reaction.
(Reproduced with permission from Gunn, S. R. *Rev. Sci. Instr.* **1958**, 29, 377.)

pure solvent. For example, for 0.25 mol of KCl in 50 g of H_2O with $V^g = 100 \text{ cm}^3$, $\delta\Delta H$, though only about -1 J in about 3800 J at 300 K, is about -12 J at 355 K.

Enthalpies of dilution can be obtained from differences of enthalpies of dissolution, but can also be obtained more accurately by direct measurements.

Accounts of the calorimetry of dissolution and dilution can be found in the literature.†

§ 4.11   Measurements of enthalpies of mixing

The mixing of two liquids is merely another kind of chemical reaction and so again not much more need be said, except that the two problems discussed for enthalpies of dissolution arise again. For two reasons the second problem is now much more acute: the magnitude of the enthalpy of mixing is likely to be much smaller and the correction for a vapour space is likely to be much larger. There has also been more interest in measurements of enthalpies of mixing over wide ranges of temperature; at temperatures approaching or even greater than the normal boiling temperatures of the liquids, especially if one liquid is much more volatile than the other, the correction for even a minute vapour space becomes large and sometimes greater than the enthalpy of mixing itself. No modern enthalpy-of-mixing calorimeter allows any vapour space whatever, and the liquids must be degassed lest they form gas bubbles in the calorimeter especially when its temperature is raised. An example of a suitable calorimeter is shown in figure 4.12. Accounts of other calorimeters can be found,‡

† *ETc(II)*, chapters 9 and 14.
‡ McGlashan, M. L. *ETc(II)*, chapter 15.
Marsh, K. N. *SPRCT(II)*, chapter 1, pp. 28–38.

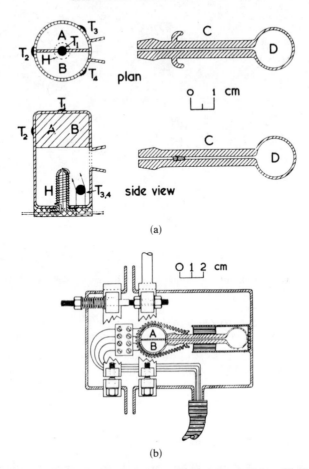

Figure 4.12. A calorimeter suitable for measurements of enthalpies of mixing. The liquids A and B are separately confined by mercury and mixed by rotation of the calorimeter. The side arm C, partly filled with mercury, allows expansion or contraction against an air space D, never in contact with the liquids, so that the mixing is at virtually constant pressure. (a), Plan and side view of the calorimeter: T, thermistors; H, electrical heater. (b), The calorimeter in its evacuable jacket arranged so that it can be rotated backwards and forwards through an angle of π clockwise (so as to keep the liquids away from the side arm) about a vertical axis in the plane of the diagram.

(Reproduced with permission from Larkin, J. A.; McGlashan, M. L. *J. Chem. Soc.* **1961**, 3425.)

including accounts of 'continuous-dilution' calorimeters in which many compositions can be studied in one experiment.

The correction for gas bubbles now usually (see the first of the references) has the approximate form:

$$\delta\Delta H \approx 10\{(1-x)V_i^{g,B} - xV_i^{g,A}\}(p_A^* - p_B^*),\qquad(4.11.1)$$

where x is the mole fraction of B in the liquid mixture, $V_i^{g,A}$ and $V_i^{g,B}$ are the volumes of vapour spaces present initially over the liquid A and over the liquid B, p_A^*

and p_B^* are the vapour pressures of pure A and of pure B, and we have used Trouton's rule: $\Delta_l^g H_m^* / RT^b \approx 10$ where T^b denotes the normal boiling temperature (see § 13.4).

As an example we consider the sign and magnitude of the molar enthalpy of mixing of $(0.5C_6H_{14} + 0.5C_{16}H_{34})$ at 373.15 K, about which at one time there was controversy. One group argued that it 'should' be about -100 J mol^{-1} and the other that it 'should' be about $+10$ J mol^{-1}. Each group measured it, in the absence of vapour spaces, and each found that it was about -100 J mol^{-1}.[†] Now suppose that a vapour space with a volume up to 0.25 cm^3 had been allowed over the hexane or over the hexadecane. For this mixture at 373.15 K equation (1) becomes

$$\delta \Delta H \approx 1.2\{(V_i^{g,B} - V_i^{g,A})/\text{cm}^3\} \text{ J} , \qquad (4.11.2)$$

so that $-0.3 < \delta \Delta H/\text{J} < +0.3$. The total amount of substance in the calorimeter was about 0.003 mol, so that $-100 < \delta \Delta H_m/\text{J mol}^{-1} < +100$ where $\delta \Delta H_m$ is the error in the molar enthalpy of mixing. Vapour spaces of up to 0.25 cm^3 could have led to either of the predicted results!

[†] A wager of a "jolly good dinner" made on the sign of this quantity was handsomely settled in 1963 at the restaurant 'Au Jardin des Gourmets', 5 Greek Street, London W1. Those present were Th. Holleman and J. Hijmans, of the Royal Dutch Shell Laboratory in Amsterdam, the author, and his wife.

Chapter 5

Thermodynamics of a phase

§ 5.1 Introduction and scope

In chapters 1, 3, and 4 we provided ourselves, in addition to all the mechanical and electrical instruments needed to measure work, and in addition to the instruments of chemical analysis, with two new characteristically thermodynamic kinds of measuring instruments or meters, namely thermometers which are used to measure temperature, and calorimeters which are used to measure differences of energy or enthalpy. In chapter 10 we shall find a third new characteristically thermodynamic kind of measuring instrument, namely chemical potentiometers,† which are used to measure differences of chemical potential (or ratios of absolute activity). Thermodynamic equations inter-relate the quantities that can be measured with these instruments.

In this chapter we shall declare as an axiom‡ an equation for any infinitesimal change in the energy of a phase. This chapter and the next one will then be concerned entirely with the properties of a single phase. We remind the reader that the characteristic of a phase is that all the intensive properties are uniform throughout the phase. Having in the next section introduced our axiom in the form of an equation containing some new quantities, we shall devote the rest of this chapter to the introduction of some auxiliary quantities and to the algebraic manipulation of the equations involving those quantities. Those auxiliary quantities are introduced only for convenience and contain nothing new; what is new is contained in the axiom. In the following chapter we shall explore the consequences of our new axiom (and of the three previously declared) for a phase of fixed composition.

§ 5.2 The fundamental equation for a change of the state of a phase

For any infinitesimal change in the state of a phase α we write

$$dU^\alpha = T^\alpha \, dS^\alpha - p^\alpha \, dV^\alpha + \sum_B \mu_B^\alpha \, dn_B^\alpha . \tag{5.2.1}$$

We regard equation (1) as an axiom and call it the fundamental equation for a change

---

† This name has not been widely used by other authors.

‡ An axiom is one of the 'rules of the game'. We remind the reader that we have already declared three axioms: the '−1th law' in § 1.10, the zeroth law in § 1.13, and the first law in § 1.18. The present axiom is one of two, together called the *second law of thermodynamics*. The other axiom of the second law, an inequality, will be declared in chapter 7. Our axiomatic foundations will then be complete and we shall then be able to play the whole of the game according to the rules, except for the third law of thermodynamics which has rather a different status. The third law will be discussed in chapter 15.

of the state of a phase α. It is one half of the second law of thermodynamics. We do not ask where it comes from. Indeed we do not admit the existence of any more fundamental relations from which it might have been derived. Nor shall we here enquire into the history of its formulation, though that is a subject of great interest to the historian of science. It is a starting point; it must be learnt by heart. It may be allowed to stand as an axiom until any single one of the host of equations that can be derived from it (with the help of the other axioms of thermodynamics) has been shown experimentally to be false. It is a synthesis into a single equation of this host of equations, none of which has in fact ever been shown to be false.

We must now discuss in turn each of the quantities that occurs in equation (1).

On the left-hand side dU^α is any infinitesimal change in the energy U^α of the phase α. We have already learnt in chapter 1 how to use a calorimeter to measure any energy difference:

$$\Delta U = U_2 - U_1 = \int_1^2 dU, \qquad (5.2.2)$$

and so need say no more here about dU^α.

In the first term on the right-hand side of equation (1) T^α is a positive intensive quantity called the *thermodynamic temperature* of the phase α, and S^α is an extensive quantity called the *entropy* of the phase α.

With the help of the other axiom of the second law we shall show in chapter 7 that the quantity T^α defined by equation (1) is a temperature; in the meantime we shall assume that it is. In chapter 6 we shall devise a recipe for the measurement of the ratio T_2/T_1 of any two thermodynamic temperatures and shall verify the conclusion, anticipated in § 1.16, that the thermodynamic temperature defined by equation (1) is proportional to the perfect-gas temperature. Accepting for the time being that this is so, then we have already learnt in § 1.16 one way of measuring the *ratio* of two such thermodynamic temperatures by use of a perfect-gas thermometer.

About the entropy S^α we as yet know nothing. Here we shall say only that we shall find, as we did for energy, that it is only differences of entropy that can be measured, and so it is only such differences that are of any physical significance.

In the second term of equation (1) p^α is the pressure of the phase α and V^α is the volume of the phase α. The reader was warned in chapter 1 that it would be taken for granted that he understands how to measure such mechanical quantities as pressure and volume, and so we shall say no more about them. Before leaving the second term, however, we shall in the next two paragraphs explore two aspects of its relation to the work W done on the phase α by its surroundings. For this purpose we shall assume that the reader remembers what was said about work in § 1.9.

In equation (1) we have assumed that the only kind of work done on the phase is pressure-volume work. That will be the case in most, though not all, of our applications of thermodynamics to chemistry. If other kinds of work are involved then we have merely to include the appropriate terms $x\,dX$ along with $-p\,dV$. For example if a change of the height of a phase has an appreciable effect on its properties then the term $-p\,dV$ must be replaced by $(-p\,dV + mg\,dh)$ where m is the mass of the phase, g is the local acceleration of free fall, and h is the height of the phase above some arbitrary zero. Again, if a change of the surface area $\mathscr{A}$ of a fluid phase has an appreciable effect on its properties then the term $-p\,dV$ must be replaced, as it will be in chapter 21, by $(-p\,dV + \gamma\,d\mathscr{A})$ where γ is the surface tension.

Returning now to situations where the only kind of work done on the phase α is pressure-volume work, we note that the second term, $-p^{\alpha}\,dV^{\alpha}$, of equation (1) is *not necessarily* equal to the work W done on the phase α by its surroundings. As was explained in § 1.9 the work W done on the phase α is, in the absence of friction,

$$W = -p_{e}^{\alpha}\,dV^{\alpha}, \tag{5.2.3}$$

where p_{e}^{α} is the *external* pressure acting on the phase α while the p^{α} of equation (1) is the pressure *of* the phase α. It is only when $p_{e}^{\alpha} = p^{\alpha}$ that we have

$$-p^{\alpha}\,dV^{\alpha} = W, \quad (p_{e}^{\alpha} = p^{\alpha}). \tag{5.2.4}$$

In that case we describe the volume change as *reversible*; if p_{e}^{α} is infinitesimally greater than p^{α} then the phase α will contract ($dV^{\alpha} < 0$), while if p_{e}^{α} is infinitesimally less than p^{α} then the phase α will expand ($dV^{\alpha} > 0$), so that the direction of change can be reversed by an infinitesimal change in p_{e}^{α}. In this book the word 'reversible' will be used only with the meaning expressed by equation (4).

In the third term of equation (1) μ_{B}^{α} is an intensive quantity called the *chemical potential* of the substance B in the phase α and n_{B}^{α} is the amount of substance of B in the phase. As far as the n's are concerned we merely recall the definition given in § 2.2 and that we can measure them by chemical analysis or synthesis of the phase. About the μ's we as yet know nothing. Here we shall say only that we shall find that it is only differences of chemical potential at the same temperature that can be measured, and so it is only such differences that are of any physical significance.

We have thus left ourselves with three tasks that must be completed before we can make full use of equation (1). These are (i) to prove that the T^{α} of equation (1) is after all identical to the perfect-gas temperature defined in § 1.16, (ii) to find out how to measure differences of entropy consequent on changes of temperature, pressure, and composition, and (iii) to find out how to measure differences of the chemical potential of a substance consequent on changes of pressure and composition. These three tasks will be completed in the course of the next five chapters, some of them with the help of the other axiom of the second law that we shall introduce in chapter 7.

In the rest of the present chapter we shall introduce some auxiliary quantities, but we emphasise that these are introduced only for convenience and contain nothing new; all that is new is contained in equation (1).

§ 5.3   Helmholtz function and Gibbs function

We have already seen in chapter 1 how the quantity $(U^{\alpha} + p^{\alpha}V^{\alpha})$ cropped up so naturally in connexion with constant-pressure calorimetry that it was convenient to give it a symbol and a name of its own. We denoted it by H^α and called it the enthalpy:

$$H^\alpha \overset{\text{def}}{=} U^\alpha + p^\alpha V^\alpha .$$ (5.3.1)

Similarly it turns out that the quantities $(U^\alpha - T^\alpha S^\alpha)$ and $(U^\alpha + p^\alpha V^\alpha - T^\alpha S^\alpha)$ crop up so naturally that it is convenient to give them each a symbol and a name. We accordingly make the definitions:

$$A^\alpha \overset{\text{def}}{=} U^\alpha - T^\alpha S^\alpha ,$$ (5.3.2)

$$G^\alpha \overset{\text{def}}{=} U^\alpha + p^\alpha V^\alpha - T^\alpha S^\alpha ,$$ (5.3.3)

and call A^α the *Helmholtz function* of the phase α and G^α the *Gibbs function* of the phase α.[†] These definitions have of course to be learnt by heart; it might be helpful to remember that U stands in the same relation to A:

$$U = A + TS ,$$ (5.3.4)

as H does to G:

$$H = G + TS .$$ (5.3.5)

By substitution for U^α from (1) or (2) or (3) into our fundamental equation (5.2.1) for any change in the state of a phase α we obtain

$$dU^\alpha = T^\alpha \, dS^\alpha - p^\alpha \, dV^\alpha + \sum_B \mu_B^\alpha \, dn_B^\alpha ,$$ (5.3.6)

$$dH^\alpha = T^\alpha \, dS^\alpha + V^\alpha \, dp^\alpha + \sum_B \mu_B^\alpha \, dn_B^\alpha ,$$ (5.3.7)

$$dA^\alpha = -S^\alpha \, dT^\alpha - p^\alpha \, dV^\alpha + \sum_B \mu_B^\alpha \, dn_B^\alpha ,$$ (5.3.8)

$$dG^\alpha = -S^\alpha \, dT^\alpha + V^\alpha \, dp^\alpha + \sum_B \mu_B^\alpha \, dn_B^\alpha ,$$ (5.3.9)

where for the sake of completeness we have begun by repeating equation (5.2.1) itself. The reader should verify for himself that equations (7) to (9) do indeed follow from equation (6) by use of the definitions of H, A, and G.

The four equations (6) to (9) are equivalent, so that we may regard any of them as the fundamental equation for a phase. Sometimes it is convenient to start with one of them and sometimes with another. The energy U may be described as the *characteristic function* for the set of independent variables S, V, and the n's. Similarly H is the characteristic function for the variables S, p, and the n's; A for T, V, and the n's;

† The names *Helmholtz free energy* for A and *Gibbs free energy* for G have often been used by other authors.

and G for T, p, and the n's. In physical chemistry by far the most generally useful of these sets of independent variables is the set: T, p, and the n's. To chemists therefore the Gibbs function G is an especially important quantity. The energy U and the enthalpy H are, as we have seen in chapter 1, the quantities measured in calorimetric experiments and derive their importance from that. The set of variables T, V, and the n's is not in general a convenient one for experimentalists. In this book much less use will be made of the Helmholtz function A than of the Gibbs function G.

Recalling the definition of a partial molar quantity in § 2.7 we discover from equation (9) that the partial molar Gibbs function is identical with the chemical potential:†

$$G_B^\alpha \stackrel{\text{def}}{=} (\partial G^\alpha / \partial n_B^\alpha)_{T^\alpha, p^\alpha, n_{A \neq B}^\alpha} = \mu_B^\alpha . \tag{5.3.10}$$

§ 5.4 Integrated forms of the fundamental equations

According to Euler's theorem (see appendix II) and in view of the fact that T, p, and the μ's are intensive quantities, while S, V, and the n's, as well as U, H, A, and G, are extensive quantities, we may integrate equations (5.3.6) to (5.3.9) and obtain

$$U^\alpha = T^\alpha S^\alpha - p^\alpha V^\alpha + \sum_B n_B^\alpha \mu_B^\alpha , \tag{5.4.1}$$

$$H^\alpha = T^\alpha S^\alpha + \sum_B n_B^\alpha \mu_B^\alpha , \tag{5.4.2}$$

$$A^\alpha = -p^\alpha V^\alpha + \sum_B n_B^\alpha \mu_B^\alpha , \tag{5.4.3}$$

$$G^\alpha = \sum_B n_B^\alpha \mu_B^\alpha . \tag{5.4.4}$$

The reader should verify that equations (1) to (4) are consistent with the definitions (5.3.1) to (5.3.3) of H, A, and G. In view of the identification of the partial molar Gibbs function with the chemical potential in equation (5.3.10) we note that equation (4) is a special case of equation (2.7.4).

§ 5.5 Gibbs–Duhem equation for a phase

Differentiation of equation (5.4.4) yields

$$dG^\alpha = \sum_B \mu_B^\alpha \, dn_B^\alpha + \sum_B n_B^\alpha \, d\mu_B^\alpha . \tag{5.5.1}$$

---

† Some authors still prefer to use the symbol G_B (or more usually $\bar{G}_B$) and the name 'partial molar Gibbs free energy' rather than the symbol μ_B and the name chemical potential.

Equation (5.3.9) is

$$dG^\alpha = -S^\alpha \, dT^\alpha + V^\alpha \, dp^\alpha + \sum_B \mu_B^\alpha \, dn_B^\alpha .$$
(5.5.2)

By subtraction of (2) from (1) we obtain

$$0 = S^\alpha \, dT^\alpha - V^\alpha \, dp^\alpha + \sum_B n_B^\alpha \, d\mu_B^\alpha ,$$
(5.5.3)

which is called the *Gibbs–Duhem equation* for the phase α. We now see that the number 0 is the characteristic thermodynamic function for the set of independent variables: T, p, and the μ's, all of which are intensive quantities. The Gibbs–Duhem equation is a fifth fundamental equation for a phase and we shall make a good deal of use of it. We note that it is a special case of equation (2.7.8).

§ 5.6   Absolute activity

It turns out to be convenient to introduce yet another auxiliary quantity λ_B^α called the *absolute activity*† of a substance B in the phase α and defined by

$$\lambda_B^\alpha \overset{\text{def}}{=} \exp(\mu_B^\alpha / RT^\alpha) ,$$
(5.6.1)

or of course

$$\mu_B^\alpha \overset{\text{def}}{=} RT^\alpha \ln \lambda_B^\alpha .$$
(5.6.2)

The Gibbs–Duhem equation (5.5.3) for a phase α can be cast into another convenient form by substitution of $RT^\alpha \ln \lambda_B^\alpha$ for μ_B^α and use of (5.4.4) and (5.3.5) when we obtain

$$0 = (H/T)^\alpha \, dT^\alpha - V^\alpha \, dp^\alpha + RT^\alpha \sum_B n_B^\alpha \, d \ln \lambda_B^\alpha .$$
(5.6.3)

The derivation of equation (3) is left as an exercise for the reader.

§ 5.7   Gibbs–Duhem equation for unit amount of substance of a phase

When we divide the Gibbs–Duhem equation (5.5.3) for a phase α by the total amount of substance $\sum_B n_B^\alpha$ in the phase we obtain the particularly convenient form:

$$0 = S_m^\alpha \, dT^\alpha - V_m^\alpha \, dp^\alpha + \sum_B x_B^\alpha \, d\mu_B^\alpha ,$$
(5.7.1)

where x_B^α denotes the mole fraction of the substance B in the phase α.

---

† The quantity λ_B would no doubt be called simply 'activity' were it not that other authors have used the name 'activity' (and symbol a_B) for variously defined ratios λ_B'/λ_B'', which should properly be called *relative activities*.

Similarly we obtain from equation (5.6.3)

$$0 = (H_m/RT^2)^\alpha \, dT^\alpha - (V_m/RT)^\alpha \, dp^\alpha + \sum_B x_B^\alpha \, d \ln \lambda_B^\alpha . \qquad (5.7.2)$$

§ 5.8 Compression factor of a phase

The reader should derive for himself from equation (5.7.2) the equation:

$$dZ_m^\alpha = (U_m/RT^2)^\alpha \, dT^\alpha + (p/RT)^\alpha \, dV_m^\alpha + \sum_B x_B^\alpha \, d \ln \lambda_B^\alpha , \qquad (5.8.1)$$

where Z_m^α is defined by

$$Z_m^\alpha \overset{\text{def}}{=} p^\alpha V_m^\alpha / RT^\alpha , \qquad (5.8.2)$$

and is called the (molar) *compression factor* of the phase α. We see that the compression factor is the characteristic thermodynamic function for the set of independent variables: T, V, and the λ's.

§ 5.9 Affinity of a chemical reaction

There are only two ways in which the amount of substance n_B^α in a phase α can change, namely by gain or loss of the substance B from or to adjacent phases in the system (or from or to the surroundings), or by changes in the extents of one or more chemical reactions. For a closed phase (see § 1.7) only the second of these possibilities remains.

For the general chemical reaction (see §§ 2.5 and 2.6) expressed by the equation: $0 = \Sigma_B \nu_B B$, the extent of reaction ξ^α in a closed phase α is defined by the relation:

$$dn_B^\alpha \overset{\text{def}}{=} \nu_B \, d\xi^\alpha , \quad \text{(for each substance B)} . \qquad (5.9.1)$$

It follows that for a closed phase α equations (5.3.6) to (5.3.9) can be written in the forms:

$$dU^\alpha = T^\alpha \, dS^\alpha - p^\alpha \, dV^\alpha - A^\alpha \, d\xi^\alpha , \qquad (5.9.2)$$

$$dH^\alpha = T^\alpha \, dS^\alpha + V^\alpha \, dp^\alpha - A^\alpha \, d\xi^\alpha , \qquad (5.9.3)$$

$$dA^\alpha = -S^\alpha \, dT^\alpha - p^\alpha \, dV^\alpha - A^\alpha \, d\xi^\alpha , \qquad (5.9.4)$$

$$dG^\alpha = -S^\alpha \, dT^\alpha + V^\alpha \, dp^\alpha - A^\alpha \, d\xi^\alpha , \qquad (5.9.5)$$

where we have introduced the *affinity* A of the chemical reaction $0 = \Sigma_B \nu_B B$ defined by

$$A^\alpha \overset{\text{def}}{=} -\sum_B \nu_B^\alpha \mu_B^\alpha . \qquad (5.9.6)$$

The affinity A can be expressed in the form of a sum of differences in chemical potential:

$$A^\alpha = -\sum_B \nu_B(\mu_B - \mu_B^{eq})^\alpha , \qquad (5.9.7)$$

where μ_B^{eq} denotes the chemical potential of the substance B at chemical equilibrium, and where we have used the relation $\Sigma_B \nu_B \mu_B^{eq} = 0$, which we shall derive in chapter 7 as the criterion of chemical equilibrium. Since each of the differences $(\mu_B - \mu_B^{eq})$ is measurable, so is the affinity A.

We conclude this chapter by noting the following series of entirely equivalent expressions for A that can be derived from equations (2) to (5):

$$A^\alpha = -\sum_B \nu_B \mu_B^\alpha = T^\alpha(\partial S^\alpha/\partial\xi^\alpha)_{U^\alpha, V^\alpha} = T^\alpha(\partial S^\alpha/\partial\xi^\alpha)_{H^\alpha, p^\alpha}$$

$$. = -(\partial U^\alpha/\partial\xi^\alpha)_{S^\alpha, V^\alpha} = -(\partial H^\alpha/\partial\xi^\alpha)_{S^\alpha, p^\alpha}$$

$$= -(\partial A^\alpha/\partial\xi^\alpha)_{T^\alpha, V^\alpha} = -(\partial G^\alpha/\partial\xi^\alpha)_{T^\alpha, p^\alpha} . \qquad (5.9.8)$$

The last of these equalities is the most important from our point of view because chemical reactions are most often studied at (more or less) constant temperature and pressure. We note that whatever 'interpretation' might be placed on the sign and magnitude of $T^\alpha(\partial S^\alpha/\partial\xi^\alpha)_{U^\alpha, V^\alpha}$ for a chemical reaction in an *isolated* system, exactly the same 'interpretation' must be placed on $-(\partial G^\alpha/\partial\xi^\alpha)_{T^\alpha, p^\alpha}$ for the same chemical reaction carried out *at constant temperature and pressure*. The same 'interpretation' can *not* be placed on $T^\alpha(\partial S^\alpha/\partial\xi^\alpha)_{T^\alpha, p^\alpha}$ which is a quite different quantity. So much for 'interpretations' of entropy changes for chemical reactions carried out at constant temperature and pressure.

Chapter 6

Change of state of a phase of fixed composition

§ 6.1   Introduction and scope

We recall the fundamental equation (5.2.1), namely

$$dU^\alpha = T^\alpha\, dS^\alpha - p^\alpha\, dV^\alpha + \sum_B \mu_B^\alpha\, dn_B^\alpha, \tag{6.1.1}$$

for any change of the state of a phase α. For a closed phase α in the absence of any change in the extent of any chemical reaction we have

$$dn_B^\alpha = 0, \quad \text{(all B)}, \tag{6.1.2}$$

so that equation (1) becomes

$$dU^\alpha = T^\alpha\, dS^\alpha - p^\alpha\, dV^\alpha, \quad (n_B^\alpha \text{ constant for all B}). \tag{6.1.3}$$

That is what we mean by a phase of fixed composition, and it is with changes of the state of such a phase that we shall be concerned in this chapter.

By far the most important equations for a phase α of fixed composition are those corresponding to the choice of T^α and p^α as independent variables. We accordingly choose as our starting point for the derivation of the equations for the dependence of the properties of a phase of fixed composition on the temperature and on the pressure, not equation (3) but the corresponding equation for dG^α, which in view of equation (5.3.9) is

$$dG^\alpha = -S^\alpha\, dT^\alpha + V^\alpha\, dp^\alpha, \quad (n_B^\alpha \text{ constant for all B}). \tag{6.1.4}$$

It will for the most part be left as an exercise for the reader to derive the corresponding equations for T^α and V^α as independent variables.

Among other things, we shall find out in §§ 6.3 to 6.6 how to measure differences of entropy, and in § 6.9 how to measure differences of chemical potential of a substance, for changes of the state of a phase of fixed composition. We shall thus dispose of part of the second and part of the third of the tasks that we set ourselves towards the end of § 5.2. These two tasks will not be completed until we find out in chapter 10 how to measure the effect of a change of composition on the entropy of a phase and on the chemical potential of a substance in a phase.

In §§ 6.16 and 6.17 we shall at last complete the first of the tasks that we set ourselves towards the end of § 5.2, namely to prove that the thermodynamic temperature T defined by equation (5.2.1) is in fact proportional to the perfect-gas temperature defined in § 1.16.

The restriction 'n_B^α constant for all B' will apply to every equation in this chapter; in the rest of the chapter it will be taken for granted. We shall also omit the superscript α from the independent variables T^α and p^α.

Our first step is to write down the two partial derivatives of G^α. These follow directly from equation (4) and are

$$(\partial G^\alpha/\partial T)_p = -S^\alpha,$$

(6.1.5)

and

$$(\partial G^\alpha/\partial p)_T = V^\alpha.$$

(6.1.6)

§ 6.2   The Gibbs–Helmholtz equation

In view of the definition:

$$G^\alpha \overset{\text{def}}{=} H^\alpha - TS^\alpha,$$

(6.2.1)

we obtain from equation (6.1.5)

$$H^\alpha = G^\alpha - T(\partial G^\alpha/\partial T)_p.$$

(6.2.2)

Equation (2) is called the *Gibbs–Helmholtz equation*.† Three other useful ways of writing it are

$$H^\alpha = G^\alpha - (\partial G^\alpha/\partial \ln T)_p,$$

(6.2.3)

$$H^\alpha = \{\partial(G^\alpha/T)/\partial(1/T)\}_p,$$

(6.2.4)

$$H^\alpha = -T^2\{\partial(G^\alpha/T)/\partial T\}_p.$$

(6.2.5)

§ 6.3   The dependence on temperature of the entropy of a phase of fixed composition

By differentiating the Gibbs–Helmholtz equation (6.2.2) with respect to T at constant p we obtain

$$(\partial H^\alpha/\partial T)_p = -T(\partial^2 G^\alpha/\partial T^2)_p,$$

(6.3.1)

which after use of equation (6.1.5) and rearrangement becomes

$$(\partial S^\alpha/\partial T)_p = (\partial H^\alpha/\partial T)_p/T = C_p^\alpha/T,$$

(6.3.2)

where we have used the definition:

$$C_p^\alpha \overset{\text{def}}{=} (\partial H^\alpha/\partial T)_p.$$

(6.3.3)

† The analogous equation $U^\alpha = A^\alpha - T(\partial A^\alpha/\partial T)_{V^\alpha}$ is also sometimes called a Gibbs–Helmholtz equation.

Equation (2) supplies us with a calorimetric method for the measurement of the entropy difference:

$$\{S^\alpha(T_2, p_1) - S^\alpha(T_1, p_1)\}, \tag{6.3.4}$$

for a phase α of fixed composition. By integration of equation (2) we obtain

$$S^\alpha(T_2, p_1) - S^\alpha(T_1, p_1) = \int_{H^\alpha(T_1, p_1)}^{H^\alpha(T_2, p_1)} T^{-1} \, dH^\alpha$$

$$= \int_{T_1}^{T_2} (C_p^\alpha / T) \, dT$$

$$= \int_{\ln T_1}^{\ln T_2} C_p^\alpha \, d\ln T. \tag{6.3.5}$$

We have already learnt in §§ 1.22 and 4.5 how to measure the enthalpy difference $\{H^\alpha(T, p_1) - H^\alpha(T_0, p_1)\}$ calorimetrically as a function of T for any arbitrary choice of T_0. It follows from equation (5) that the desired entropy difference $\{S^\alpha(T_2, p_1) - S^\alpha(T_1, p_1)\}$ is equal to the area under a curve of $1/T$ plotted as ordinate, against $\{H^\alpha(T, p_1) - H^\alpha(T_0, p_1)\}$ as abscissa, between the values $\{H^\alpha(T_1, p_1) - H^\alpha(T_0, p_1)\}$ and $\{H^\alpha(T_2, p_1) - H^\alpha(T_0, p_1)\}$. An example of such a plot is shown in figure 6.1.

We have seen in § 1.22 that the heat capacity at constant pressure C_p^α is not directly measured. What is measured is the quantity $\{H^\alpha(T + \Delta T, p) - H^\alpha(T, p)\}/\Delta T$ where ΔT is finite, whereas C_p^α is the limiting value of that quantity as $\Delta T \to 0$.

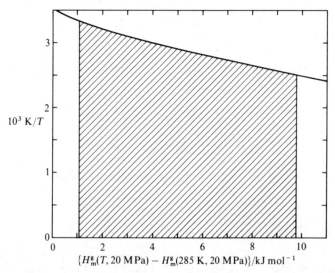

Figure 6.1. Values of $1/T$ plotted against $\{H_m^g(T, p) - H_m^g(T_0, p)\}$ at $p = 20$ MPa for gaseous C_2H_4 (ethene) with T_0 chosen arbitrarily as 285 K. The molar entropy difference: $\{S_m^g (400 \text{ K}, 20 \text{ MPa}) - S_m^g (300 \text{ K}, 20 \text{ MPa})\} = 25.10$ J K^{-1} mol^{-1}, was evaluated by measurement of the area under the curve from $T = 300$ K to $T = 400$ K.

Nevertheless the results of such measurements are usually† evaluated from plots of C_p^α/T against T. Figure 6.2 shows such a plot for the same experimental results as were plotted in figure 6.1. A plot of C_p^α against $\ln(T/K)$ would of course have served equally well.

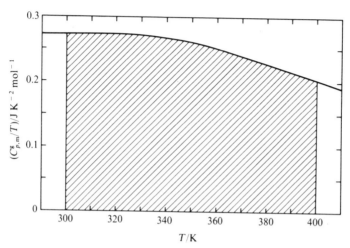

Figure 6.2. Values of $C_{p,m}^g/T$, where $C_{p,m}^g$ is the molar heat capacity at constant pressure, plotted against T at $p = 20$ MPa for gaseous C_2H_4 (ethene). The molar entropy difference: $\{S_m^g(400\text{ K}, 20\text{ MPa}) - S_m^g(300\text{ K}, 20\text{ MPa})\} = 25.10\text{ J K}^{-1}\text{ mol}^{-1}$, was evaluated by measurement of the area under the curve from $T = 300$ K to $T = 400$ K.

§ 6.4   Maxwell's equations

We differentiate equation (6.1.5) with respect to p at constant T:

$$\{\partial(\partial G^\alpha/\partial T)_p/\partial p\}_T = -(\partial S^\alpha/\partial p)_T, \tag{6.4.1}$$

and equation (6.1.6) with respect to T at constant p:

$$\{\partial(\partial G^\alpha/\partial p)_T/\partial T\}_p = (\partial V^\alpha/\partial T)_p. \tag{6.4.2}$$

By the 'cross-differentiation rule' (appendix I) we then have

$$(\partial S^\alpha/\partial p)_T = -(\partial V^\alpha/\partial T)_p. \tag{6.4.3}$$

Equation (3) is called a *Maxwell equation*.

By writing $(\partial S^\alpha/\partial p)_T$ in the form $(\partial S^\alpha/\partial V^\alpha)_T/(\partial p/\partial V^\alpha)_T$ we obtain

$$(\partial S^\alpha/\partial V^\alpha)_T = -(\partial V^\alpha/\partial T)_p(\partial p/\partial V^\alpha)_T = (\partial p/\partial T)_{V^\alpha}, \tag{6.4.4}$$

where we have used the '−1 rule' (appendix I). Equation (4) is also called a Maxwell equation; it can alternatively be derived by a route parallel to that used above to

† But why? Why not record the results of measurements of enthalpy increments, obtain the entropy by integration according to the first of equations (5), and then obtain the heat capacity, if it is needed, by differentiation?

obtain equation (3) but starting from $dA^\alpha = -S^\alpha\,dT - p\,dV^\alpha$ instead of $dG^\alpha = -S^\alpha\,dT + V^\alpha\,dp$.

§ 6.5   The dependence on pressure of the entropy of a phase of fixed composition

Maxwell's equation (6.4.3) supplies us with an experimental method for the measurement of the difference:

$$\{S^\alpha(T_1, p_2) - S^\alpha(T_1, p_1)\}, \tag{6.5.1}$$

for a phase α of fixed composition. By integration of equation (6.4.3) we obtain

$$S^\alpha(T_1, p_2) - S^\alpha(T_1, p_1) = -\int_{p_1}^{p_2} (\partial V^\alpha/\partial T)_p\,dp. \tag{6.5.2}$$

Measurements of V^α as a function of T over a range of values around T_1 at constant pressure p_1 lead to a value of $(\partial V^\alpha/\partial T)_p$ at $T = T_1$ and $p = p_1$. Similar sets of measurements around the same temperature T_1 but at other pressures in the range p_1 to p_2 lead to values of $(\partial V^\alpha/\partial T)_p$ at $T = T_1$ as a function of p over the range p_1 to p_2. The entropy difference (1) is then equal to the area under the curve drawn through those values of $(\partial V^\alpha/\partial T)_p$ plotted as ordinate against p as abscissa from p_1 to p_2. An example of such a plot is shown in figure 6.3.

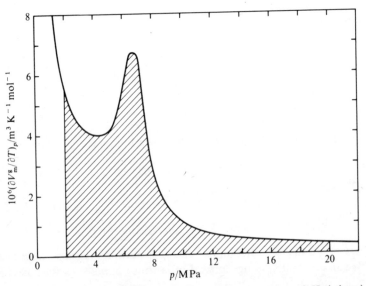

Figure 6.3. Values of $(\partial V_m^g/\partial T)_p$ at $T = 300$ K plotted against p for gaseous C_2H_4 (ethene). The molar entropy difference: $\{S_m^g(300\text{ K}, 20\text{ MPa}) - S_m^g(300\text{ K}, 2\text{ MPa})\} = -37.82\text{ J K}^{-1}\text{ mol}^{-1}$, was evaluated by measurement of the area under the curve from $p = 2$ MPa to $p = 20$ MPa.

§ 6.6 The dependence on temperature and on pressure of the entropy of a phase of fixed composition

We have now found out, in § 6.3 how to measure the entropy difference: $\{S^\alpha(T_2, p_1) - S^\alpha(T_1, p_1)\}$, and in § 6.5 how to measure the entropy difference: $\{S^\alpha(T_1, p_2) - S^\alpha(T_1, p_1)\}$, for a phase of fixed composition. Measurement of the entropy difference $\{S^\alpha(T_2, p_2) - S^\alpha(T_1, p_1)\}$ for a change of both temperature and pressure presents no new problem since

$$\{S^\alpha(T_2, p_2) - S^\alpha(T_1, p_1)\} = \{S^\alpha(T_2, p_2) - S^\alpha(T_1, p_2)\}$$
$$+ \{S^\alpha(T_1, p_2) - S^\alpha(T_1, p_1)\}. \qquad (6.6.1)$$

For example when we use the quantities evaluated from figures 6.1 (or 6.2) and 6.3 for gaseous C_2H_4 (ethene) we obtain

$$S_m^g(400 \text{ K}, 20 \text{ MPa}) - S_m^g(300 \text{ K}, 2 \text{ MPa})$$

$$= \{S_m^g(400 \text{ K}, 20 \text{ MPa}) - S_m^g(300 \text{ K}, 20 \text{ MPa})\}$$

$$+ \{S_m^g(300 \text{ K}, 20 \text{ MPa}) - S_m^g(300 \text{ K}, 2 \text{ MPa})\}$$

$$= (25.10 - 37.82) \text{ J K}^{-1} \text{ mol}^{-1}$$

$$= -12.72 \text{ J K}^{-1} \text{ mol}^{-1}. \qquad (6.6.2)$$

Thus we know now how to measure the difference between the entropies of any two states of a phase of fixed composition.

§ 6.7 The dependence on temperature and on pressure of the enthalpy of a phase of fixed composition

We already know how to measure the difference between the enthalpies of any two states of a phase of fixed composition by means of calorimetric experiments. No more need be said about the enthalpy difference $\{H^\alpha(T_2, p_1) - H^\alpha(T_1, p_1)\}$; it can be measured calorimetrically but in no other way. The difference $\{H^\alpha(T_1, p_2) - H^\alpha(T_1, p_1)\}$ can also be measured calorimetrically, usually in a flow calorimeter, as described in § 1.23; now, however, there is also another way.

Recalling that according to the definitions of H and G

$$H^\alpha = G^\alpha + TS^\alpha, \qquad (6.7.1)$$

we obtain on differentiation with respect to p at constant T

$$(\partial H^\alpha / \partial p)_T = (\partial G^\alpha / \partial p)_T + T(\partial S^\alpha / \partial p)_T. \qquad (6.7.2)$$

When we use equations (6.1.6) and (6.4.3) equation (2) becomes

$$(\partial H^\alpha / \partial p)_T = V^\alpha - T(\partial V^\alpha / \partial T)_p. \qquad (6.7.3)$$

On integration from p_1 to p_2 at constant temperature T_1 we then obtain

$$H^\alpha(T_1, p_2) - H^\alpha(T_1, p_1) = \int_{p_1}^{p_2} \{V^\alpha - T(\partial V^\alpha/\partial T)_p\}\, dp. \qquad (6.7.4)$$

Until now we have been dealing with the recipes for measuring quantities. Now at last we have produced in (3) or its integrated form (4) a thermodynamic equation, the first of many, in which the quantities occurring on each side can be measured independently. In particular the quantity on the left of equation (4) can be measured in a flow calorimeter while the quantity on the right can be evaluated from measurements of p, V, and T. If we have confidence in our experiments then we need not measure both these quantities but only that which is the easier. If we are less confident then we can measure both quantities and use equation (4) as a check on the *thermodynamic consistency* of our results.

It is convenient at this point to derive a closely related equation that we shall also need. We differentiate equation (3) with respect to T at constant p, use the 'cross-differentiation rule', and obtain

$$(\partial C_p^\alpha/\partial p)_T = -T(\partial^2 V^\alpha/\partial T^2)_p, \qquad (6.7.5)$$

or in integrated form:

$$C_p^\alpha(T_1, p_2) - C_p^\alpha(T_1, p_1) = -T_1 \int_{p_1}^{p_2} (\partial^2 V^\alpha/\partial T^2)_p\, dp. \qquad (6.7.6)$$

§ 6.8   The dependence on pressure and on temperature of the Gibbs function of a phase of fixed composition

We now turn to the measurement of the difference between the Gibbs functions of two states of a phase of fixed composition. The difference $\{G^\alpha(T_1, p_2) - G^\alpha(T_1, p_1)\}$ presents no difficulty so we shall deal with it first. By integration of equation (6.1.6) we obtain immediately

$$G^\alpha(T_1, p_2) - G^\alpha(T_1, p_1) = \int_{p_1}^{p_2} V^\alpha\, dp. \qquad (6.8.1)$$

Alternatively we might have dealt with $\{G^\alpha(T_1, p_2) - G^\alpha(T_1, p_1)\}$ simply by noting that it follows from the definition $G = H - TS$ that *at constant temperature* $\Delta G = \Delta H - T\,\Delta S$, so that the recipes we already have for the measurement of ΔS and of ΔH at constant temperature, that is equations (6.5.2) and (6.7.4), suffice also for the measurement of ΔG.

The difference $\{G^\alpha(T_2, p_1) - G^\alpha(T_1, p_1)\}$ on the other hand *cannot* be measured. Using equation (6.1.5) we obtain formally

$$G^\alpha(T_2, p_1) - G^\alpha(T_1, p_1) = -\int_{T_1}^{T_2} S^\alpha\, dT. \qquad (6.8.2)$$

Since the entropy S^α of a phase cannot be measured neither can the difference $\{G^\alpha(T_2, p_1) - G^\alpha(T_1, p_1)\}$.

§ 6.9   The dependence on pressure and on temperature of the chemical potential and of the absolute activity of a substance in a phase of fixed composition

We now turn to the measurement of the difference between the chemical potentials of a substance in two states of a phase of fixed composition. In view of relation (5.4.4):

$$G^\alpha = \sum_B n_B^\alpha \mu_B^\alpha, \tag{6.9.1}$$

and (5.3.10):

$$\mu_B^\alpha = (\partial G^\alpha / \partial n_B^\alpha)_{T,p,n_{A \neq B}^\alpha}, \tag{6.9.2}$$

between the chemical potentials μ_B^α and the Gibbs function G^α it will be no surprise to find that our conclusions about the measurability of the chemical-potential differences $\{\mu_B^\alpha(T_1, p_2) - \mu_B^\alpha(T_1, p_1)\}$ and of $\{\mu_B^\alpha(T_2, p_1) - \mu_B^\alpha(T_1, p_1)\}$ are exactly parallel to the conclusions that we reached in the previous section about differences of the Gibbs function, namely that the first of these differences is measurable but that the second is not.

Making use of equations (2), (6.1.6), and (2.7.1), and of the 'cross-differentiation rule', we obtain for $(\partial \mu_B^\alpha / \partial p)_T$

$$(\partial \mu_B^\alpha / \partial p)_T = \{\partial(\partial G^\alpha / \partial n_B^\alpha)_{T,p,n_{A \neq B}^\alpha} / \partial p\}_T = \{\partial(\partial G^\alpha / \partial p)_T / \partial n_B^\alpha\}_{T,p,n_{A \neq B}^\alpha}$$

$$= (\partial V^\alpha / \partial n_B^\alpha)_{T,p,n_{A \neq B}^\alpha}$$

$$= V_B^\alpha, \tag{6.9.3}$$

where V_B^α is the partial molar volume of the substance B in the phase α of fixed composition, a quantity that we learnt how to measure in chapter 2. Integration of equation (3) then gives

$$\mu_B^\alpha(T_1, p_2) - \mu_B^\alpha(T_1, p_1) = \int_{p_1}^{p_2} V_B^\alpha \, dp, \tag{6.9.4}$$

which tells us how to determine the chemical potential difference $\{\mu_B^\alpha(T_1, p_2) - \mu_B^\alpha(T_1, p_1)\}$ for a phase of fixed composition from measurements of the partial molar volume V_B^α made over a range of pressures from p_1 to p_2.

We recall that the absolute activity λ_B^α is defined by the relation (5.6.2):

$$\mu_B^\alpha \stackrel{\text{def}}{=} RT \ln \lambda_B^\alpha, \tag{6.9.5}$$

and now recast the previous paragraph in terms of *ratios* of absolute activities rather than *differences* of chemical potentials. It follows immediately from equations (3) and (5) that

$$(\partial \ln \lambda_B^\alpha / \partial p)_T = V_B^\alpha / RT. \tag{6.9.6}$$

Integration of equation (6) then gives

$$\ln\{\lambda_B^\alpha(T_1, p_2)/\lambda_B^\alpha(T_1, p_1)\} = \int_{p_1}^{p_2} (V_B^\alpha/RT_1)\, dp, \qquad (6.9.7)$$

which tells us how to determine the ratio of absolute activities $\{\lambda_B^\alpha(T_1, p_2)/\lambda_B^\alpha(T_1, p_1)\}$ for a phase of fixed composition from measurements of the partial molar volume V_B^α made over a range of pressures from p_1 to p_2.

By far the most important application of partial molar volumes is in connexion with equations (4) and (7).

Using a similar method to that used to derive equation (3) we obtain for $(\partial\mu_B^\alpha/\partial T)_p$

$$(\partial\mu_B^\alpha/\partial T)_p = -S_B^\alpha, \qquad (6.9.8)$$

where S_B^α is the partial molar entropy of the substance B in the phase α. Formal integration of equation (8) then gives

$$\mu_B^\alpha(T_2, p_1) - \mu_B^\alpha(T_1, p_1) = -\int_{T_1}^{T_2} S_B^\alpha\, dT. \qquad (6.9.9)$$

Since the partial molar entropy S_B^α of a substance B in a phase α cannot be measured neither can the chemical-potential difference $\{\mu_B^\alpha(T_2, p_1) - \mu_B^\alpha(T_1, p_1)\}$. We shall nevertheless find considerable use for equations (8) and (9) in later chapters.

By use of the equation:

$$H_B^\alpha = G_B^\alpha + TS_B^\alpha = \mu_B^\alpha + TS_B^\alpha, \qquad (6.9.10)$$

and the definition (5) of λ_B^α we obtain

$$(\partial \ln \lambda_B^\alpha/\partial T)_p = -H_B^\alpha/RT^2, \qquad (6.9.11)$$

where H_B^α is the partial molar enthalpy of the substance B in the phase α. Formal integration of equation (11) gives

$$\ln\{\lambda_B^\alpha(T_2, p_1)/\lambda_B^\alpha(T_1, p_1)\} = -\int_{T_1}^{T_2} (H_B^\alpha/RT^2)\, dT. \qquad (6.9.12)$$

Since the partial molar enthalpy H_B^α of a substance B in a phase α cannot be measured neither can the ratio of absolute activities $\lambda_B^\alpha(T_2, p_1)/\lambda_B^\alpha(T_1, p_1)$. We shall nevertheless find considerable use for equations (11) and (12) in later chapters.

§ 6.10 Expansivity and compressibility

In §§ 6.11 and 6.12 we shall derive two important thermodynamic equations which, like equation (6.7.4), inter-relate independently measurable quantities for a phase of fixed composition. The methods used in these two derivations should be studied carefully as models for the derivation of other such thermodynamic equations. In the rest of this chapter we shall for the sake of simplicity omit the superscript $^\alpha$; we are still, however, discussing a single phase.

Before beginning those two derivations it is convenient to introduce two new auxiliary quantities. These are† the *isobaric expansivity* (or *expansivity* for short; formerly called 'coefficient of thermal expansion') α defined by

$$\alpha \overset{\text{def}}{=} V^{-1}(\partial V/\partial T)_p = (\partial \ln V/\partial T)_p,\qquad(6.10.1)$$

and the *isothermal compressibility* κ_T defined by

$$\kappa_T \overset{\text{def}}{=} -V^{-1}(\partial V/\partial p)_T = -(\partial \ln V/\partial p)_T.\qquad(6.10.2)$$

The expansivity α of a phase is *usually* positive (it is negative for example for liquid water at temperatures between 273.15 K and 277.13 K). The isothermal compressibility κ_T is *always* positive.‡ The two quantities α and κ_T are related by the equation:

$$\alpha/\kappa_T = (\partial p/\partial T)_V,\qquad(6.10.3)$$

which follows from the '−1 rule'.

§ 6.11   Relation between heat capacities at constant pressure and at constant volume

We now seek an expression for the difference $(C_p - C_V)$ between the heat capacities at constant pressure and at constant volume as far as possible in terms of familiar derivatives $(\partial X/\partial T)_p$ and $(\partial X/\partial p)_T$. From the definitions of C_p, C_V, and H we obtain

$$C_p - C_V = (\partial H/\partial T)_p - (\partial U/\partial T)_V$$
$$= (\partial H/\partial T)_p - (\partial H/\partial T)_V + V(\partial p/\partial T)_V.\qquad(6.11.1)$$

We now use the 'rule for changing the variable held constant' (see appendix I) on $(\partial H/\partial T)_V$ and obtain

$$(\partial H/\partial T)_V = (\partial H/\partial T)_p + (\partial H/\partial p)_T(\partial p/\partial T)_V.\qquad(6.11.2)$$

In view of (2) equation (1) can be written in the form:

$$C_p - C_V = \{V - (\partial H/\partial p)_T\}(\partial p/\partial T)_V,\qquad(6.11.3)$$

which on use of equation (6.7.3) and of the '−1 rule' becomes

$$C_p - C_V = -T\{(\partial V/\partial T)_p\}^2/(\partial V/\partial p)_T,\qquad(6.11.4)$$

which is the sought-for expression. When we use the definitions (6.10.1) and (6.10.2)

---

† *Isobaric* means 'at constant pressure'; *isothermal* means 'at constant temperature'. Similarly *isochoric* means 'at constant volume', *isentropic* 'at constant entropy', *isenthalpic* 'at constant enthalpy', *isoenergetic* 'at constant energy'.

‡ This is one of the thermodynamic inequalities that we shall prove in chapter 7; $\kappa_T > 0$ is the criterion for the mechanical stability of a phase, that is to say the condition that prevents an isolated fluid phase from splitting into two phases with different pressures.

of α and κ_T equation (4) can be rewritten in the form:

$$C_p - C_V = T\alpha^2 V/\kappa_T. \tag{6.11.5}$$

Since T, α^2, V, and κ_T are all necessarily positive or zero, it follows from equation (5) that C_p is never less than C_V. Since C_V is necessarily positive† so is C_p. Both C_p and C_V can be measured calorimetrically; all the quantities on the right of equation (5) can be measured by use only of a thermometer, a dilatometer, and a manometer. It is therefore unnecessary to measure all the quantities occurring in equation (5). Of these quantities C_V is incomparably the most difficult to measure directly. Equation (5) is therefore usually used to calculate C_V from measured values of C_p, α, κ_T, T, and V.

§ 6.12   Isentropic compressibility

We now seek an expression for the difference between $(\partial V/\partial p)_T$ and $(\partial V/\partial p)_S$ in terms of familiar derivatives $(\partial X/\partial T)_p$ and $(\partial X/\partial p)_T$. By use of the 'rule for changing the variable held constant' (see appendix I) we obtain

$$(\partial V/\partial p)_T - (\partial V/\partial p)_S = -(\partial V/\partial T)_p (\partial T/\partial p)_S, \tag{6.12.1}$$

which on use of the '−1 rule' becomes

$$(\partial V/\partial p)_T - (\partial V/\partial p)_S = (\partial V/\partial T)_p (\partial S/\partial p)_T / (\partial S/\partial T)_p. \tag{6.12.2}$$

We now use equations (6.4.3) and (6.3.2) in (2) and obtain

$$(\partial V/\partial p)_T - (\partial V/\partial p)_S = -T\{(\partial V/\partial T)_p\}^2 / C_p, \tag{6.12.3}$$

which is the sought-for expression. When we use the definitions (6.10.1) and (6.10.2) of α and κ_T equation (3) can be rewritten in the form:

$$\kappa_T - \kappa_S = T\alpha^2 V/C_p, \tag{6.12.4}$$

where κ_S defined by

$$\kappa_S \overset{\text{def}}{=} -V^{-1}(\partial V/\partial p)_S, \tag{6.12.5}$$

is called the *isentropic compressibility*.‡ All the quantities occurring in equation (4) except κ_S are obviously measurable; we shall now find that κ_S is measurable too.

The fundamental equation for any change of state of a phase of fixed composition is according to (5.1.3)

$$dU = T\,dS - p\,dV. \tag{6.12.6}$$

---

† This is one of the thermodynamic inequalities that we shall prove in chapter 7; $C_V > 0$ is the criterion for the thermal stability of a phase, that is to say the condition that prevents an isolated phase from splitting into two phases with different temperatures.

‡ κ_S is often loosely called the 'adiabatic compressibility'. *Isentropic* does not mean *adiabatic* but (*adiabatic* + *reversible*).

Combining equation (6) with equation (1.18.9) for the first law:

$$dU = W + Q, \tag{6.12.7}$$

we obtain

$$T \, dS = Q + (W + p \, dV). \tag{6.12.8}$$

It follows from equation (8) that any change of state that takes place adiabatically ($Q = 0$), *and* reversibly ($W = -p \, dV$), is also isentropic ($dS = 0$). We can carry out an expansion or compression adiabatically by thermally insulating the apparatus as completely as possible and then quickly changing the pressure and remeasuring the volume and pressure. We can carry out an expansion or compression reversibly by changing the pressure slowly. If the thermal insulation is good enough then we can carry out an expansion or compression both adiabatically and reversibly, that is to say isentropically. In practice it is easy to carry out an expansion isentropically, and many direct measurements of κ_S have been made, especially for liquids.

Another important method for the determination of κ_S depends on measurements of the speed of sound u in the phase. In the absence of dispersion the isentropic compressibility κ_S of a fluid† is related to the speed of sound u by the formula:

$$\kappa_S = 1/u^2\rho = V_m/u^2 M, \tag{6.12.9}$$

where ρ is the density, V_m the molar volume, and M the molar mass, of the phase.

We now return to equation (4). If all the quantities κ_T, κ_S, α, T, V, and C_p have been measured for a phase then equation (4) can be used to test the measurements for thermodynamic consistency. If all save one of the quantities have been measured then equation (4) can be used to calculate that one. For fluids no one of the quantities is much harder to measure than the others. For crystalline solids, however, the isothermal compressibility κ_T is difficult to measure, has seldom been measured with high accuracy, and is much needed in connexion with theories of solids. The isentropic compressibility κ_S (via the speed of sound), α (via the temperature dependence of the lattice constant from X-ray diffraction measurements), V, and C_p, are all easier to measure and have more often been measured with high accuracy. Equation (4) is therefore especially useful for obtaining values of κ_T for solids from measured values of κ_S, α, T, V, and C_p.

Since T, α^2, V, and C_p are all necessarily positive or zero, it follows from equation (4) that κ_T is never less than κ_S.

§ 6.13   Isentropic expansion

We have seen in § 6.12 that a reasonably fast‡ expansion or compression in a thermally insulated vessel is isentropic. We shall now study the variation of

---

† For crystalline solids the velocities u_l of longitudinal waves and u_t of transverse waves must both be measured and the relation of these to κ_S is somewhat more complicated.

‡ But not too fast and especially not explosively fast; in an explosion the pressure of the system is neither uniform nor even well defined and the expansion is not reversible.

temperature with pressure in an isentropic expansion or compression. By use of the '−1 rule' we obtain

$$(\partial T/\partial p)_S = -(\partial S/\partial p)_T/(\partial S/\partial T)_p, \tag{6.13.1}$$

which by use of equations (6.4.3), (6.3.2), and (6.10.1) becomes

$$(\partial T/\partial p)_S = T(\partial V/\partial T)_p/C_p = T\alpha V/C_p = T\alpha V_m/C_{p,m}. \tag{6.13.2}$$

Equation (2) is exact for any phase. For a perfect ($pV_m = RT$) gas equation (2) becomes

$$(\partial T/\partial p)_S = RT/pC_{p,m}. \tag{6.13.3}$$

Integrating equation (3), assuming constant $C_{p,m}$, we obtain

$$\ln(T_2/T_1) = (R/C_{p,m})\ln(p_2/p_1), \quad (S \text{ constant}), \tag{6.13.4}$$

or

$$T_2 = T_1(p_2/p_1)^{R/C_{p,m}}, \quad (S \text{ constant}). \tag{6.13.5}$$

By substitution from $pV_m = RT$ we can express equation (5) for the isentropic expansion of a perfect gas in the alternative forms:

$$p_2 = p_1(V_{m,1}/V_{m,2})^{C_{p,m}/(C_{p,m}-R)}, \quad (S \text{ constant}), \tag{6.13.6}$$

$$T_2 = T_1(V_{m,1}/V_{m,2})^{R/(C_{p,m}-R)}, \quad (S \text{ constant}). \tag{6.13.7}$$

As an example let us calculate T_2 when $T_1 = 300 \text{ K}$, $p_1 = 3.2 \text{ MPa}$, and $p_2 = 0.1 \text{ MPa}$ for a perfect monatomic gas, for which $C_{p,m} = 5R/2$, that is to say the final temperature of a perfect monatomic gas when it expands isentropically at room temperature from an initial pressure of 3.2 MPa to atmospheric pressure. We have

$$T_2 = 300 \text{ K} (32)^{-2/5} = 300 \text{ K}/4 = 75 \text{ K}! \tag{6.13.8}$$

It is no wonder that isentropic expansion is used to produce low temperatures and to liquefy 'permanent' gases. The dramatic effect is of course familiar to anyone who has opened a cylinder of carbon dioxide and the converse effect to anyone who has used a bicycle pump.

§ 6.14 Isoenergetic (Joule) expansion

We shall now discuss in modern terms the famous 'free-expansion' experiment that Joule used in 1845, before the first law of thermodynamics had been properly formulated, to study the dependence of the temperature of a gas on its volume at constant energy. He used the apparatus reproduced from his paper[†] in figure 6.4 to

[†] Joule, J. P. *Phil. Mag. (Series 3)* **1845**, 26, 369.

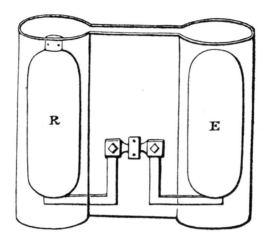

Figure 6.4. Joule's apparatus for his 'free-expansion' experiment.

measure $(\partial T/\partial V)_U$, describing the experiment in these words:

"Having filled the receiver R with about 22 atmospheres [2.2 MPa] of dry air, and having exhausted the receiver E by means of an air-pump, I screwed them together, and then put them into a tin can containing $16\frac{1}{2}$ lbs. [7.5 kg] of water. The water was first thoroughly stirred and its temperature taken by the same delicate thermometer which was made use of in former experiments. The stop-cocks were then opened, and the air allowed to pass from the full into the empty receiver until equilibrium was established between the two. Lastly, the water was again stirred and its temperature carefully noted. The following Table contains the results of a series of [six] experiments, interpolated with others to eliminate the effects of stirring, evaporation, &c. The difference between the means of the experiments and interpolations being exactly such as was found to be due to the increased effect of the temperature of the room in the latter case, we arrive at the conclusion, that no *change of temperature occurs when air is allowed to expand in such a manner as not to develope mechanical power*."

Since the bulbs in Joule's apparatus were rigid and the water bath was thermally insulated from its surroundings (or, rather, corrections were made for imperfect insulation), the system was isolated and anything happening in it was isoenergetic: U was constant. By the methods of §§ 6.11 to 6.13 we derive the equation:†

$$(\partial T/\partial V)_U = \{p - T(\partial p/\partial T)_V\}/C_V$$
$$= (p - T\alpha/\kappa_T)/(C_p - T\alpha^2 V/\kappa_T). \qquad (6.14.1)$$

For a perfect ($pV_m = RT$) gas the numerator of equation (1) vanishes while the denominator does not: for a perfect gas $(\partial T/\partial V)_U = 0$. It is left to the reader to prove that for a perfect gas it follows that $(\partial U/\partial V)_T = 0$, $(\partial U/\partial p)_T = 0$, $(\partial H/\partial p)_T = 0$, and $(\partial H/\partial V)_T = 0$: the energy and the enthalpy of a perfect gas depend only on its temperature.

Joule's thermometer had a sensitivity of about ±0.01 K. We now know, from rough estimates of the total heat capacity of his calorimeter (air + copper vessels +

---

† The complete derivation of equation (6.14.1) is left as an exercise for the reader.

water + tin can: $C_{p,\text{total}} \approx 37000\,\text{J K}^{-1}$) and from what we now know of α, κ_T, and $C_{V,\text{m}}$ for air, that with a modern thermometer he would have observed a fall of about 0.003 K in temperature. Had the heat capacity of the whole of the calorimeter been that of only the air that he was studying, the temperature would have fallen by about 3 K.†

The reader should prove for himself that

$$(\partial T/\partial p)_U = (T\alpha V - p\kappa_T V)/(C_p - p\alpha V). \qquad (6.14.2)$$

For a real gas at low pressures $(\partial T/\partial p)_U$ is positive (except at extremely high temperatures where it becomes negative; see § 12.3). For nitrogen at room temperature and pressure $(\partial T/\partial p)_U \approx 0.003\,\text{K kPa}^{-1}$ while for 2,2-dimethylpropane under the same conditions $(\partial T/\partial p)_U \approx 0.017\,\text{K kPa}^{-1}$. The expected effects are so small, in view of the inevitably large heat capacity of the rigid container, and the experiments are so difficult, that the method has seldom if ever been used for any modern study of the deviations of the properties of real gases from those of a perfect gas.

§ 6.15   Isenthalpic (Joule–Thomson) expansion

We learnt in § 1.23 that an adiabatic throttling experiment is isenthalpic. In a Joule–Thomson experiment one measures $\{T(H, p_2) - T(H, p_1)\}/(p_2 - p_1)$, which in the limit $(p_2 - p_1) \to 0$ becomes the Joule–Thomson coefficient $\mu_{JT} = (\partial T/\partial p)_H$, given according to equation (6.7.3) and the '−1 rule' by

$$\mu_{JT} = (\partial T/\partial p)_H = -\{V - T(\partial V/\partial T)_p\}/C_p. \qquad (6.15.1)$$

The Joule–Thomson coefficient is zero for a perfect gas. For a real gas at low pressures $(\partial T/\partial p)_H$ is positive (except at very high temperatures where it becomes negative; see § 12.3). For nitrogen at room temperature and pressure $(\partial T/\partial p)_H \approx 0.002\,\text{K kPa}^{-1}$ while for 2,2-dimethylpropane under the same conditions $(\partial T/\partial p)_H \approx 0.025\,\text{K kPa}^{-1}$.

§ 6.16   Isothermal Joule–Thomson expansion

In an isothermal Joule–Thomson experiment, power is supplied to the flowing fluid on the downstream side of the throttle so as just to restore the outlet temperature to the inlet temperature. In such an experiment one measures $\{H(T, p_2) - H(T, p_1)\}/(p_2 - p_1)$, which in the limit $(p_2 - p_1) \to 0$ becomes the isothermal Joule–Thomson coefficient $\phi_{JT} = (\partial H/\partial p)_T$, given by equation (6.7.3):

$$\phi_{JT} = (\partial H/\partial p)_T = V - T(\partial V/\partial T)_p. \qquad (6.16.1)$$

† It is paradoxical that had Joule observed such changes then the progress of thermodynamics might have been slower.

Like the (isenthalpic) Joule–Thomson coefficient, the isothermal Joule–Thomson coefficient is zero for a perfect gas. For a real gas at low pressures $(\partial H/\partial p)_T$ is negative (except at very high temperatures where it becomes positive; see § 12.3). For N_2 at room temperature and pressure $(\partial H_m/\partial p)_T \approx -60 \text{ cm}^3 \text{ mol}^{-1}$ while for $C(CH_3)_4$ under the same conditions $(\partial H_m/\partial p)_T \approx -3000 \text{ cm}^3 \text{ mol}^{-1}$.

§ 6.17   Measurement of the ratio of two thermodynamic temperatures

We are now ready to prove that the thermodynamic temperature T in equation (5.2.1) is completely defined, that is to say that the ratio T_2/T_1 of two thermodynamic temperatures can be unambiguously measured. The reader should carefully satisfy himself that we do not anywhere in the present section assume what we are trying to prove. We assume only that temperature exists (§ 1.13), and that T is defined by equation (5.2.1). We shall prove in § 7.3 that T is a temperature; in the meantime we assume that to be so. We pretend to know nothing about any relation between thermodynamic temperature T defined by equation (5.2.1) and the perfect-gas temperature discussed in § 1.16.

The Gibbs–Helmholtz equation (6.2.3):

$$H = G - (\partial G/\partial \ln T)_p, \qquad (6.17.1)$$

supplies us with a general[†] method for the measurement of the ratio $T(\theta_2)/T(\theta_1)$ of two thermodynamic temperatures corresponding to two empirical temperatures θ_2 and θ_1. Rearranging and integrating equation (1) we obtain

$$\ln\{T(\theta_2)/T(\theta_1)\} = \int_{\theta_1}^{\theta_2} \{\Delta G(\theta) - \Delta H(\theta)\}^{-1} \, \mathrm{d}\Delta G(\theta), \qquad (6.17.2)$$

where we have used $\Delta X(\theta)$, with $X = H$ or $X = G$, to denote the difference $\{X(\theta, p_2, \xi_2) - X(\theta, p_1, \xi_1)\}$ between H or G for a closed phase observed at different pressures, or at different extents of a chemical reaction, or both, but at the same temperature. Measurements of $\Delta G(\theta)$ and of $\Delta H(\theta)$ for the same isothermal process at each of a series of temperatures covering the range θ_1 to θ_2, followed by integration according to the right-hand side of equation (2), then lead to a value for the ratio $T(\theta_2)/T(\theta_1)$.

The most important case is that for a phase of fixed composition (no chemical reaction) for which

$$\Delta X(\theta) = X(\theta, p_2) - X(\theta, p_1), \qquad (6.17.3)$$

and especially that for a gas at a low pressure. By use of equation (6.1.6), $\Delta G(\theta)$ for such a phase can be expressed in the form:

$$\Delta G(\theta) = \int_{p_1}^{p_2} V(\theta, p) \, \mathrm{d}p, \qquad (6.17.4)$$

† If magnetic work is involved then we must use an analogue of equation (6.17.1) in which the magnetic work is included.

and so can be evaluated from measurements of the volume of the phase over a range of pressures from p_1 to p_2 at the temperature θ. The corresponding $\Delta H(\theta)$ is $\{H(\theta, p_2) - H(\theta, p_1)\}$ and can be measured flow-calorimetrically as was explained in § 1.23. We shall show in the following section how such measurements on dilute gases lead unequivocally to the identification of the ratio of two thermodynamic temperatures with the ratio of two perfect-gas temperatures defined in § 1.16. In the meantime we have now learnt how to measure $T(\theta_2)/T(\theta_1)$.

We shall discover in chapter 10 how to measure the change of Gibbs function which accompanies a chemical reaction at constant temperature and pressure:

$$\Delta G(\theta) = G(\theta, p, \xi_2) - G(\theta, p, \xi_1) . \tag{6.17.5}$$

The corresponding $\Delta H(\theta)$ can be measured calorimetrically. It follows that such measurements can also be used to determine the ratio $T(\theta_2)/T(\theta_1)$ of two thermodynamic temperatures. So can measurements of the corresponding quantities for the mixing of two phases (for example two miscible liquids) to form a single mixed phase. In practice, however, these methods have not been used for thermometry.

§ 6.18 The proportionality of thermodynamic temperature to perfect-gas temperature

We are now at last ready to verify the conclusion that we anticipated in § 1.16 that the thermodynamic temperature defined by equation (5.2.1) is proportional to the perfect-gas temperature discussed in § 1.16.

We recall (§ 1.16) that it is an experimental fact established by a vast array of experiments on a wide variety of gases that the results of measurements of pressure p and molar volume V_m at given empirical temperature θ can be expressed by the series:

$$pV_m = \Theta + Bp + C'p^2 + \cdots , \tag{6.18.1}$$

which converges rapidly at low pressures and in which the coefficient Θ depends only on θ, while $B, C', \cdots$, depend on θ and also on the nature of the gas. Our object is to find by experiment how Θ depends on the thermodynamic temperature T.

After differentiation of (1) with respect to T we obtain

$$V_m - T(\partial V_m/\partial T)_p = (\Theta - T \, d\Theta/dT)/p + (B - T \, dB/dT)$$
$$+ (C' - T \, dC'/dT)p + \cdots . \tag{6.18.2}$$

We shall have achieved our object if we can measure the quantity $(\Theta - T \, d\Theta/dT)$ that occurs in equation (2).

According to equation (6.7.3) we have

$$(\partial H_m/\partial p)_T = V_m - T(\partial V_m/\partial T)_p , \tag{6.18.3}$$

which, since we know that Θ is *some* function of T though not yet *what* function, we may rewrite in the form:

$$(\partial H_m/\partial p)_\Theta = V_m - T(\partial V_m/\partial T)_p$$
$$= (\Theta - T\,d\Theta/dT)/p + (B - T\,dB/dT)$$
$$+ (C' - T\,dC'/dT)p + \cdots. \qquad (6.18.4)$$

The quantity $(\partial H_m/\partial p)_\Theta$ can be measured as we have seen in § 1.23 by means of a flow calorimeter fitted with a throttle and an electric heater.† The dependence of $(\partial H_m/\partial p)_\Theta$ on pressure p then leads according to (4) to information about $(\Theta - T\,d\Theta/dT)$, $(B - T\,dB/dT)$, $\cdots$. Here we are interested only in the first of these.

If $(\Theta - T\,d\Theta/dT)$ had any value other than zero then according to equation (4) a plot of $(\partial H_m/\partial p)_\Theta$ against p at given Θ would go to infinity as $p \to 0$. It is an experimental fact established by a wide variety of experiments‡ on a wide variety of gases that a plot of $(\partial H_m/\partial p)_\Theta$ against p at given Θ becomes more and more accurately linear at smaller and smaller pressures; no 'infinity catastrophe' has ever been found. This is illustrated in figure 6.5 for CO_2 at three temperatures.§ The

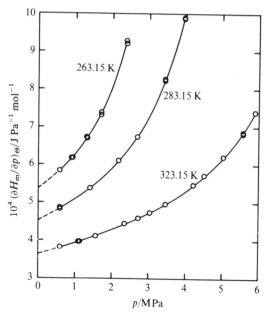

Figure 6.5. Calorimetric values of $(\partial H_m/\partial p)_\Theta$ plotted against pressure p for CO_2 at three temperatures.

---

† Strictly, what is measured is the quantity $\{H(\Theta, p_2) - H(\Theta, p_1)\}/(p_2 - p_1)$. The quantity $(\partial H/\partial p)_\Theta$ is the limiting value as $(p_2 - p_1) \to 0$.

‡ The isenthalpic Joule–Thomson coefficient $(\partial \Theta/\partial p)_H$ can be used as well for this purpose since it is known that the heat capacity at constant pressure $(\partial H/\partial \Theta)_p$ remains finite as $p \to 0$.

§ The measurements plotted in figure 6.5 are recent ones that were not made to confirm equation (5) but rather, equation (5) being assumed, to determine $(B - T\,dB/dT)$ and higher coefficients of equation (4).

inescapable conclusion is

$$\Theta - T \, d\Theta/dT = 0 . \tag{6.18.5}$$

It follows from equation (5) that

$$\Theta \propto T , \tag{6.18.6}$$

so that

$$\Theta_2/\Theta_1 = T_2/T_1 , \tag{6.18.7}$$

which is what we set out to demonstrate.

Problems for chapter 6

Problem 6.1

Calculate

(a) $\{S_m(400 \text{ K}, 2 \text{ MPa}) - S_m(300 \text{ K}, 0.1 \text{ MPa})\}$,

and

(b) $\{H_m(400 \text{ K}, 2 \text{ MPa}) - H_m(300 \text{ K}, 0.1 \text{ MPa})\}$,

for a perfect $(pV_m = RT)$ monatomic $(C_{p,m} = 5R/2)$ gas.

Problem 6.2

The following are values of the molar heat capacity at constant pressure $C_{p,m}$ for gaseous CO_2 at the temperatures T and the pressure 9 MPa. Calculated values of $C_{p,m}/T$ are also given.

T/K	340	350	370	390	410	430	450	460
$C_{p,m}/\text{J K}^{-1} \text{mol}^{-1}$	88.1	75.1	63.2	57.7	54.8	53.0	52.0	51.6
$(C_{p,m}/T)/\text{J K}^{-2} \text{mol}^{-1}$	0.259	0.215	0.171	0.148	0.134	0.123	0.116	0.112

Evaluate

(a) $\{S_m(450 \text{ K}, 9 \text{ MPa}) - S_m(350 \text{ K}, 9 \text{ MPa})\}$,

and

(b) $\{H_m(350 \text{ K}, 9 \text{ MPa}) - H_m(350 \text{ K}, 9 \text{ MPa})\}$.

Hints: (i) A convenient and quite accurate way of evaluating the area under a curve is by cutting out and weighing on an analytical balance. (ii) Remember to measure the area from zero on the ordinate scale.

Problem 6.3

The following are some values of the molar volume V_m and of $(\partial V_m/\partial T)_p$ for gaseous CO_2 at the pressures p and the temperature 350 K.

p/MPa	0.75	1	2	3	5	7	9	10
$V_m/cm^3\,mol^{-1}$	3794	2823	1367	881	490	321	226	192
$(\partial V_m/\partial T)_p/cm^3\,K^{-1}\,mol^{-1}$ 11.70	8.94	4.82	3.48	2.47	2.13	2.04	2.05

Evaluate

(a) $\{G_m(350\ K, 9\ MPa) - G_m(350\ K, 1\ MPa)\}$,

(b) $\{S_m(350\ K, 9\ MPa) - S_m(350\ K, 1\ MPa)\}$,

and

(c) $\{H_m(350\ K, 9\ MPa) - H_m(350\ K, 1\ MPa)\}$.

Problem 6.4

From the results of problems 6.2 and 6.3 calculate

(a) $\{S_m(450\ K, 9\ MPa) - S_m(350\ K, 1\ MPa)\}$,

and

(b) $\{H_m(450\ K, 9\ MPa) - H_m(350\ K, 1\ MPa)\}$.

Why was ΔG_m asked for in problem 6.3 but not in problem 6.2 or in this problem?

Problem 6.5

Calculate the specific heat capacity at constant volume c_V for liquid hexane at 298.15 K and 100 kPa given that $c_p = 2.27\ J\,g^{-1}\,K^{-1}$, $\rho = 0.655\ g\,cm^{-3}$, $\alpha = 1.38 \times 10^{-3}\ K^{-1}$, and $\kappa_T = 1.67 \times 10^{-3}\ MPa^{-1}$, where c_p denotes specific heat capacity at constant pressure, ρ density, α isobaric expansivity, and κ_T isothermal compressibility.

Problem 6.6

For liquid cyclohexane at 298.15 K and 100 kPa $\kappa_T = (1.130 \pm 0.002)\ GPa^{-1}$, $u = (1253 \pm 1)\ m\,s^{-1}$, $c_p = (1.861 \pm 0.001)\ J\,K^{-1}\,g^{-1}$, $\rho = (0.77387 \pm 0.00001)\ g\,cm^{-3}$, and $\alpha = (1.215 \pm 0.001)\ kK^{-1}$, where u denotes the speed of sound and the other symbols have the same meanings as in problem 6.5. Are these results thermodynamically consistent within the admitted experimental errors?

Problem 6.7

Verify the statements made in § 6.14 about Joule's free-expansion experiment. The volume of each of the brass receivers R and E was about 2.2 dm^3 and the total heat capacity of the (air + brass receivers + water + can) was about 37 kJ K^{-1}. Assume that air, present initially in R at a pressure of 2.2 MPa, has a molar heat capacity at constant volume of $5R/2$. Assume also that air follows the equation of state $pV_m = RT + Bp$, and that $dB/dT = +0.2$ cm^3 mol^{-1} K^{-1} at $T = 288$ K.

Problem 6.8

The results of measurements of p, V_m, and T on gaseous C_6H_6 (benzene) at temperatures between 330 and 420 K and at pressures below atmospheric can be fitted within the experimental uncertainty of about ± 20 cm^3 mol^{-1} by the formula: $pV_m = RT + Bp$, with B given by

$$B/cm^3 \ mol^{-1} = -77 - 66.15 \ exp(906.7 \ K/T) .$$

Measurements of $\{H_m(T, p_2) - H_m(T, p_1)\}/(p_2 - p_1)$ made in a flow calorimeter fitted with a throttle and an electric heater lead to the result at 373.5 K and $p \to 0$:

$$(\partial H_m/\partial p)_T = -(2630 \pm 20) \ cm^3 \ mol^{-1} .$$

Measurements of the molar heat capacity $C_{p,m}$ made as a function of pressure in a flow calorimeter fitted with an electric heater lead to the result at 402.3 K and $p \to 0$:

$$(\partial C_{p,m}/\partial p)_T = (15.3 \pm 0.4) \ cm^3 \ K^{-1} \ mol^{-1} .$$

Are these results thermodynamically consistent within the admitted experimental errors?

Chapter 7

Thermodynamic inequalities and their consequences

§ 7.1 Scope

We shall begin the chapter by completing our statement of the second law by declaring as an axiom one inequality for the entropy change that is occurring if anything is happening in an isolated system. We shall then have the five axioms (the '−1th law' in § 1.10, the zeroth law in § 1.13, the first law in § 1.18, the equality of the second law in § 5.2, and now the inequality of the second law) which (together with Nernst's heat theorem, already mentioned in a footnote on p. 59) are all the rules that we shall need to play out the whole game of thermodynamics.

In § 7.2 we shall express the inequality in alternative forms by use of the auxiliary quantities H, A, and G. We shall then derive the conditions for anything to be happening, and the conditions for nothing to be happening – that is to say for equilibrium: in § 7.3 for an isolated system of two phases at different temperatures (where we shall incidentally prove at last that T defined by equation (5.2.1) *is* a temperature); in § 7.4 for an isolated system of two phases at the same temperature but at different pressures; in § 7.5 for an isolated system of two phases at the same temperature, but not necessarily at the same pressure, with different chemical potentials of one of the substances; and in § 7.6 for an isolated system of one phase in which the possibility exists of a change in the extent of a chemical reaction. We shall then derive in §§ 7.7 to 7.9 the conditions for the thermal stability of a phase, the hydrostatic stability of a phase, and the diffusional (or 'material') stability of a phase. In § 7.10 we shall briefly discuss the consequences of the inequality for heat engines working in cycles. Finally in § 7.11 we shall indulge ourselves in a philosophical digression on the 'interpretation' of entropy changes.

§ 7.2 The fundamental inequality

If anything is happening in an *isolated*† system Σ composed of one or more phases α, β, $\cdots$, that is to say if any measurable property of the system is changing perceptibly, then the entropy S^Σ of the system is increasing. In symbols:

$$(\partial S^\Sigma/\partial t)_{U^\Sigma, V^\Sigma, N^\Sigma} > 0 , \tag{7.2.1}$$

---

† We remind the reader that an isolated system is one of constant energy, volume, and content. The fundamental inequality applies *only* to an isolated system.

where t denotes time and where N^Σ denotes total material content of the system without regard to states of chemical or physical aggregation.[†]

If nothing is happening in the isolated system then of course nothing is happening to the entropy and

$$(\partial S^\Sigma / \partial t)_{U^\Sigma, V^\Sigma, N^\Sigma} = 0 . \tag{7.2.2}$$

The isolated system is then said to be in a state of equilibrium. Any system that has been isolated for long enough to ensure that anything that was happening has stopped happening is in a state of equilibrium.

The inequality (1) forms the other part of the second law of thermodynamics; equation (5.2.1) and inequality (1) together form a complete statement of the law. We shall regard inequality (1), like equation (5.2.1), as an axiom that stands or falls by its observable consequences.

It will be noticed that the definition of an equilibrium state given above includes states that we know to be *metastable* with respect to some change to a more stable state (for example a mixture of hydrogen and oxygen at room temperature in the absence of a catalyst), provided that anything that is or might be happening is happening so slowly as to be undetected with the measuring instruments available to us during the time of our observation. We may thus regard a particular system as being at equilibrium when we study it with a particular set of measuring instruments and over a particular time interval, in spite of our knowing (or not knowing) that it is not at equilibrium when we study it with more sensitive measuring instruments or over a longer time interval.

§ 7.3 Derived inequalities

It follows from equation (7.2.1) by use of the '−1 rule' for a system of constant uniform temperature T, since $(\partial U^\Sigma / \partial S^\Sigma)_{t, V^\Sigma, N^\Sigma} = T$, that

$$(\partial U^\Sigma / \partial t)_{S^\Sigma, V^\Sigma, N^\Sigma} < 0 , \tag{7.3.1}$$

or in words that if anything is happening in a system of constant entropy, volume, and content, then the energy is decreasing.

It follows similarly for a system of constant uniform pressure p that

$$(\partial S^\Sigma / \partial t)_{H^\Sigma, p, N^\Sigma} > 0 , \tag{7.3.2}$$

and that

$$(\partial H^\Sigma / \partial t)_{S^\Sigma, p, N^\Sigma} < 0 , \tag{7.3.3}$$

or in words that if anything is happening in a system of constant enthalpy, pressure, and content, then the entropy is increasing; and that if anything is happening in a system of contant entropy, pressure, and content, then the enthalpy is decreasing.

[†] We shall be using time t as one of our variables in this chapter. There are those who say that time has no place in thermodynamics. They are wrong.

It follows similarly for a system of constant uniform temperature and constant volume and content that,

$$(\partial A^{\Sigma}/\partial t)_{T,V^{\Sigma},N^{\Sigma}} < 0 , \qquad (7.3.4)$$

and for a system of constant uniform temperature and pressure and constant content that

$$(\partial G^{\Sigma}/\partial t)_{T,p,N^{\Sigma}} < 0 , \qquad (7.3.5)$$

or in words that if anything is happening in a system of constant temperature, volume, and content, then the Helmholtz function is decreasing, and that if anything is happening in a system of constant temperature, pressure, and content, then the Gibbs function is decreasing.

Since most chemical reactions are carried out at (more or less) constant temperature and pressure, equation (5) is a particularly important one for chemists. We note that whatever 'interpretation' might be placed on the increasing nature of the entropy in an isolated system (or in an isenthalpic and isobaric closed system), exactly the same 'interpretation' must be placed on the decreasing nature of the Gibbs function divided by temperature for the same process (such as a chemical reaction) carried out in a closed system at constant temperature and pressure, and that the same 'interpretation' can *not* be placed on $(\partial S^{\Sigma}/\partial t)_{T,p,N^{\Sigma}}$, which if anything is happening in the system may have either sign.

§ 7.4   Thermal equilibrium

We consider the exchange of energy between two phases α and β separated by a rigid impermeable diathermic partition in the isolated system shown diagrammatically in figure 7.1. Let the thermodynamic temperatures of the phases be T^{α} and T^{β}. Since the system is isolated and the partition is rigid and impermeable, we have

$$dU^{\Sigma} = dU^{\alpha} + dU^{\beta} = 0 , \qquad (7.4.1)$$

$$dV^{\alpha} = 0 , \quad \text{and} \quad dV^{\beta} = 0 , \qquad (7.4.2)$$

$$dn_{B}^{\alpha} = 0 , \quad \text{and} \quad dn_{B}^{\beta} = 0 , \quad \text{(all B)} . \qquad (7.4.3)$$

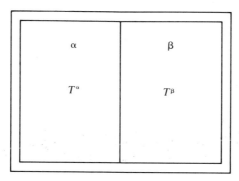

Figure 7.1. Isolated system composed of two phases α and β separated by a rigid impermeable diathermic partition.

Application of equation (5.2.1) yields

$$dS^\alpha = dU^\alpha/T^\alpha, \quad \text{and} \quad dS^\beta = dU^\beta/T^\beta, \tag{7.4.4}$$

so that

$$dS^\Sigma = (dU^\alpha/T^\alpha) + (dU^\beta/T^\beta) = dU^\alpha(T^\beta - T^\alpha)/T^\alpha T^\beta. \tag{7.4.5}$$

We now use the fundamental inequality (7.2.1). If $T^\beta > T^\alpha$, and if anything is happening in the isolated system, then it follows from the inequality (7.2.1) that $dU^\alpha/dt > 0$ and $dU^\beta/dt < 0$, that is to say energy is flowing from the phase with the higher temperature T^β to that with the lower temperature T^α. If $T^\beta = T^\alpha$ then $dS^\Sigma/dt = 0$ and nothing is happening; the system is in a state of *thermal equilibrium*.

It thus follows from the fundamental inequality (7.2.1) that the quantity T defined by equation (7.2.1), which we have called the *thermodynamic temperature, is* a temperature in the sense described in § 1.13.

As promised in § 1.13 we shall henceforth use the word 'system' only for a system that is in internal thermal equilibrium and that therefore has a uniform thermodynamic temperature T.

We shall have no further occasion to use any temperature other than the thermodynamic temperature and shall accordingly henceforth usually omit the adjective 'thermodynamic'.

§ 7.5   Hydrostatic equilibrium

We consider the exchange of volume between two phases α and β separated by a movable impermeable diathermic partition in the isolated system of uniform temperature T shown diagrammatically in figure 7.2. Let the pressures of the phases be p^α and p^β. Since the system is isolated and the partition is impermeable we have

$$dU^\Sigma = dU^\alpha + dU^\beta = 0, \tag{7.5.1}$$

$$dV^\Sigma = dV^\alpha + dV^\beta = 0, \tag{7.5.2}$$

$$dn_B^\alpha = 0, \quad \text{and} \quad dn_B^\alpha = 0, \quad \text{(all B)}. \tag{7.5.3}$$

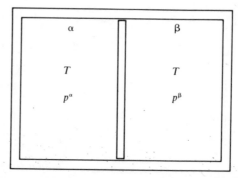

Figure 7.2. Isolated system composed of two phases α and β separated by a movable impermeable diathermic partition.

Application of equation (5.2.1) yields

$$T\,dS^\alpha = dU^\alpha + p^\alpha\,dV^\alpha, \quad \text{and} \quad T\,dS^\beta = dU^\beta + p^\beta\,dV^\beta, \tag{7.5.4}$$

so that

$$T\,dS^\Sigma = (dU^\alpha + dU^\beta) + (p^\alpha\,dV^\alpha + p^\beta\,dV^\beta)$$
$$= (p^\alpha - p^\beta)\,dV^\alpha. \tag{7.5.5}$$

We now use the inequality (7.2.1). If $p^\beta > p^\alpha$, and if anything is happening in the isolated system, then it follows from the inequality that $dV^\alpha/dt < 0$ and $dV^\beta/dt > 0$, that is to say the volume of the phase having the lower pressure p^α is decreasing and that of the phase having the higher pressure p^β is increasing, or in other words the partition shown in figure 7.2 is moving to the left. If $p^\beta = p^\alpha$ then $dS^\Sigma/dt = 0$ and nothing is happening; the system is in a state of *hydrostatic equilibrium*.

We should have been entitled to assume this purely mechanical result but it is satisfactory to have derived it from our statement of the second law of thermodynamics.

Most, *but not all*, of the systems with which we shall have to deal will be in internal hydrostatic equilibrium and will therefore have a uniform pressure p.

§ 7.6   Diffusive equilibrium

We consider the exchange of a substance A between two phases α and β separated by a rigid diathermic partition permeable to the substance A in the isolated system of uniform temperature T shown diagrammatically in figure 7.3. Let the chemical potentials of the substance A be μ_A^α and μ_A^β. Since the system is isolated and the partition is rigid and impermeable to substances other than A we have

$$dU^\Sigma = dU^\alpha + dU^\beta = 0, \tag{7.6.1}$$

$$dV^\alpha = 0, \quad \text{and} \quad dV^\beta = 0, \tag{7.6.2}$$

$$dn_A^\Sigma = dn_A^\alpha + dn_A^\beta = 0, \tag{7.6.3}$$

$$dn_B^\alpha = 0, \quad \text{and} \quad dn_B^\beta = 0, \quad \text{(all B} \neq \text{A)}. \tag{7.6.4}$$

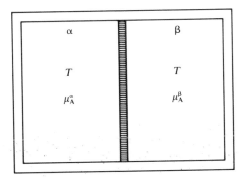

Figure 7.3. Isolated system composed of two phases α and β separated by a rigid diathermic partition permeable to the substance A.

Application of equation (5.2.1) yields

$$T \, dS^\alpha = dU^\alpha + p^\alpha \, dV^\alpha - \mu_A^\alpha \, dn_A^\alpha,$$

and

$$T \, dS^\beta = dU^\beta + p^\beta \, dV^\beta - \mu_A^\beta \, dn_A^\beta, \tag{7.6.5}$$

so that

$$T \, dS^\Sigma = (\mu_A^\beta - \mu_A^\alpha) \, dn_A^\alpha. \tag{7.6.6}$$

We now use the inequality (7.2.1). If $\mu_A^\beta > \mu_A^\alpha$, and if something is happening in the isolated system, then $dn_A^\alpha/dt > 0$ and $dn_A^\beta/dt < 0$, that is to say the substance A is flowing from the phase in which it has the higher chemical potential μ_A^β to that in which it has the lower chemical potential μ_A^α. If $\mu_A^\alpha = \mu_A^\beta$ then $dS^\Sigma/dt = 0$ and nothing is happening; the system is in a state of *diffusive equilibrium*.

We note that we did *not* assume above that the system was in a state of hydrostatic equilibrium. Diffusive equilibrium ($\mu_A^\alpha = \mu_A^\beta$) for some but not all of the substances B in the absence of hydrostatic equilibrium ($p^\alpha \neq p^\beta$) is called *osmotic equilibrium* and will be extensively discussed in chapter 10.

Had we assumed instead that the partition was movable, then since $p^\alpha = p^\beta$ and $dV^\alpha = -dV^\beta$, the same result would have been obtained.

We note that we did, however, assume that the system was in a state of thermal equilibrium. Indeed we shall find that the chemical-potential difference ($\mu_A^\alpha - \mu_A^\beta$) is defined (that is to say is measurable) only when $T^\alpha = T^\beta$.

§ 7.7   Chemical equilibrium

In §§ 7.4 to 7.6 we implicitly excluded the possibility of any change in the extent of any chemical reaction in either of the phases α and β. We now consider a change in the extent of reaction ξ for the chemical reaction according to the equation:

$$0 = \sum_B \nu_B B, \tag{7.7.1}$$

in the isolated single-phase system shown diagrammatically in figure 7.4. Since the system is isolated we have

$$dU = 0, \tag{7.7.2}$$

$$dV = 0, \tag{7.7.3}$$

$$dn_B = \nu_B \, d\xi, \quad \text{(all B)}. \tag{7.7.4}$$

Application of equation (5.2.1) yields

$$T \, dS = dU + p \, dV - \sum_B \mu_B \, dn_B$$

$$= -\left(\sum_B \nu_B \mu_B\right) d\xi = A \, d\xi, \tag{7.7.5}$$

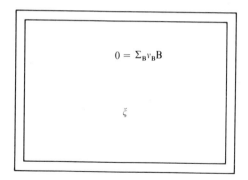

Figure 7.4. Isolated single-phase system containing a chemically reacting mixture.

where we have used the affinity A introduced in § 5.9. We now use the inequality (7.2.1). If $\sum_B \nu_B \mu_B < 0$ so that $A > 0$, and if anything is happening in the isolated system, then $d\xi/dt > 0$, that is to say the extent of the reaction is increasing. If $\sum_B \nu_B \mu_B > 0$ so that $A < 0$, and if anything is happening in the isolated system, then $d\xi/dt < 0$, that is to say the extent of reaction is decreasing; the reaction defined by the chemical equation (1) is 'going backwards'. If $\sum_B \nu_B \mu_B = 0$ so that $A = 0$ then $dS/dt = 0$ and nothing is happening; the system is in a state of chemical equilibrium.

The extension of the results of this section to more than one chemical reaction in a phase α is straightforward. An independent equation of the form $\sum_B \nu_B \mu_B^\alpha = 0$ is satisfied for each independent chemical reaction. One or more of the substances can of course be common to two or more of the reactions.

The extension of the results of this section to one or more chemical reactions in a system Σ consisting of two or more phases $\alpha, \beta, \cdots$, is likewise straightforward. Provided that any membrane separating a pair of phases is impermeable to not more than one of the reacting substances then, when the system has reached diffusive and chemical equilibrium, $\mu_B^\alpha = \mu_B^\beta$ for each of the substances and $\sum_B \nu_B \mu_B^\alpha = \sum_B \nu_B \mu_B^\beta = 0$.

§ 7.8   Thermal stability of a phase

We shall now derive the condition that prevents an isolated phase of fixed composition from acquiring different temperatures in different parts. We suppose that the phase, initially of energy $2U$ and volume $2V$, can split into two halves, one half having the energy $(U + \delta U)$ and volume V and the other half having the energy $(U - \delta U)$ and volume V. If that were to happen then the entropy change would be

$$\delta S = S(U + \delta U, V) + S(U - \delta U, V) - S(2U, 2V), \tag{7.8.1}$$

which on expansion by Taylor's theorem (see appendix III) gives

$$\delta S = (\partial^2 S / \partial U^2)_V (\delta U)^2 + O(\delta U)^4, \tag{7.8.2}$$

where $O(\delta U)^4$ means terms in $(\delta U)^4$ and higher powers of δU. Since $(\partial S/\partial U)_V = 1/T$, we have

$$(\partial^2 S/\partial U^2)_V = (\partial T^{-1}/\partial U)_V = -1/T^2 C_V. \tag{7.8.3}$$

From equations (2) and (3) we deduce that δS would be positive, so that the contemplated change would be happening, if C_V were negative. The condition that determines the thermal stability of an isolated phase is therefore

$$C_V > 0. \tag{7.8.4}$$

That is the condition for example that prevents an isolated bar of metal in thermal equilibrium from becoming hot at one end and cold at the other end.

If energy is added to a phase of fixed composition and constant volume then its temperature always increases.

§ 7.9 Hydrostatic stability of a phase

We shall now derive for an isolated phase of fixed composition the condition that prevents movement of a partition, which would cause the pressures of the two parts to become different. We suppose that the partition between the two parts of the phase, initially each of temperature T and volume V, can move so that one part has the volume $(V + \delta V)$ and temperature T and the other part has volume $(V - \delta V)$ and temperature T. The partition may of course be imaginary. For the variables T and V the appropriate thermodynamic function is the Helmholtz function A. If the contemplated change were to happen then

$$\delta A = A(T, V + \delta V) + A(T, V - \delta V) - A(T, 2V), \tag{7.9.1}$$

which on expansion by Taylor's theorem gives

$$\delta A = (\partial^2 A/\partial V^2)_T (\delta V)^2 + O(\delta V)^4. \tag{7.9.2}$$

Since $(\partial A/\partial V)_T = -p$, we have

$$(\partial^2 A/\partial V^2)_T = -(\partial p/\partial V)_T. \tag{7.9.3}$$

From equations (2) and (3) we deduce that δA would be negative, so that the contemplated change would be happening, if $(\partial p/\partial V)_T$ were positive. The condition that determines the hydrostatic stability of an isolated phase in thermal equilibrium is therefore

$$(\partial p/\partial V)_T < 0, \tag{7.9.4}$$

or

$$-V^{-1}(\partial V/\partial p)_T = \kappa_T > 0. \tag{7.9.5}$$

If the pressure of a phase of fixed composition is increased at constant temperature then its volume always decreases.

§ 7.10   Diffusional stability of a phase

We shall now derive for an isolated phase the condition that prevents a substance from becoming concentrated in one part and depleted in another part. We suppose that the phase, initially of temperature T, pressure p, and containing amounts of substance $2n_A$, $2n_B$, $2n_C$, $\cdots$, of A, B, C, $\cdots$, can split into two halves, one half having the temperature T, pressure p, amount of substance $(n_A + \delta n_A)$ of A, and amounts of substance n_B, n_C, $\cdots$, of B, C, $\cdots$; and the other half having temperature T, pressure p, amount of substance $(n_A - \delta n_A)$ of A, and amounts of substance n_B, n_C, $\cdots$, of B, C, $\cdots$. For the variables T, p, and the n's the appropriate thermodynamic function is the Gibbs function G. If the contemplated change were to happen then

$$\delta G = G(T,\ p,\ n_A + \delta n_A, n_B, n_C, \cdots) + G(T,\ p,\ n_A - \delta n_A, n_B, n_C, \cdots)$$
$$- G(T, p, 2n_A, 2n_B, 2n_C, \cdots), \tag{7.10.1}$$

which on expansion by Taylor's theorem gives

$$\delta G = (\partial^2 G/\partial n_A^2)_{T,p,n_B,n_C,\cdots}(\delta n_A)^2 + O(\delta n_A)^4. \tag{7.10.2}$$

Since $(\partial G/\partial n_A)_{T,p,n_B,n_C,\cdots} = \mu_A$, we have

$$(\partial^2 G/\partial n_A^2)_{T,p,n_B,n_C,\cdots} = (\partial \mu_A/\partial n_A)_{T,p,n_B,n_C,\cdots}. \tag{7.10.3}$$

From equations (2) and (3) we deduce that δG would be negative, so that the contemplated change would be happening, if $(\partial \mu_A/\partial n_A)_{T,p,n_B,n_C,\cdots}$ were negative. The condition that determines the diffusional stability of an isolated phase in thermal and hydrostatic equilibrium is therefore

$$(\partial \mu_A/\partial n_A)_{T,p,n_B,n_C,\cdots} > 0. \tag{7.10.4}$$

If the substance A is added to a phase of otherwise fixed composition at constant temperature and pressure then the chemical potential of A always increases.

§ 7.11   The efficiency of a heat engine

For a complete cycle carried out on the working substance of a heat engine we have

$$\Delta U = 0, \tag{7.11.1}$$

so that the total work $-W$ done on the surroundings may be related to the several quantities of heat Q_1, Q_2, $\cdots$, by

$$-W = \sum_i Q_i. \tag{7.11.2}$$

If we distinguish between positive quantities of heat Q_j (that is to say quantities of heat absorbed by the working substance from its surroundings) and negative quantities of heat Q_k (that is to say quantities of heat absorbed by the surroundings from the working substance), we may rewrite equation (2) in the forms:

$$-W = \sum_j Q_j + \sum_k Q_k = \sum_j |Q_j| - \sum_k |Q_k| . \qquad (7.11.3)$$

The quantity η defined by

$$\eta \overset{\text{def}}{=} -W \Big/ \sum_j |Q_j| = \Big(\sum_j |Q_j| - \sum_k |Q_k| \Big) \Big/ \sum_j |Q_j| , \qquad (7.11.4)$$

is called the *efficiency* of the heat engine. It is the ratio of the work done by the system on its surroundings to the heat done by the surroundings on the system. Obviously the efficiency would be greatest, that is have the value 1, if no heat were 'wasted' by flowing from the system to the surroundings. The first law, all that we have so far used in this section, does not forbid that. We shall see that the second law does.

For a complete cycle on the working substance of a heat engine we also have

$$\Delta S = 0 . \qquad (7.11.5)$$

Making use of the second-law equation for a phase of fixed composition we obtain for any step i in the cycle

$$\Delta_i S = \Big(\Delta U + \int p \, dV \Big)_i \Big/ T_i$$

$$= Q_i / T_i + \Big(W + \int p \, dV \Big)_i \Big/ T_i . \qquad (7.11.6)$$

If the expansion or compression is reversible then $(W + \int p \, dV)_i = 0$; otherwise $(W + \int p \, dV)_i > 0$. Thus

$$\Delta_i S \geqslant Q_i / T_i , \qquad (7.11.7)$$

the equality applying for a reversible step and the inequality for an irreversible step. Summing over all the steps of a complete cycle and making use of equation (5) we obtain

$$\sum_i Q_i / T_i \leqslant 0 . \qquad (7.11.8)$$

Since all the T_i's are positive and since some of the Q_i's must be positive if $-W > 0$, equation (8) implies that at least one of the Q_i's must be negative. Thus it is impossible to construct a heat engine that operates in cycles and produces work as the result of only cooling its surroundings. A heat engine must absorb some heat $\Sigma_j |Q_j|$ (from a 'source') and must reject some heat $\Sigma_k |Q_k|$ (into a 'sink'). The useful work $-W$ is the difference between those two quantities of heat.

§ 7.12 Carnot's cycle

We now consider a cycle in which heat is absorbed at only one temperature T_1, in an isothermal step, and is rejected at only one temperature T_2, in another isothermal step. The other steps in the cycle must be adiabatic ($Q = 0$). Thus

$$Q_1/T_1 + Q_2/T_2 \leqslant 0, \tag{7.12.1}$$

or

$$|Q_1|/T_1 - |Q_2|/T_2 \leqslant 0, \tag{7.12.2}$$

so that

$$1 - |Q_2|/|Q_1| \leqslant (T_1 - T_2)/T_1. \tag{7.12.3}$$

The efficiency of this particular cycle, called Carnot's cycle, is given by

$$\eta = 1 - |Q_2|/|Q_1|, \tag{7.12.4}$$

which on substitution from (3) becomes

$$\eta \leqslant (T_1 - T_2)/T_1. \tag{7.12.5}$$

The efficiency of a Carnot's cycle cannot be greater than $(T_1 - T_2)/T_1$, and can be equal to $(T_1 - T_2)/T_1$ only if the expansion and compression are each done reversibly.

It can be shown that no cycle can be as efficient as a reversible Carnot's cycle.

Some authors base their introduction of the second law on an analysis of Carnot's cycle. If that is done rigorously then it is inevitably laborious.[†] More often it is done improperly by the implicit assumption of that which is to be proved or by seeming to deduce the second law from the first.[‡] Arguments based on heat engines have little appeal to chemists. We have preferred to use Gibbs's method,[§] followed by Guggenheim,[¶] namely to state the second law as an algebraic equation (5.2.1) and an inequality (7.2.1), to treat those relations as axioms, to deduce their consequences, and to compare their consequences with experiment.

§ 7.13   Philosophical digression

Thermodynamics is an experimental science, and not a branch of metaphysics. It consists of a collection of equations, and also some inequalities, which inter-relate certain kinds of measurable physical quantities. In any thermodynamic equation every quantity is independently measurable. What can such an equation 'tell one'

[†] One author for example (Denbigh, K. G. *The Principles of Chemical Equilibrium*. Cambridge University Press. Third edition. 1971.) takes about 15 pages rigorously to introduce the second law *via* Carnot's cycle.

[‡] Briefly, $\Delta U = W + Q$; *assume* that a substance exists for which simultaneously (i) $pV = nRT$ and (ii) $dU = C_V\, dT$. Assumption (ii) implies that $(\partial U/\partial V)_V = 0$ and $(\partial U/\partial T)_V = C_V(T)$. Then $Q/T = dU/T - W/T = C_V\, dT/T + nR\, dV/V + (p_e - p)\, dV/T$, so that when $p_e = p$, that is to say when the volume change is reversible, $Q/T = C_V(T)\, dT/T + nR\, dV/V$. Therefore, for a reversible change on a substance defined by both (i) and (ii), $\Sigma(Q/T)$ depends only on the initial and final temperatures and on the initial and final volumes, and so for a complete reversible cycle on such a substance $\Sigma(Q/T) = 0$. But the assumption that $(\partial U/\partial V)_T = 0$ for a perfect gas is a *consequence* of the second law.

[§] Gibbs, J. W. *The Collected Works*. Volume I. *Thermodynamics*. Longmans Green: New York. **1928**.

[¶] Guggenheim, E. A. *Thermodynamics*. North-Holland: Amsterdam. Fifth edition. **1967**.

about one's system or process? Or, in other words, what can we learn from such an equation about the microscopic explanation of a macroscopic change? Nothing whatever. What then is the use of thermodynamic equations? They are useful because some quantities are easier to measure than others.

For some physical quantities, like length and temperature, we arrive into the world already equipped with meters, albeit crude ones. The human foot can be used to measure the length of a room or the temperature of a bath. For other physical quantities, such as entropy, we have no built-in physiological 'yardstick'. Presumably, because we 'understand' such questions as 'How long is it?' and 'How hot is it?', we are lured into trying to provide ourselves with similar familiarity with thermodynamic functions. In particular, the ordinary physical quantity called the entropy has become the subject of almost mystical veneration. Popular songs, even, have been written about it. It has been surrounded and obscured by a cloud of home-spun philosophy, made up of statements having no meaning, of dangerously misleading half-truths, and of downright lies, none of which has any place in physical science.

An understanding of entropy changes is coextensive with an understanding of how to measure them and of how to relate them to other measured quantities. Yet many of those who pontificate about the so-called meaning of entropy changes are quite unable to answer such an ordinary question as 'How would you measure the entropy change that accompanies the mixing of two gases at room temperature and pressure?'.

Thermodynamics reveals nothing of any microscopic or molecular meaning for its functions. The science of statistical mechanics (which will be introduced in chapter 15) does provide a severely quantum-mechanical microscopic interpretation for the thermodynamic functions, but leads to no simply geometrical 'picture' of them. Nevertheless it is widely assumed by chemists and others that increase of entropy implies increase of 'disorder' or of 'randomness' or of 'mixed-upness'.†

When, if ever, is this assumption justified?

---

† The word 'probability' is also used. If it is used to mean 'the number of accessible eigenstates' then it is formally correct at least for isolated systems. But it is a curious use of the word. Whereas statistical probabilities have values that lie between 0 and 1, these 'probabilities' can have values enormously greater than 1. More often the word is used in such a way that the words 'entropy' and 'probability' have the same relation to 'something that is happening'. One of the words is then redundant, and the statement 'The entropy of an isolated system is increasing because the probability is increasing' is equivalent to the statement 'the entropy of an isolated system is increasing because something is happening', a statement that adds no 'interpretation' to the inequality (7.2.1).

In the words of a distinguished philosopher of science (Margenau, H. *The Nature of Physical Reality*. McGraw-Hill: New York. **1950**, p. 215): "Entropy is a quantity which is both a proper variable of state and measurable. To many students of science entropy is something very abstract, something closely related to probabilities. This attitude represents an improper approach to thermodynamics, where entropy enjoys the same status as pressure and temperature. The concept would have perfectly good meaning even if probabilities had never been invented – and we beseech the philosophic reader to dismiss at this point all associations with bags of mixed-up marbles and decks of shuffled cards if he desires a correct view of things. Entropy is as definite and clear a thing as other thermodynamic quantities. With this in mind we proceed to state the Second Law as follows: The entropy of a closed system never decreases."

When something is happening *in an isolated system*, the entropy is increasing. By the methods of statistical mechanics the entropy increase in such an isolated system can be simply related to the increase in the number Ω of eigenstates to which the system has access:

$$S(U, V, N, t_2) - S(U, V, N, t_1) = k \ln\{\Omega(U, V, N, t_2)/\Omega(U, V, N, t_1)\}. \quad (7.13.1)$$

When, if ever, has that number Ω anything to do with any of those words like 'mixed-upness'?

It does, but only for two very special cases. These are mixtures of perfect gases, and crystals at temperatures close to zero, neither of which is commonly studied in the ordinary chemical laboratory. In each of those two special cases the number Ω of accessible eigenstates is simply related to the purely geometrical or spatial 'disorder' or 'randomness' or 'mixed-upness',[†] and so therefore is the entropy. But that is so only in those two special cases.

In all other, and all ordinary chemical processes, the entropy change is capable of no simply geometrical interpretation *even for changes in isolated systems*. One may not wantonly extrapolate statistical-mechanical conclusions for systems of non-interacting particles (that is to say, perfect gases) or for very cold crystals, to beakers of liquids at ordinary temperatures.

Consider what is happening in the following experiment. Just before a super-saturated solution of aqueous $Na_2S_2O_3$ (a perfectly definite equilibrium state) was isolated, for example in a Dewar flask, a crystal of $Na_2S_2O_3$ had been added. Something is happening. The system being an isolated one the entropy is therefore increasing. But what is happening is the partial unmixing of the solution. The 'mixed-upness' is decreasing. There is spatial sorting of the solute from the solvent.[‡] Many other examples can be found. One more must suffice. Consider an isolated system consisting of a mixture of hydrogen + argon separated by a palladium membrane (permeable to hydrogen but not to argon) from an evacuated space. The process of sorting or unmixing that is happening is accompanied by an increase of entropy, but can hardly be described as an increase of 'mixed-upness'.

But the situation is even worse than this, for chemists commonly behave not only as though increases of entropy in *isolated* systems were a measure of increasing disorder, but also as though that were true of increases of entropy in systems *at constant temperature and pressure*, under which conditions, very different from those of isolation, most chemical reactions are actually carried out. Even if the entropy were a measure of disorder in an isolated system (and we have seen that it is not), the

---

† When the partition is broken between two different perfect gases in an isolated container the volume available to each kind of molecule is increased. For each gas the number of combinations of translational energy eigenvalues that can be used to make up its energy is thereby increased. In crystals at temperatures close to zero the geometrical orientations of the molecules on the lattice sites may be regular or irregular, may be ordered or disordered.

‡ Those who are confused by the rise in temperature in tnat experiment (it is irrelevant) might think instead about the similar experiment with a supersaturated solution of Na_2SO_4 for which in isolation the temperature falls as the solid anhydrous salt precipitates.

corresponding measure in an isothermal and isobaric experiment would be the Gibbs function, and not the entropy:

$$(\partial S/\partial t)_{U,V,N} = -T^{-1}(\partial G/\partial t)_{T,p,N} \geqslant 0 . \tag{7.13.2}$$

Thus, several well known text-books of physical chemistry assert that the entropy of mixing of liquids as ordinarily measured at constant temperature and pressure is necessarily positive. One counter-example is enough; many are known. The molar entropy of mixing of $\{0.5H_2O + 0.5(C_2H_5)_2NH\}$ at 322.25 K is[†] -8.78 J K^{-1} mol^{-1}.

Worse still, many chemists interpret *standard* entropy changes $\Delta S^{\ominus}$ (see chapter 12) in the same way. Values of $\Delta S^{\ominus}$ can change sign for a different conventional choice of the standard pressure $p^{\ominus}$ or of the standard molality $m^{\ominus}$ which they usually contain.

Some of this kind of misunderstanding has even rubbed off on the layman. I suspect that Lord Snow's choice of the second law of thermodynamics as a test of literacy for the humanist (he said[‡] it was "about the scientific equivalent of: *Have you read a work of Shakespeare's?*") derives from just such a misunderstanding about the meaning of entropy, and especially from its supposed implications with respect to the Fate of the Universe.[§]

Since I regard those supposed implications as nonsensical I can hardly be expected to give a coherent account of them, but I think that the argument runs something like this. Any process that actually takes place in the universe increases the entropy of the universe. [Untestable.] Increase of entropy implies increase of mixed-upness. [Untrue, even in an isolated system.] Therefore, the inevitable fate of the universe is a state of more and more mixed-upness. In the end there would be neither sea nor dry land, but I suppose a kind of porridge. But then there would be no earth and no heavens but just some species of structureless particle distributed uniformly throughout space. Then the universe would have suffered the so-called 'heat-death', and chaos would be come again.

Now I suppose that if thermodynamics could be shown (but how?) to be applicable to the universe, and if the universe were known to be a bounded and isolated system (whatever that might mean in the case of the universe), then we might deduce that the universe would eventually reach a state of complete thermal, hydrostatic, diffusional, chemical, $\cdots$, equilibrium. But there is no scientific reason to suppose that the universe is a bounded isolated system. Isolated? Isolated from what? And even if it were, there is no reason to suppose that thermodynamics, which is an experimental

---

[†] Copp, J. L.; Everett, D. H. *Discuss. Faraday Soc.* **1953**, 15, 174.

[‡] Snow, C. P. *The Two Cultures and a Second Look.* Cambridge University Press. **1964**, p. 15.

[§] "Finally, a word of warning against the tendency to drag 'the universe' into thermodynamic theory when to do so is neither required nor justifiable", quoted from Buchdahl, H. A. *The Concepts of Classical Thermodynamics.* Cambridge University Press. **1966**, p. 17.

"The third misconception is that thermodynamics, and in particular the concept of entropy, can without further enquiry be applied to the whole universe", quoted from Landsberg, P. T. *Thermodynamics.* Interscience: New York. **1961**, p. 391.

science, can be applied to systems as large as the universe. Clausius's famous aphorism: "Die Entropie der Welt strebt einem Maximum zu"† is, to say the least, misleading. The pessimistic idea that the fate of the universe is chaos no doubt has an appeal to a certain kind of mind, but it is a matter of faith which has no support from the science of thermodynamics.

† "The entropy of the universe tends towards a maximum."

Chapter 8

Thermodynamics of a heterogeneous system

§ 8.1 Scope

Chapters 5 and 6 were devoted to the thermodynamics of a phase, that is to say of a *homogeneous* system. In the present chapter we shall extend our equations to a system composed of more than one phase, that is to say to a *heterogeneous* system. Using the results of §§ 7.4 to 7.7 we shall show how those equations can be simplified for the several kinds of phase equilibrium. We shall complete the chapter by deriving Gibbs's *phase rule*.

§ 8.2 Change of the state of a heterogeneous system

The equations of chapter 5 for a system composed of just one phase α can be readily extended to a system Σ composed of several phases α, β, $\cdots$. We recall the fundamental equation (5.3.9):

$$dG^\alpha = -S^\alpha \, dT^\alpha + V^\alpha \, dp^\alpha + \sum_B \mu_B^\alpha \, dn_B^\alpha, \tag{8.2.1}$$

for a change of the state of a phase α. Now, since

$$dG^\Sigma = d(G^\alpha + G^\beta + \cdots) = d\sum_\alpha G^\alpha = \sum_\alpha dG^\alpha, \tag{8.2.2}$$

it follows that the extension of equation (1) to a heterogeneous system Σ is simply

$$dG^\Sigma = -\sum_\alpha S^\alpha \, dT^\alpha + \sum_\alpha V^\alpha \, dp^\alpha + \sum_\alpha \sum_B \mu_B^\alpha \, dn_B^\alpha. \tag{8.2.3}$$

Any other of the equations of chapter 5 for a phase α can similarly be extended to a heterogeneous system $\Sigma = \alpha + \beta + \cdots$.

§ 8.3 Thermal equilibrium in a heterogeneous system

We shall deal in this book only with systems Σ that are in complete thermal equilibrium so that

$$T^\alpha = T^\beta = \cdots = T^\Sigma = T, \tag{8.3.1}$$

where here and henceforth we omit the superscript $^\Sigma$ from T, and from any other intensive variable that has a uniform value throughout the system.

For such systems in complete thermal equilibrium equation (8.2.3) becomes

$$dG^\Sigma = -S^\Sigma dT + \sum_\alpha V^\alpha dp^\alpha + \sum_\alpha \sum_B \mu_B^\alpha dn_B^\alpha. \qquad (8.3.2)$$

§ 8.4   Hydrostatic equilibrium in a heterogeneous system

With a few important exceptions we shall deal in this book with systems that are not only in complete thermal equilibrium but also in complete hydrostatic equilibrium, so that

$$\left. \begin{aligned} T^\alpha = T^\beta = \cdots = T^\Sigma = T, \\ p^\alpha = p^\beta = \cdots = p^\Sigma = p. \end{aligned} \right\} \qquad (8.4.1)$$

For such systems in complete thermal and hydrostatic equilibrium equation (8.2.3) becomes

$$dG^\Sigma = -S^\Sigma dT + V^\Sigma dp + \sum_\alpha \sum_B \mu_B^\alpha dn_B^\alpha. \qquad (8.4.2)$$

§ 8.5   Diffusive equilibrium in a heterogeneous system

With a few important exceptions we shall deal in this book with systems that are in complete diffusive equilibrium with respect to every substance present. When the partitions separating the phases are permeable to *all* the substances present then the substances will diffuse between the phases until not only complete diffusive equilibrium, and of course thermal equilibrium, but also hydrostatic equilibrium have all been established. For such systems we then have

$$\left. \begin{aligned} T^\alpha = T^\beta = \cdots = T^\Sigma = T, \\ p^\alpha = p^\beta = \cdots = p^\Sigma = p, \\ \mu_B^\alpha = \mu_B^\beta = \cdots = \mu_B^\Sigma = \mu_B, \quad \text{(all B)}, \end{aligned} \right\} \qquad (8.5.1)$$

so that equation (8.2.3) becomes

$$dG^\Sigma = -S^\Sigma dT + V^\Sigma dp + \sum_B \mu_B dn_B^\Sigma, \qquad (8.5.2)$$

which we see is the same as equation (8.2.1) for a phase α, except that the superscript $^\alpha$ for a phase has been replaced by $^\Sigma$ for the system and that superscripts have been omitted from the symbols for uniform intensive quantities.

§ 8.6   Osmotic equilibrium

The few but important exceptions to systems described by equation (8.5.2) arise when two phases are separated by a *semi-permeable* partition or *membrane*, that is to say a partition permeable to some but not to all of the substances present. In general, equilibrium is possible in such systems only when the pressures of the two phases are different. We then have instead of equations (8.5.1)

$$\left. \begin{aligned} T^\alpha &= T^\beta, \\ \mu_B^\alpha &= \mu_B^\beta, \quad \text{(some but not all B)}, \\ p^\alpha &\neq p^\beta. \end{aligned} \right\} \qquad (8.6.1)$$

Such a partial equilibrium is called an *osmotic equilibrium* and the difference of pressures $|p^\alpha - p^\beta|$ is called the *osmotic pressure*. Osmotic equilibria will be extensively discussed in chapter 10.

§ 8.7   Chemical equilibrium in a heterogeneous system

In § 7.7 we learnt that if the chemical reaction:

$$0 = \sum_B \nu_B B, \qquad (8.7.1)$$

has reached chemical equilibrium in the phase α, then

$$\sum_B \nu_B \mu_B^\alpha = 0. \qquad (8.7.2)$$

The extension of equation (2) to a heterogeneous system is trivial when the system is in complete diffusive equilibrium with respect to all the substances present and so is also in complete thermal and hydrostatic equilibrium. This will be the case in all the applications dealt with in this book. For such systems we then have

$$\left. \begin{aligned} T^\alpha &= T^\beta = \cdots = T^\Sigma = T, \\ p^\alpha &= p^\beta = \cdots = p^\Sigma = p, \\ \mu_B^\alpha &= \mu_B^\beta = \cdots = \mu_B^\Sigma = \mu_B, \quad \text{(all B)}, \\ \sum_B \nu_B \mu_B &= 0. \end{aligned} \right\} \qquad (8.7.3)$$

Two points may be noted. First, if some of the substances present in the system play no part in the chemical reaction (1) they are still included among the B's but with their ν_B's put equal to zero. Second, there is no difficulty in the extension of equations (3) to systems in which more than one chemical reaction is taking place.

§ 8.8   Closed heterogeneous systems

For a system in complete thermal, hydrostatic, and diffusive equilibrium we have according to equation (8.5.2):

$$dG^\Sigma = -S^\Sigma \, dT + V^\Sigma \, dp + \sum_B \mu_B \, dn_B^\Sigma. \qquad (8.8.1)$$

If the system is also one of *fixed composition* so that $dn_B^\Sigma = 0$ for all B then equation (1) becomes

$$dG^\Sigma = -S^\Sigma \, dT + V^\Sigma \, dp, \qquad (8.8.2)$$

which is the same as equation (6.1.4) for a phase α of fixed composition except for the substitution of $^\Sigma$ for $^\alpha$.

If the system is *closed* then it is also one of fixed composition unless the extent ξ^Σ of a chemical reaction is changing, in which case we have, for the reaction $0 = \Sigma_B \, \nu_B B$,

$$dn_B^\Sigma = \nu_B \, d\xi^\Sigma, \quad \text{(all B)}, \qquad (8.8.3)$$

so that for a *closed system* equation (1) becomes

$$dG^\Sigma = -S^\Sigma \, dT + V^\Sigma \, dp + \left(\sum_B \nu_B \mu_B \right) d\xi^\Sigma$$

$$= -S^\Sigma \, dT + V^\Sigma \, dp - A \, d\xi^\Sigma, \qquad (8.8.4)$$

where we remind the reader that $A \overset{\text{def}}{=} -(\Sigma_B \, \nu_B \mu_B)$ is called the affinity of the chemical reaction.

Equation (4) reduces to equation (2) under either of two conditions. If no catalyst for the reaction is present then the reaction is *frozen* and $d\xi^\Sigma = 0$. If an efficient catalyst is present then chemical equilibrium is maintained and $A = 0$.

In the rest of this book we shall be concerned only with *closed systems* in complete *thermal equilibrium*. Most, but not all, of our systems will also be in complete *diffusive* and *hydrostatic equilibrium* so that equation (4) applies. Some of them will also be of fixed composition, or be maintained at chemical equilibrium, so that equation (2) applies.

§ 8.9   The phase rule

We recall the Gibbs–Duhem equation (5.7.1) for a phase α:

$$0 = S_m^\alpha \, dT - V_m^\alpha \, dp + \sum_B x_B^\alpha \, d\mu_B, \qquad (8.9.1)$$

where we have omitted the superscript $^\alpha$ from $T, p,$ and the μ_B's because we propose to discuss systems that are in complete thermal, hydrostatic, and diffusive equilibrium.

We see from equation (1) that the number of independent intensive variables for a phase is just $(\mathscr{C} + 1)$ where $\mathscr{C}$ is the number of *components* in the phase, that is to say the number of terms in the sum Σ_B. We might choose these $(\mathscr{C} + 1)$ independent intensive variables to be $T, p,$ and $(\mathscr{C} - 1)$ of the $\mathscr{C}$ chemical potentials μ_B, or we might choose them to be $T, p,$ and the $(\mathscr{C} - 1)$ independent mole fractions x_B^α, or in many other ways. Whichever choice we make, the number of independent intensive variables is always $(\mathscr{C} + 1)$.

Now let us consider a system of $\mathscr{P}$ phases in complete thermal, hydrostatic, and diffusive equilibrium. In place of equation (1) we now have a set of $\mathscr{P}$ equations, one for each phase:

$$
\left.
\begin{aligned}
0 &= S_m^\alpha \, dT - V_m^\alpha \, dp + \sum_B x_B^\alpha \, d\mu_B \, , \\[1em]
0 &= S_m^\beta \, dT - V_m^\beta \, dp + \sum_B x_B^\beta \, d\mu_B \, , \\[1em]
\cdot &= \cdots \qquad \cdots \qquad \cdots \quad , \\
\cdot &= \cdots \qquad \cdots \qquad \cdots \quad .
\end{aligned}
\right\}
\tag{8.9.2}
$$

The $(\mathscr{P}-1)$ additional equations impose $(\mathscr{P}-1)$ additional restrictions so that the number $\mathscr{F}$ of independent intensive variables is now given by

$$
\mathscr{F} = (\mathscr{C}+1)-(\mathscr{P}-1) = \mathscr{C}+2-\mathscr{P} \, . \tag{8.9.3}
$$

This number $\mathscr{F}$ is called the *variance* or the number of degrees of freedom of the system. Equation (3) is called the *phase rule*.

So far we have tacitly excluded chemical reactions. If the system is also in equilibrium with respect to the chemical reaction $0 = \Sigma_B \nu_B B$, then in addition to the $\mathscr{P}$ Gibbs–Duhem equations (2) we have also the equation:

$$
\sum_B \nu_B \mu_B = 0 \, , \tag{8.9.4}
$$

which reduces the variance $\mathscr{F}$ by one. If $\mathscr{R}$ *independent* chemical reactions are all in equilibrium then equation (3) must be replaced by

$$
\mathscr{F} = \mathscr{C}+2-\mathscr{P}-\mathscr{R} \, . \tag{8.9.5}
$$

Equation (5) is the phase rule extended to systems in which chemical reactions occur.

As an example of the application of equation (5) consider a gaseous mixture of nitrogen, hydrogen, and ammonia in the presence of a catalyst. We note that in view of the relation:

$$
x(N_2)+x(H_2)+x(NH_3) = 1 \, , \tag{8.9.6}
$$

only two of the mole fractions are independent.† When we apply equation (5) we obtain

$$
\mathscr{F} = 3+2-1-1 = 3 \, . \tag{8.9.7}
$$

We may choose these three independent variables to be two of the mole fractions, $x(NH_3)$ and $x(N_2)$ say, and the temperature T. Then the pressure p is a dependent variable. If instead we choose T, p, and $x(NH_3)$ as independent variables, then $x(N_2)$ or $x(H_2)$ is a dependent variable. We cannot simultaneously impose arbitrarily

---

† We shall often as here write a particular example of an x_B in the form $x(H_2O)$, or of an x_B^g in the form $x(H_2O, g)$, if only to avoid multiple subscripts.

chosen values of T, p, $x(NH_3)$, and $x(N_2)$ on a mixture of nitrogen, hydrogen, and ammonia *in the presence of a catalyst*. In the absence of a catalyst, however, we have $\mathcal{R} = 0$ so that $\mathcal{F} = 4$ and T, p, $x(NH_3)$, and $x(N_2)$ can all be varied independently.

Much fruitless argument has gone on in the past about the number $\mathcal{C}$ of components in a system of chemically reacting substances. For example, if we consider a system consisting *initially* of pure ammonia one might argue that $\mathcal{C} = 1$. Indeed in the absence of a catalyst that is correct. In the presence of a catalyst, however, there are three substances present and $\mathcal{C} = 3$. Blind use of equation (5) now leads to $\mathcal{F} = 3$, but that is wrong. We have forgotten that we have imposed the restriction:

$$x(H_2) = 3x(N_2),\tag{8.9.8}$$

by assuming that the system consisted initially of pure ammonia. That restriction reduces $\mathcal{F}$ by 1 from 3 to 2. A restriction such as (8) is no different in principle from any other restriction, such as for example a statement that the temperature is fixed at 500 K. Any such restriction of course reduces $\mathcal{F}$ by 1.

It will be obvious that if other 'work' terms than $V\,dp$ are needed in equation (1), as they might be for example for a system in a variable electromagnetic field, then the number of such extra terms must be added to the right-hand side of equation (5).

Problems for chapter 8

Problem 8.1

A system consists of two sub-systems each containing some nitrogen, some hydrogen, and some ammonia, in the presence of a catalyst. The two sub-systems are separated by a palladium membrane permeable to hydrogen but not to nitrogen or ammonia. What are the equilibrium conditions (a) for the separate sub-systems and (b) for the system as a whole. How would the results differ if the membrane were replaced by one permeable to two of the susbstances but not to the third?

Problem 8.2

Prove that for the equilibrium of two phases α and β of a system consisting of two components A and B that if $x_B^\alpha = x_B^\beta = x$ then $dT/dx = 0$ when p is constant and $dp/dx = 0$ when T is constant. What is the physical significance of those results?

Problem 8.3

Try to anticipate what we shall do in chapter 10 by devising a method for the experimental measurement of $\{\mu_B(T, p, x^\beta) - \mu_B(T, p, x^\alpha)\}$ where x is the mole fraction of B in a mixture of A and B.

Hint: You are allowed a membrane permeable to B but impermeable to A.

Chapter 9

Phase equilibria for pure substances

§ 9.1   Introduction

For the equilibrium of two phases α and β of a pure substance we have $\mathscr{C} = 1$, $\mathscr{P} = 2$, and $\mathscr{R} = 0$, so that according to the phase rule $\mathscr{F} = 1$. Thus if the temperature T is chosen as the independent variable then the pressure becomes a dependent variable and is then denoted by $p^{\alpha+\beta}$ and called the *equilibrium pressure* or, if one of the two phases is a gas, the *vapour pressure* or 'saturated vapour pressure'. If the pressure p is chosen as the independent variable then the temperature becomes a dependent variable and is then denoted by $T^{\alpha+\beta}$ and called the *boiling temperature*, or *melting temperature*, or *sublimation temperature*, or *transition temperature*, according as the two phases are liquid + gas, or solid + liquid, or solid + gas, or solid(α) + solid(β). The molar volumes V_m^α and V_m^β of coexisting phases at given temperature or pressure are called *orthobaric* molar volumes.

For the equilibrium of three phases α, β, and γ of a pure substance we have $\mathscr{C} = 1$, $\mathscr{P} = 3$, and $\mathscr{R} = 0$, so that $\mathscr{F} = 0$. The system is *invariant*; there are no independent intensive variables. Thus the temperature $T^{\alpha+\beta+\gamma}$, called the *triple-point temperature*, and the pressure $p^{\alpha+\beta+\gamma}$, called the *triple-point pressure*, are both fixed by the nature of the pure substance, as are also the triple-point molar volumes (or triple-point densities) of the three phases.

A projection of the three-dimensional plot of p, T, and V_m on the (p, T) plane is shown in figure 9.1 for the two-phase and three-phase equilibria of the substance CO_2. The curves show how the equilibrium pressures $p^{\alpha+\beta}$ depend on the temperature T. The three curves meet at the triple point (s + l + g). The (s + g) curve approaches zero at lower pressures. The (s + l) curve has a large and positive slope. (The melting line has a large and positive slope for most substances, but for water it has a large and negative slope.) The (s + l) curve continues upwards, as far as we know indefinitely. The (l + g) curve terminates at the *critical point*. There is continuity between the liquid and gaseous states; that can be seen by considering a pathway round the critical point such as that shown in figure 9.1. The temperature of a 'liquid' is lowered from the critical temperature, then its pressure is raised above the critical pressure, then its temperature is raised above the critical temperature, then its pressure is lowered to the original value, and then its temperature is lowered towards the original value. The 'liquid' condenses to a 'liquid'. Where is the 'gas'? Alternatively we may proceed in the other direction round the cycle, beginning with a 'gas' which then evaporates to a 'gas'. Where is the 'liquid'?

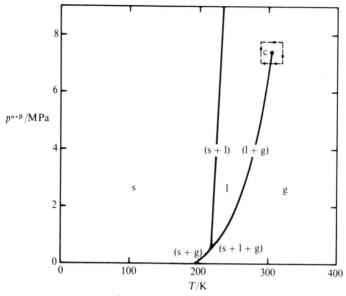

Figure 9.1. Equilibrium pressures $p^{\alpha+\beta}$ plotted against temperature T for carbon dioxide. When the temperature of solid carbon dioxide is increased at atmospheric pressure (0.1 MPa) the solid sublimes; for most other substances the triple-point pressure is below atmospheric pressure, so that the substance first melts and then boils. The path round the critical point c, illustrating continuity of the liquid and gaseous states, is described in the text.

The critical point c in figure 9.1 is the projection on to the (p, T) plane of a maximum of the projection on to the (T, V_m) plane, shown in figure 9.2(a), or of the projection on to the (p, V_m) plane, shown in figure 9.2(b). We now see that the

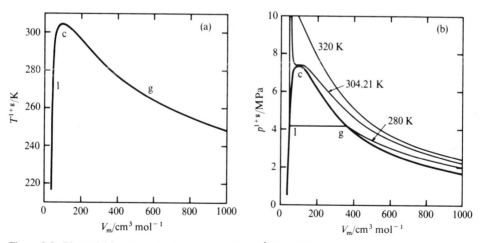

Figure 9.2. Plots of (a), the equilibrium temperature T^{l+g}, and (b), the equilibrium pressure (vapour pressure) p^{l+g}, against the molar volume V_m for the coexistence of two fluid phases, 'liquid' and 'gas', of CO_2. In (b) three isotherms are also shown; they are described in the text.

critical temperature is the highest temperature at which two fluid phases, 'liquid' and 'gas', of a pure substance can coexist. In figure 9.2(b) three isothermal sections are also shown. At 320 K, which is greater than the critical temperature, no condensation occurs whatever the pressure; there is no region in which two fluid phases can coexist. At 280 K, which is less than the critical temperature, two fluid phases coexist at the pressure 4.16 MPa for molar volumes between 49.8 and 361.8 cm^3 mol^{-1}. One fluid phase (the 'gas') exists at pressures less than 4.16 MPa and one fluid phase (the 'liquid') exists at pressures greater than 4.16 MPa. At the critical temperature, 304.21 K, the *critical isotherm* has a horizontal point of inflexion at the critical point:

$$(\partial p/\partial V_m)_T = 0, \quad (\partial^2 p/\partial V_m^2)_T = 0, \quad (T = T^c). \tag{9.1.1}$$

Equation (1) may be regarded as defining the critical point of a pure fluid. We shall discover some other kinds of critical point when we deal with fluid mixtures in chapters 16 and 17.

In figure 9.3 $T^{\alpha+\beta}$ is plotted against the reciprocal of the molar volume (or the

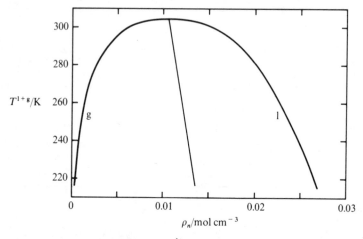

Figure 9.3. Plot of the equilibrium temperatures T^{l+g} against the amount-of-substance density ρ_n for the coexistence of two fluid phases, 'liquid' and 'gas', of CO$_2$. The mean values $(\rho_n^l + \rho_n^g)/2$ of ρ_n^l and ρ_n^g lie on a straight line, confirming the law of the rectilinear diameter.

amount-of-substance density) $V_m^{-1} = \rho_n$, instead of against the molar volume itself as in figure 9.2(a). The mean values $(\rho_n^l + \rho_n^g)/2$ lie on a straight line, an example of the *law of the rectilinear diameter* which is obeyed with high accuracy by all pure substances.

There is no formal distinction between a 'liquid' and a 'gas'. When two fluid phases of a pure substance coexist we may call the denser phase the liquid and the less dense phase the gas.† When there is only one fluid phase any distinction is arbitrary.

---

† Even that distinction sometimes fails for a fluid mixture when a small change of temperature or pressure causes two coexisting fluid phases to change places in the earth's gravitational field (the *barotropic effect*; see chapter 17).

Nevertheless the properties of a dilute gas (such as low density, high compressibility, and positive temperature coefficient of viscosity) are very different from those of a liquid at temperatures well below the critical temperature (high density, low compressibility, and negative temperature coefficient of viscosity). They are usually studied in separate experiments if only because of the large differences in the molar volumes and compressibilities. The experimental results are usually analysed differently. A different definition of the standard chemical potential is used for each of them. For those reasons we shall deal separately with gases and gaseous mixtures in chapter 12 and with liquid mixtures in chapter 16 and shall return to fluid mixtures in chapter 17.

A substance having more than one solid phase will have more than one triple point. An example is shown in figure 9.4 where $p^{\alpha+\beta}$ is plotted against T for sulphur. There are four triple points; three of them are stable: (rhombic + monoclinic + g), (monoclinic + l + g), and (rhombic + monoclinic + l), and one of them is metastable: (rhombic + l + g). The metastable curves, and hence the metastable triple point (rhombic + l + g), can be realized by quickly raising the temperature of rhombic sulphur at a pressure greater than p_1.

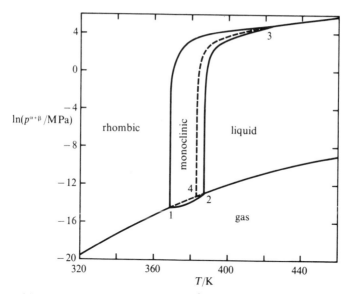

Figure 9.4. Logarithm of the equilibrium pressure $p^{\alpha+\beta}$ plotted against temperature T for sulphur. Metastable lines are broken. There are three stable triple points: 1, (rhombic + monoclinic + g); 2, (monoclinic + l + g); and 3, (rhombic + monoclinic + l); and one metastable triple point: 4, (rhombic + l + g).

§ 9.2   Entropy change for a two-phase transition

For a pure substance we have according to equations (5.4.4) and (5.3.5)

$$\mu^* = G_m^* = H_m^* - TS_m^* . \tag{9.2.1}$$

For the equilibrium of two phases α and β of a pure substance we have according to (8.5.1)

$$\mu^{*\alpha} = \mu^{*\beta} . \tag{9.2.2}$$

By substitution of (1) into (2) we then obtain for two-phase equilibrium of a pure substance

$$G_m^{*\beta} - G_m^{*\alpha} = 0 , \tag{9.2.3}$$

and

$$H_m^{*\beta} - H_m^{*\alpha} = T^{\alpha+\beta}(S_m^{*\beta} - S_m^{*\alpha}) . \tag{9.2.4}$$

We rewrite equations (3) and (4) in the forms:

$$\Delta_\alpha^\beta G_m^* = 0 , \tag{9.2.5}$$

$$\Delta_\alpha^\beta H_m^* = T^{\alpha+\beta} \Delta_\alpha^\beta S_m^* . \tag{9.2.6}$$

In specific applications we shall replace α and β by l and g (evaporation), by s and l (melting), or by s and g (sublimation).

We already know (§ 4.6) how to measure $\Delta_\alpha^\beta H_m^*$ calorimetrically. Equation (6) therefore provides us with a method of measuring another kind of entropy change, namely the molar entropy of transition $\Delta_\alpha^\beta S_m^*$ which is equal to $\Delta_\alpha^\beta H_m^*/T^{\alpha+\beta}$.

We may of course include the method of § 6.6 to obtain such a molar entropy difference for a pure substance as

$$S_m^{*g}(T_2, p_2) - S_m^{*l}(T_1, p_1) = \{S_m^{*g}(T_2, p_2) - S_m^{*g}(T_2, p^{l+g})\}$$

$$+ \{S_m^{*g}(T_2, p^{l+g}) - S_m^{*g}(T^{l+g}, p^{l+g})\}$$

$$+ \{S_m^{*g}(T^{l+g}, p^{l+g}) - S_m^{*l}(T^{l+g}, p^{l+g})\}$$

$$+ \{S_m^{*l}(T^{l+g}, p^{l+g}) - S_m^{*l}(T^{l+g}, p_1)\}$$

$$+ \{S_m^{*l}(T^{l+g}, p_1) - S_m^{*l}(T_1, p_1)\}$$

$$= - \int_{p^{l+g}}^{p_2} (\partial V_m^{*g}/\partial T)_{T=T_2,p} \, dp$$

$$+ \int_{T^{l+g}}^{T_2} \{C_{p,m}^{*g}(T, p^{l+g})/T\} \, dT + \Delta_l^g H_m^*(T^{l+g})/T^{l+g}$$

$$- \int_{p_1}^{p^{l+g}} (\partial V_m^{*l}/\partial T)_{T=T^{l+g},p} \, dp$$

$$+ \int_{T_1}^{T^{l+g}} \{C_{p,m}^{*l}(T, p_1)/T\} \, dT . \tag{9.2.7}$$

§ 9.3   Clapeyron's equation

For the equilibrium of two phases α and β of a pure substance $\mathscr{F} = 1$ and so there must exist a thermodynamic relation between the equilibrium pressure $p^{\alpha+\beta}$ and the temperature T (or between the equilibrium temperature $T^{\alpha+\beta}$ and the pressure p). For such a two-phase equilibrium the set of Gibbs–Duhem equations (8.9.2) takes the simple form:

$$0 = S_m^{*\alpha}\, dT - V_m^{*\alpha}\, dp + d\mu^*, \left.\begin{array}{c}\\ \\ \end{array}\right\}$$
$$0 = S_m^{*\beta}\, dT - V_m^{*\beta}\, dp + d\mu^*. \tag{9.3.1}$$

Eliminating $d\mu^*$ from equations (1) and rearranging we obtain

$$dp^{\alpha+\beta}/dT = (S_m^{*\beta} - S_m^{*\alpha})/(V_m^{*\beta} - V_m^{*\alpha}) = \Delta_\alpha^\beta S_m^* / \Delta_\alpha^\beta V_m^*, \tag{9.3.2}$$

or when we use equation (9.2.6)

$$dp^{\alpha+\beta}/dT = \Delta_\alpha^\beta H_m^* / T\Delta_\alpha^\beta V_m^*. \tag{9.3.3}$$

Either of equations (2) or (3) is called *Clapeyron's equation*. (We have chosen to regard the temperature as the independent variable and so have labelled the symbol for the equilibrium pressure, but we could equally well have made the inverse choice.)

If β is the phase stable at higher temperatures and α that stable at lower temperatures then $\Delta_\alpha^\beta H_m^*$ is positive (or zero).[†] The sign of $dp^{\alpha+\beta}/dT$ thus depends on the sign of $\Delta_\alpha^\beta V_m^*$. When β is a gas $\Delta_l^g V_m^*$ or $\Delta_s^g V_m^*$ is always positive so that dp^{l+g}/dT or dp^{s+g}/dT is positive; the vapour pressure of a liquid or solid increases with increasing temperature, and the boiling temperature or sublimation temperature increases with increasing pressure. When neither α nor β is a gas but β is the higher-temperature phase, $\Delta_\alpha^\beta V_m^*$ is usually but not always positive so that $dp^{\alpha+\beta}/dT$ is usually but not always positive; the melting temperature or transition temperature of a solid usually increases with increasing pressure. The best known exception is the melting of ice for which $\Delta_s^l V_m^*$ is negative so that dp^{s+l}/dT is negative; the melting temperature of ice decreases with increasing pressure.

§ 9.4   Approximate form of Clapeyron's equation applicable when one of the phases is a dilute gas

When one of the two phases is a dilute gas we may derive an approximate form of Clapeyron's equation by assuming (i) that the gas is perfect ($pV_m^{*g} = RT$) and (ii) that the molar volume V_m^{*l} or V_m^{*s} is negligible compared with V_m^{*g}. Equation (9.3.3)

[†] The molar enthalpy of evaporation $\Delta_l^g H_m$ decreases with increasing temperature and becomes zero at the critical temperature. For 'second-order' transitions, such as that at the 'lambda temperature', 2.18 K, of liquid helium, or the order-disorder transition of the alloy CuZn (β-brass) at 742 K, and for higher-order transitions, $\Delta_\alpha^\beta H_m = 0$. For accounts of second-order and higher-order transitions see for example: Guggenheim, E. A. *Thermodynamics*. North-Holland: Amsterdam. **1967**, fifth edition, pp. 252–265; Callen, H. B. *Thermodynamics*. Wiley: New York. **1960**, chapter 9.

then becomes

$$d \ln p^{\alpha+g}/dT \approx \Delta_\alpha^g H_m^* / RT^2 ,$$ (9.4.1)

or

$$d \ln p^{\alpha+g}/d(1/T) \approx -\Delta_\alpha^g H_m^* / R .$$ (9.4.2)

At least over short ranges of temperature a plot of $\ln(p^{\alpha+g}/p^\ominus)$ against $1/T$ (where $p^\ominus$ denotes some standard pressure) is roughly linear. The further approximation may therefore be made that $\Delta_\alpha^g H_m^*$ is independent of temperature over such ranges; equation (2) may then be integrated to give

$$\ln\{p^{\alpha+g}(T_2)/p^{\alpha+g}(T_1)\} \approx \Delta_\alpha^g H_m^* (T_2 - T_1)/RT_1 T_2 .$$ (9.4.3)

Equation (3) can be used to calculate an approximate value of $\Delta_\alpha^g H_m^*$ from measurements of vapour pressure at each of two temperatures T_1 and T_2. Alternatively, if the molar enthalpy of evaporation has been measured or can be estimated, and if the vapour pressure has been measured at one temperature, then equation (3) can be used to estimate the vapour pressures at other temperatures.†

Having approximated so far we might as well go further by assuming for $(l+g)$ that $\Delta_l^g H_m^*/RT^b \approx 10$, where T^b denotes the normal boiling temperature at which $p^{l+g}(T^b) \approx 0.1$ MPa. This is called Trouton's rule and will be further discussed in chapter 13. We then obtain

$$p^{l+g}(T) \approx 0.1 \exp\{10(T - T^b)/T\} \text{ MPa} .$$ (9.4.4)

Accurate values of $\Delta_\alpha^g H_m^*$ should, however, be calculated by first fitting the vapour pressures to an empirical formula in T, and then differentiating and using equation (9.3.3), which may be written in the form:

$$dp^{\alpha+g}/dT = \Delta_\alpha^g H_m^*/T(RT/p^{\alpha+g} + B + C'p^{\alpha+g} + \cdots - V_m^{*\alpha}) .$$ (9.4.5)

Equation (5) has sometimes been used to obtain values of the second virial coefficient B from measurements of $\Delta_l^g H_m^*$, V_m^{*l}, and p^{l+g} as a function of T for values of p^{l+g} low enough to ensure that the term $C'p^{l+g}$ and higher terms are negligible.

§9.5 Measurement of vapour pressure

Here we shall very briefly describe methods for the measurement of vapour pressure as a function of temperature, and methods for the analysis of the results of such measurements. Full accounts can be found in the literature.‡

---

† For evaporation $\ln\{p^{l+g}(T)/p^{l+g}(T^\ominus)\}$ is a much more nearly linear function of $1/T$ (see for example figure 13.4) throughout the range of existence of the liquid than we have any obvious reason to expect. Errors from the neglect of the temperature dependence of $\Delta_l^g H_m$ and errors from the neglect of the difference between $\Delta_l^g V_m$ and RT/p^{l+g} evidently nearly compensate each other.

‡ Thomson, G. W.; Douslin, D. R. *Techniques of Chemistry*, Volume I, *Physical Methods of Chemistry*, Part V. Weissberger, A.; Rossiter, B.W.: Editors. Wiley-Interscience: New York. **1971**, chapter II, pp. 23–104.

Ambrose, D. *SPRCT(I)*, chapter 7.

Ambrose, D. *ETd(II)*, chapter 13. (Chapter 4 of that work gives an extensive account of manometry.)

Mercury manometers are used for the measurement of vapour pressures in the range from 0.001 MPa or less to 0.2 MPa or more. They are sometimes isolated from the two phases in equilibrium by a transducer such as a glass 'spoon' gauge, or a hollow glass helix, or a metal bellows, or a deformable plate and a fixed plate forming an electric capacitor. At higher pressures, up to the critical pressure (which seldom exceeds 5 MPa), pressure balances (piston gauges) are used. In these a piston rotating freely in a cylindrical hole is loaded with weights until the pressure is just balanced. The piston is 'sealed' with oil or, in low-pressure balances, with air. At lower pressures the Knudsen-effusion method is used: the loss Δm of mass in time t through a small orifice of area $\mathscr{A}$ from an evacuated vessel containing the liquid or solid is related to the vapour pressure $p^{\alpha+g}$ by the equation:

$$p^{\alpha+g} = C(\Delta m/\mathscr{A}t)(2\pi RT/M)^{1/2},\tag{9.5.1}$$

where M is the molar mass of the gas and C is called the *Clausing factor* (1 for $l/r \to 0$, 1.95 for $l/r = 2$, where l is the length and r the radius of the orifice).

In the range from 0.002 MPa to 0.2 MPa two kinds of method are used. In 'static' or equilibrium methods the liquid and gaseous (or s+g) phases of a rigorously degassed sample of the substance are brought to equilibrium and the pressure is measured. In 'ebulliometric' methods the liquid phase is continuously boiled and the vapour condensed and returned to the boiler; the pressure is measured when the apparatus has reached a steady state.† It is assumed that the steady-state pressure is negligibly different from the equilibrium pressure. At pressures higher than 0.2 MPa static methods are nearly always used.

Over short ranges of temperature values of the vapour pressure may be fitted to equation (9.4.2) in the form:

$$\ln(p/p^{\ominus}) = a + b/T.\tag{9.5.2}$$

The Antoine equation:

$$\ln(p/p^{\ominus}) = a + b/(c + T),\tag{9.5.3}$$

gives a more accurate fit over wider ranges of temperature. Highly complex equations are needed, however, to fit vapour pressures within experimental error from the triple-point temperature to the critical temperature. The use of orthogonal (Chebyshev) polynomials for this purpose has found a good deal of favour.

§ 9.6   Heat capacities of a pure substance at two-phase equilibrium

We define the heat capacity $C^{*\alpha}_{\alpha+\beta}$ of a phase α of a pure substance at two-phase equilibrium ($\mu^{*\alpha} = \mu^{*\beta}$) by the relation:

$$C^{*\alpha}_{\alpha+\beta} \overset{\text{def}}{=} T(\partial S^{*\alpha}/\partial T)_{\alpha+\beta}.\tag{9.6.1}$$

---

† In a system at equilibrium nothing is happening. In a system at a steady state none of the intensive properties of any phase is changing but something is nevertheless happening.

By use of the 'rule for changing the variable held constant' (appendix I), treating $(\alpha + \beta)$ as such a variable, we obtain

$$(\partial S^{*\alpha}/\partial T)_{\alpha+\beta} = (\partial S^{*\alpha}/\partial T)_p + (\partial S^{*\alpha}/\partial p)_T (\partial p/\partial T)_{\alpha+\beta}, \qquad (9.6.2)$$

which on use of Maxwell's relation (6.4.3) for $(\partial S/\partial p)_T$ and of equation (9.3.3) for $(\partial p/\partial T)_{\alpha+\beta}$ gives, after multiplication by T,

$$C^{*\alpha}_{\alpha+\beta} = C^{*\alpha}_p - (\partial V^{*\alpha}/\partial T)_p \, \Delta^\beta_\alpha H^*_m / \Delta^\beta_\alpha V^*_m. \qquad (9.6.3)$$

Equation (3) relates the heat capacity of a pure phase at two-phase equilibrium to its heat capacity at constant pressure. The quantity $C^{*\alpha}_{\alpha+\beta}$ is often called the 'heat capacity of the phase α at saturation'.

In view of equation (6.12.8) for any phase of fixed composition, namely

$$Q = T \, dS - (p \, dV + W), \qquad (9.6.4)$$

we may interpret $C^{*\alpha}_{\alpha+\beta}$ or $T(\partial S^{*\alpha}/\partial T)_{\alpha+\beta}$ as the limiting value as $\delta T \to 0$ of $Q/\delta T$ where Q is the heat flow into a phase needed to cause a change δT of temperature subject to the phase α being and remaining in equilibrium with a vanishingly small amount of the phase β (for example: 'wet steam'), any work being done reversibly so that $(p \, dV + W) = 0$.

We shall now see that $C_{\alpha+\beta}$, unlike C_p or C_V, can have *either* sign; in particular it is sometimes necessary to cause a flow of heat out of a phase $(Q < 0)$ in order to raise its temperature while maintaining two-phase equilibrium.

For $\alpha = g$ and $\beta = l$, provided that the pressure is not too far above atmospheric pressure, we may use the crude approximations: $(\partial V^{*g}_m/\partial T)_p \approx R/p$, $\Delta^g_l H^*_m \approx 10RT$, and $\Delta^g_l V^*_m \approx V^{*g}_m \approx RT/p$. Substituting these approximations into equation (3) we obtain

$$C^{*g}_{g+l,m} \approx C^{*g}_{p,m} - 10R. \qquad (9.6.5)$$

Since $C^{*g}_{p,m}$ is often less than $10R$, the molar heat capacity $C^{*g}_{g+l,m}$ of a pure substance in the gaseous phase at (gas + liquid) equilibrium is often negative.

For $\alpha = l$ and $\beta = g$ on the other hand we may under the same conditions use the crude approximation: $(\partial V^{*l}_m/\partial T)_p \approx 10^{-3} V^{*l}_m \, \mathrm{K}^{-1}$. We then obtain

$$C^{*l}_{g+l,m} \approx C^{*l}_{p,m} - 10^{-2} V^{*l}_m p \, \mathrm{K}^{-1}, \qquad (9.6.6)$$

which for a typical liquid having a molar volume of say $50 \, \mathrm{cm}^3 \, \mathrm{mol}^{-1}$ and at atmospheric pressure becomes

$$C^{*l}_{g+l,m} \approx C^{*l}_{p,m} - 10^{-2} \times 50 \times 10^{-6} \, \mathrm{m}^3 \, \mathrm{mol}^{-1} \, 10^5 \, \mathrm{Pa} \, \mathrm{K}^{-1}$$

$$\approx C^{*l}_{p,m} - 0.05 \, \mathrm{J} \, \mathrm{K}^{-1} \, \mathrm{mol}^{-1}$$

$$\approx C^{*l}_{p,m}. \qquad (9.6.7)$$

The last step in (7) depends on the fact that the molar heat capacity at constant pressure is always much greater than $0.05 \, \mathrm{J} \, \mathrm{K}^{-1} \, \mathrm{mol}^{-1}$ for a pure liquid. The molar heat capacity $C^{*l}_{g+l,m}$ of a pure substance in the liquid phase at (gas + liquid) equilibrium is positive, and approximately equal to the molar heat capacity at constant pressure.

§ 9.7   The dependence of the enthalpy of a phase transition on temperature

The temperature dependence $(\partial \Delta_\alpha^\beta H_m^* / \partial T)_p$ of the molar enthalpy of a phase transition of a pure substance at constant pressure is in general difficult to measure directly, because $\Delta_\alpha^\beta H_m^*$ is difficult to measure except at the equilibrium temperature $T^{\alpha+\beta}(p)$. At any other temperature the possibility of $\Delta_\alpha^\beta H_m^*$ being measured depends on the realization of metastable phases, such as a 'supercooled' liquid at temperatures below the melting temperature. Such measurements, though always possible in principle, are seldom feasible in practice.

The temperature dependence $(\partial \Delta_\alpha^\beta H_m^* / \partial T)_{\alpha+\beta}$ of the molar enthalpy of a phase transition of a pure substance at two-phase equilibrium is easily obtained from measurements of the molar enthalpy of phase transition at each of a series of equilibrium temperatures $T^{\alpha+\beta}$. We accordingly use a method similar to that of § 9.6 to derive *Planck's equation*:

$$(\partial \Delta_\alpha^\beta H_m^* / \partial T)_{\alpha+\beta} = (\partial \Delta_\alpha^\beta H_m^* / \partial T)_p + (\partial \Delta_\alpha^\beta H_m^* / \partial p)_T (\partial p / \partial T)_{\alpha+\beta}$$

$$= \Delta_\alpha^\beta C_{p,m}^* + (\Delta_\alpha^\beta H_m^* / T^{\alpha+\beta})\{1 - T^{\alpha+\beta}(\partial \ln \Delta_\alpha^\beta V_m^* / \partial T)_p\}. \qquad (9.7.1)$$

If phase β is a gas at a low pressure we may use the approximation $\Delta_\alpha^g V_m^* \approx V_m^{*g} \approx RT/p$, so that $\{\cdots\}$ in equation (1) goes approximately to zero and we have

$$(\partial \Delta_\alpha^g H_m^* / \partial T)_{\alpha+g} \approx \Delta_\alpha^g C_{p,m}^* . \qquad (9.7.2)$$

If on the other hand $\alpha = s$ and $\beta = l$, we have in effect replaced the problem of measuring $(\partial \Delta_s^l H_m^* / \partial T)_p$ by the hardly easier problem of measuring $(\partial \Delta_s^l V_m^* / \partial T)_p$. The last term in equation (1) is now usually far from negligible, and little is usually known about $(\partial \Delta_s^l V_m^* / \partial T)_p$ except by extrapolation.

§ 9.8   The critical point of a pure substance

If we can measure the vapour pressure at temperatures up to the critical then we can measure the critical temperature and the critical pressure. The critical temperature T^c and critical pressure p^c are usually determined visually as the particular temperature T and pressure p at which the meniscus separating two fluid phases disappears (or appears) in the middle of the tube as T and p are raised (or lowered). At a lower temperature the meniscus disappears at the top of the tube when the pressure is raised or at the bottom of the fluid sample when the pressure is lowered; at a lower pressure the meniscus disappears at the bottom of the fluid sample when the temperature is raised or at the top of the tube when the temperature is lowered. At a higher temperature than T^c or a higher pressure than p^c only one fluid phase is ever present. Close to the critical point a characteristic and beautiful opalescence can be seen.† The critical molar volume V_m^c is much harder to measure accurately because of the flatness of the top of the coexistence curve of T against V_m or of p against V_m. It is usually obtained from the molar volumes V_m^l and V_m^g of the two coexisting phases measured at a series of temperatures below the critical temperature, by use of the law of the rectilinear diameter. For the same reason it is possible to determine T^c by raising the temperature of a sealed tube containing amount of substance n in the volume V if n has been adjusted so that V/n is roughly equal to the critical molar volume V_m^c. Since the compressibility becomes infinite at the critical point, the earth's gravitational field causes an unusually large gradient of density within a phase in the critical region. That effect causes changes in the shapes of the curves of $p(V_m, T)$ and of the coexistence curve in the critical region. Those shapes, in the absence of a gravitational field, are of great current interest as will be discussed below. The effect can be reduced by working with a horizontal rather than a vertical tube of fluid.

In figure 9.5 experimental values of the pressure p along isotherms at temperatures close to the critical are shown plotted against the density ρ for xenon. The critical isotherm calculated from a 'classical' equation of state is shown for

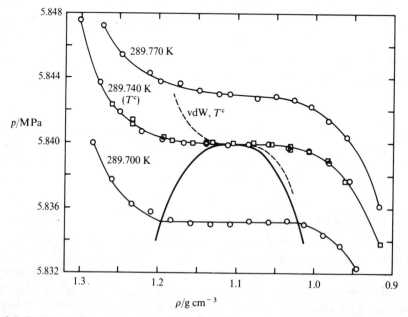

Figure 9.5. Isotherms of pressure p plotted against density ρ for xenon at temperatures close to the critical. The broken curve is the critical isotherm calculated according to van der Waals's equation.

---

† The molecular explanation of critical opalescence is that the 'correlation length', or the mean distance over which the molecules are ordered, grows from about 1 nm through the wavelength of visible light around 600 nm.

comparison as a broken line; the equation of state is that of van der Waals:

$$p = RT/(V_m - b) - a/V_m^2,\qquad(9.8.1)$$

but any other analytic† equation of state would give similar results. At the critical point, in addition to $(\partial p/\partial \rho)_T = 0$ and $(\partial^2 p/\partial \rho^2)_T = 0$, the measurements plotted in figure 9.5 and other similar measurements show that it is certain that $(\partial^4 p/\partial \rho^4)_T = 0$ and is probable that $(\partial^3 p/\partial \rho^3)_T = 0$; $(\partial^2 p/\partial \rho^2)_T$ certainly has a point of inflexion at $\rho = \rho^c$ and probably a horizontal point of inflexion. Such behaviour is not predicted by any analytic equation of state.

During the last fifteen years or so it has at last been fully realized that no analytic equation of state can fit results near the critical point. For example, any analytic equation of state, such as that of van der Waals, must give for the difference of coexisting densities as the critical temperature is approached

$$(\rho^l - \rho^g) \propto (T^c - T)^\beta,\quad (T \to T^c),\qquad(9.8.2)$$

with $\beta = \tfrac{1}{2}$ (the 'classical' value), whereas the curve drawn through the experimental results for xenon in figure 9.6 has the value $\beta = 0.351$. The corresponding curve for $\beta = \tfrac{1}{2}$ is shown for comparison as a broken curve.

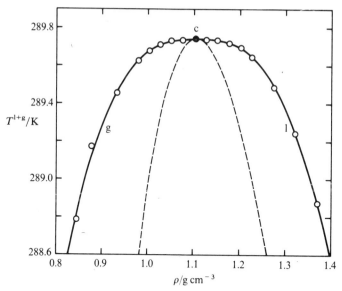

Figure 9.6. Temperatures T^{l+g} close to the critical temperature plotted against coexisting density ρ^g or ρ^l for xenon. The unbroken curve is that calculated from the equation: $(\rho^l - \rho^g)/\rho^c = 3.609\,(1 - T/T^c)^{0.351}$. The broken curve is that calculated from van der Waals's equation for which the critical exponent $\beta = \tfrac{1}{2}$.

---

† An analytic equation is one for which the expansion as a Taylor series about the point in question converges in the neighbourhood of the point.

Similarly, direct measurements of $C_V(V_m = V_m^c, T)$, made in a rigid container have a mass less than that of the fluid, have shown that it diverges at the critical temperature instead of remaining finite as predicted by any analytic equation of state. In terms of the critical exponent α

$$C_V(V_m = V_m^c, T) \propto |T - T^c|^{-\alpha}, \quad (T \to T^c). \tag{9.8.3}$$

These and some of the other critical exponents are defined in table 9.1 for the

Table 9.1. Definitions of critical exponents.

Definition	Pathway
$C_V \propto |T - T^c|^{-\alpha}, \quad (T \to T^c)$	$\begin{cases} \alpha_1^+ : \text{along the critical isochore for } T > T^c. \\ \alpha_2^- : \text{along the critical isochore for } T < T^c. \\ \alpha_1^- : \text{for either phase along the orthobaric curve.} \end{cases}$
$(V_m^g - V_m^l) \propto (T^c - T)^\beta, \quad (T \to T^c)$ $(V_m^g - V_m^c) \propto (T^c - T)^\beta, \quad (T \to T^c)$ $(V_m^c - V_m^l) \propto (T^c - T)^\beta, \quad (T \to T^c)$	β: for either phase along the orthobaric curve.
$\kappa_T \propto |T - T^c|^{-\gamma}, \quad (T \to T^c)$	$\begin{cases} \gamma_1^+ : \text{along the critical isochore for } T > T^c. \\ \gamma_1^- : \text{for either phase along the orthobaric curve.} \end{cases}$
$|p - p^c| \propto |V_m - V_m^c|^\delta, \quad (V_m \to V_m^c)$	δ: along the critical isotherm: $T = T^c$.

pathways briefly explained there and illustrated in figure 9.7. The subscripts $_1$ and $_2$ relate to the number of phases and the superscripts $^+$ and $^-$ to $T > T^c$ and $T < T^c$.

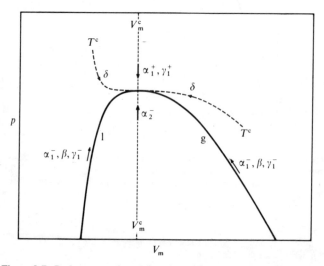

Figure 9.7. Pathways used to define the critical exponents of table 9.1.

Six plausible conjectures have been made about critical exponents.

1. We have tacitly assumed that there is symmetry about the critical point, so that we need not distinguish for example between β^l and β^g.

2. We have tacitly assumed the principle of universality according to which any critical exponent is the same for all substances; universality implies that the critical exponents arise from what all fluids have in common rather than from their differences.

3. We shall assume that we may omit all subscripts and superscripts from any critical exponent. That this may be done follows from the plausible but unproved scaling laws based on the assumption that certain thermodynamic functions are homogeneous functions of their arguments.

By purely thermodynamic analysis making use of the stability conditions $C_V > 0$ and $\kappa_T > 0$ it can be shown that the critical exponents are bound by certain inequalities:

$$\alpha + 2\beta + \gamma \geqslant 2, \tag{9.8.4}$$

$$\alpha + \beta(\delta + 1) \geqslant 2. \tag{9.8.5}$$

4. The inequalities are satisfied as equalities. This, which also follows from the scaling laws, would reduce the number of independent exponents in (4) and (5) to two.

5. When a 'dictionary' is used to translate to corresponding variables, the nature of a critical point remains the same for other kinds of critical point than those in a pure fluid. Other kinds of critical point include those in fluid mixtures (see chapter 17), those in solids like Cu_3Au that have order-disorder transitions, and the Curie point of a ferromagnet. For example, when we translate from a pure fluid to a binary fluid mixture: for α, C_V is replaced by C_p; for β, $(V_m^g - V_m^l)$ is replaced by the difference $(x^\alpha - x^\beta)$ between the mole fractions in coexisting phases; for γ, $-\kappa_T$ is replaced by $(\partial^2 G_m / \partial x^2)_{T,p}$; and for δ, $(p - p^c)$ as a function of $(V_m - V_m^c)$ at $T = T^c$ is replaced by $\{(\mu_A - \mu_A^c) - (\mu_B - \mu_B^c)\}$ as a function of $(x - x^c)$ at $T = T^c$ and $p = p^c$.

6. For any of these critical points the exponents are given correctly by the solution of the three-dimensional Ising or lattice-gas model. In the Ising model it is assumed that there are only two ways in which a lattice site can be occupied: by a molecule or by a vacancy (pure fluid); by a molecule A or by a molecule B (fluid mixture); by a 'right' atom or by a 'wrong' atom (order-disorder transition); by a spin 'up' or by a spin 'down' (ferromagnet). It is further assumed that interactions count only when they are between nearest neighbours on the lattice. Even that simple model has so far defied algebraic analysis for a lattice of three dimensions, though it is easy in one dimension and was solved for two dimensions in a famous paper by Onsager. The values of the critical exponents are now known for the three-dimensional lattice as the result of numerical work.

All these conjectures are at present the basis of many careful experiments and much analysis. All the exponents are difficult to measure accurately. For small enough values of $|T/T^c - 1|$ the experimental accuracy falls catastrophically; for larger values other terms than the leading ones given in table 9.1 might be important. Proper allowance for gravitational effects is experimentally exacting. For further details the reader should delve into the copious modern literature.† We shall conclude this section by comparing in table 9.2 the understandably approximate 'best' experimental values of the critical exponents, their 'classical' values, and the values calculated according to the methods of statistical mechanics from the three-dimensional Ising or lattice-gas model.

Table 9.2. Some values of critical exponents: 'best' experimental values for pure fluids, 'classical' values, and values calculated according to the three-dimensional Ising model.

Exponent	Experiment	Classical	Ising model
α	≈ 0.1	0	0.125
β	≈ 0.35	0.5	0.3125
γ	≈ 1.2	1	1.25
δ	≈ 4.4	3	5
$\alpha + 2\beta + \gamma$	≈ 2.0	2	2
$\alpha + \beta(\delta + 1)$	≈ 2.0	2	2

Problems for chapter 9

Problems 9.1

Calculate $\{S_m(CO_2, 230\ K, 2\ MPa) - S_m(CO_2, 300\ K, 0.1\ MPa)\}$ and $\{H_m(CO_2, 230\ K, 2\ MPa) - H_m(CO_2, 300\ K, 0.1\ MPa)\}$ from each of the two series of steps for which $\Delta_i^f S_m$ and $\Delta_i^f H_m$ are tabulated below. Draw the pathway implied by each series on a sketch of figure 9.1, and briefly describe how each quantity in each series could be measured. Comment on your results.

Series	initial state	final state	$\dfrac{\Delta_i^f S_m}{J\,K^{-1}\,mol^{-1}}$	$\dfrac{\Delta_i^f H_m}{J\,mol^{-1}}$
(a)	(g, 300 K, 0.1 MPa)	(g, 300 K, 2 MPa)	+26.89	+836
	(g, 300 K, 2 MPa)	(g, 253.77 K, 2 MPa)	+8.28	+2251
	(g, 253.77 K, 2 MPa)	(l, 253.77 K, 2 MPa)	+48.79	+12380
	(l, 253.77 K, 2 MPa)	(l, 230 K, 2 MPa)	+8.63	+2084
(b)	(g, 300 K, 0.1 MPa)	(g, 330 K, 0.1 MPa)	−3.60	−1144
	(g, 330 K, 0.1 MPa)	(g, 330 K, 10 MPa)	+50.71	+5220
	(g, 330 K, 10 MPa)	(g, 230 K, 10 MPa)	+46.42	+13387
	(g, 230 K, 10 MPa)	(g, 230 K, 2 MPa)	−1.00	+80

† Levelt Sengers, J. M. H. *ETd(II)*, chapter 14, pp. 657–724.

Stanley, H. E. *Introduction to Phase Transitions and Critical Phenomena.* Clarendon Press: Oxford. **1971**.

Wood, D. W. *Specialist Periodical Report, Statistical Mechanics*, Volume 2, chapter 2. Singer, K.: Senior Reporter. The Chemical Society: London. **1975**, pp. 55–187.

Rowlinson, J. S. *Liquids and Liquid Mixtures.* Second edition. Butterworth: London. **1969**, pp. 76–103.

Münster, A. *Statistical Thermodynamics.* Volume II. Springer-Verlag: Berlin. **1974**, pp. 181–214.

Problem 9.2

Calculate dp^{s+l}/dT for the equilibrium of ice + water at temperatures close to 273 K given that at 273.15 K and 101.325 kPa the molar enthalpy of melting $\Delta_s^l H_m$ of $H_2O(s)$ is 6007 J mol^{-1} and that the densities are 0.9168 g cm^{-3} for ice and 0.9999 g cm^{-3} for water. Hence estimate the difference between the melting temperatures of ice at atmospheric pressure and at the triple-point pressure (0.61 kPa). Account for the difference between your result and the measured difference $(T^{s+l+g} - T_{ice}) = +(0.0100 \pm 0.0002)$ K between the triple-point temperature T^{s+l+g} of water and the 'normal ice temperature' T_{ice} of an (ice + water) mixture measured in an open Dewar flask at atmospheric pressure.

Problem 9.3

The vapour pressure of thiophene C_4H_4S at 298.15 K is 10.62 kPa, and its molar enthalpy of evaporation at 318.51 K is 33606 J mol^{-1}. Estimate the vapour pressure at 340.08 K (a) by use of equation (9.4.3), and (b) given that the normal boiling temperature is 357.3 K by use of equation (9.4.4). Compare your result with the measured vapour pressure of 57.80 kPa at 340.08 K.

Problem 9.4

For 2,2-dimethylpropane (neopentane) C_5H_{12} at 298.15 K the calorimetrically measured molar enthalpy of evaporation is 21.84 kJ mol^{-1}, the vapour pressure p^{l+g} is 171.418 kPa, $dp^{l+g}/dT = 5.471$ kPa K^{-1}, and the molar volume V_m^{*l} of the liquid is 123 cm^3 mol^{-1}. Neglecting the third and higher virial coefficients calculate the second virial coefficient B. Compare your result with $-(900 \pm 25)$ cm^3 mol^{-1} obtained by extrapolation of the directly measured values.

Problem 9.5

For H_2O at 373 K and atmospheric pressure $C_{p,m}^g = 36.6$ J K^{-1} mol^{-1}, $C_{p,m}^l = 76.0$ J K^{-1} mol^{-1}, $\Delta_l^g H_m = 40660$ J mol^{-1}, $\Delta_l^g V_m = 30140$ cm^3 mol^{-1}, $(\partial V_m^g/\partial T)_p = 86.8$ cm^3 K^{-1} mol^{-1}, and $(\partial V_m^l/\partial T)_p = 0.012$ cm^3 K^{-1} mol^{-1}. Calculate the molar heat capacities $C_{l+g,m}^g$ and $C_{l+g,m}^l$ at saturation at 373 K.

Problem 9.6

Use the following experimental values of the orthobaric densities ρ^l and ρ^g of dinitrogen oxide N_2O to test the law of the rectilinear diameter. Given that the critical temperature is 309.54 K estimate the critical density.

T/K	293.15	295.15	297.15	299.15	300.15	303.15	304.15	305.15	306.15
$\rho^l/$g cm^{-3}	0.789	0.775	0.752	0.731	0.723	0.689	0.671	0.655	0.638
$\rho^g/$g cm^{-3}	0.159	0.171	0.183	0.197	0.205	0.234	0.246	0.260	0.277

Problem 9.7

Find a value for the critical exponent β for Xe from the following values of $(\rho^1 - \rho^c)$ and of $(\rho^c - \rho^g)$ for small values of $(T^c - T)$ where ρ^1 and ρ^g denote orthobaric densities and ρ^c denotes the critical density.

$(T^c - T)/\text{K}$	1.606	0.868	0.497	0.255	0.096	0.046
$(\rho^1 - \rho^c)/\text{g cm}^{-3}$	0.332	0.265	0.217	0.169	0.121	0.098
$(T^c - T)/\text{K}$	0.954	0.568	0.287	0.114	0.061	0.030
$(\rho^c - \rho^g)/\text{g cm}^{-3}$	0.260	0.227	0.172	0.125	0.101	0.077

Chapter 10

Dependence of thermodynamic functions on composition

§ 10.1 Scope

We shall begin in § 10.2 by reviewing the two kinds of change of composition for which the enthalpy change ΔH can be measured calorimetrically, namely change of composition as a result of mixing and change of composition as a result of chemical reaction.

The heart of the chapter will be an account in §§ 10.3 to 10.5 and 10.7 of the experimental determination of the difference:

$$\{\mu_B(T, p, x_C^\beta) - \mu_B(T, p, x_C^\alpha)\}, \tag{10.1.1}$$

between the chemical potentials at given temperature and pressure of any substance B in any two phases having any compositions x_C^α and x_C^β, where x_C^α is used here to denote the set of $(\mathscr{C} - 1)$ independent mole fractions x_B^α, x_C^α, $\cdots$, in the phase α, $\mathscr{C}$ being the number of components present in the phase.

Having found out how to measure the difference (1) it will be easy to show in §§ 10.10 and 10.11 that the change ΔG of Gibbs function and the change ΔS of entropy can be measured for just the same two kinds of change of composition as those for which the enthalpy change ΔH can be measured calorimetrically, namely change of composition as a result of mixing and change of composition as a result of chemical reaction.

§ 10.2 Dependence of enthalpy on composition

We recall that there are two (and only two) kinds of change of composition for which the enthalpy difference ΔH can be measured calorimetrically. These are *mixing* and *chemical reaction* which we shall now review in turn.

We can use a calorimeter to measure the enthalpy difference:

$$\Delta H = H(T, p, n_A, n_B) - H(T, p, n_A, 0) - H(T, p, 0, n_B), \tag{10.2.1}$$

which accompanies the mixing at constant temperature and pressure of an amount of substance n_A of pure A and an amount of substance n_B of pure B to give an amount of substance $(n_A + n_B)$ of a mixture of A and B containing mole fraction

$x = n_B/(n_A + n_B)$ of B. Equation (1) can be written in the alternative forms:

$$\Delta H = (n_A + n_B)\{H_m(T, p, x) - (1-x)H_m(T, p, 0) - xH_m(T, p, 1)\}$$
$$= (n_A + n_B)\Delta_{mix}H_m(T, p, x), \qquad (10.2.2)$$

where H_m denotes molar enthalpy and $\Delta_{mix}H_m$ the *molar enthalpy of mixing*.

More generally we can use a calorimeter to measure the enthalpy difference:

$$\Delta H = H\{T, p, (n_A^\alpha + n_A^\beta + \cdots), (n_B^\alpha + n_B^\beta + \cdots), \cdots\} - H(T, p, n_A^\alpha, n_B^\alpha, \cdots)$$
$$- H(T, p, n_A^\beta, n_B^\beta, \cdots) - \cdots, \qquad (10.2.3)$$

which accompanies the mixing at constant temperature and pressure of any number of phases $\alpha, \beta, \cdots$, having any compositions, to give a single phase (or for that matter to give any number of new phases). Equation (3) can be rewritten in the form:

$$\Delta H = \{(n_A^\alpha + n_A^\beta + \cdots) + (n_B^\alpha + n_B^\beta + \cdots) + \cdots\}H_m(T, p, x_C^{\alpha+\beta+\cdots})$$
$$- (n_A^\alpha + n_B^\alpha + \cdots)H_m(T, p, x_C^\alpha) - (n_A^\beta + n_B^\beta + \cdots)H_m(T, p, x_C^\beta)$$
$$- \cdots. \qquad (10.2.4)$$

Equation (3) or (4) reduces to equation (1) or (2) when just two phases are mixed, one of which is pure A and the other pure B.

If the $\Delta_{mix}H_m$ of equation (2) has been measured as a function of x then we can also determine values, as was explained in § 4.9, of the differences:

$$\{H_A(T, p, x) - H_A(T, p, 0)\}, \quad \text{and} \quad \{H_B(T, p, x) - H_B(T, p, 1)\}, \qquad (10.2.5)$$

where $H_A(T, p, x)$ and $H_B(T, p, x)$ are the partial molar enthalpies of A and of B in a mixture containing mole fraction x of B, and $H_A(T, p, 0) = H_A^*(T, p)$ and $H_B(T, p, 1) = H_B^*(T, p)$ are the partial molar (or of course the molar) enthalpies of pure A and of pure B.†

If the quantities (4) have been determined for two or more different values of x then it follows that we can also determine values of the differences:

$$\{H_A(T, p, x^\beta) - H_A(T, p, x^\alpha)\}, \quad \text{and} \quad \{H_B(T, p, x^\beta) - H_B(T, p, x^\alpha)\}, \qquad (10.2.6)$$

between the partial molar enthalpies of a substance in two binary mixtures having different compositions.

More generally we can determine from the results of calorimetric experiments the difference:

$$\{H_B(T, p, x_C^\beta) - H_B(T, p, x_C^\alpha)\}, \qquad (10.2.7)$$

between the partial molar enthalpies at given temperature and pressure of any substance B in any two phases having any compositions.

---

† According to context we shall use $X_B(T, p, x = 1)$ or $X_B^*(T, p)$ or $X_m^*(B, T, p)$ to denote a molar quantity for the *pure* substance B.

We turn now to the measurement of the enthalpy difference:

$$\Delta H = H(T, p, n''_C) - H(T, p, n'_C),\qquad(10.2.8)$$

which accompanies a change in the extent ξ of the chemical reaction:

$$0 = \sum_B \nu_B B,\qquad(10.2.9)$$

from ξ' to ξ'' at constant temperature and pressure. We recall that the amounts of substance n_B of the several substances B are related to the extent of reaction ξ by the equations:

$$n_B(\xi) = n_B(0) + \nu_B \xi, \quad \text{(all B)},\qquad(10.2.10)$$

where the $n_B(0)$'s are constant for a given closed system. Equation (8) can then be written in the alternative form:

$$\Delta H = H(T, p, \xi'') - H(T, p, \xi').\qquad(10.2.11)$$

In principle it is always possible to measure the enthalpy difference (8) or (11) in a calorimeter. In practice it is often convenient to obtain the required result by combining the results of two or more different calorimetric experiments. For example one might measure

$$\{H(T, p, \xi^{eq}) - H(T, p, \xi)\},\qquad(10.2.12)$$

where ξ^{eq} is the *equilibrium* value of ξ, first for $\xi = \xi'$ and second for $\xi = \xi''$, and then obtain the required difference (11) by subtraction. More usually one would measure values of ΔH for several chemical reactions that can be added and subtracted so as to give the required result.

All the corresponding quantities with heat capacity at constant pressure C_p in place of enthalpy H can be measured by repeating the measurement of ΔH over a range of temperatures. All the corresponding quantities with volume V in place of enthalpy H can be measured dilatometrically or densitometrically (see § 2.9).

§ 10.3   Chemical potentiometer

We are now ready to devise an experimental method for the measurement of the difference:

$$\{\mu_B(T, p, x^\beta_C) - \mu_B(T, p, x^\alpha_C)\},\qquad(10.3.1)$$

between the chemical potentials at given temperature and pressure of any substance B in any two phases having different compositions. The symbol x^α_C is again used to denote the set of $(\mathscr{C}-1)$ independent mole fractions $x^\alpha_B, x^\alpha_C, \cdots$, in the phase α where $\mathscr{C}$ is the number of components present in the phase.

We construct the apparatus shown in figure 10.1 in which a phase α consisting of a mixture of A, B, C, $\cdots$, present with mole fractions $(1 - x^\alpha_B - x^\alpha_C - \cdots), x^\alpha_B, x^\alpha_C, \cdots$,

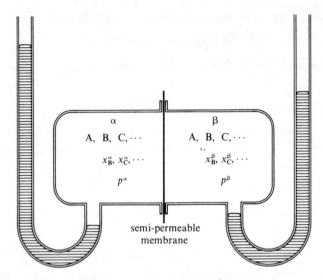

Figure 10.1. A chemical potentiometer.

is separated from a phase β consisting of a mixture of A, B, C, $\cdots$, present with mole fractions $(1 - x_B^\beta - x_C^\beta - \cdots)$, x_B^β, x_C^β, $\cdots$, by a membrane permeable to the substance B but impermeable to at least one of the other substances A, C, $\cdots$. Such a membrane is called *semi-permeable*. Let p^α and p^β be the pressures of the two phases at osmotic equilibrium. We then have

$$\mu_B(T, p^\beta, x_C^\beta) = \mu_B(T, p^\alpha, x_C^\alpha). \tag{10.3.2}$$

The sought-for difference (1) is then given by

$$\mu_B(T, p, x_C^\beta) - \mu_B(T, p, x_C^\alpha) = \{\mu_B(T, p, x_C^\beta) - \mu_B(T, p^\beta, x_C^\beta)\}$$
$$+ \{\mu_B(T, p^\beta, x_C^\beta) - \mu_B(T, p^\alpha, x_C^\alpha)\}$$
$$+ \{\mu_B(T, p^\alpha, x_C^\alpha) - \mu_B(T, p, x_C^\alpha)\}$$
$$= \int_{p^\beta}^{p} \{\partial \mu_B(T, p, x_C^\beta)/\partial p\}_{T, x_C^\beta} \, dp + 0$$
$$- \int_{p^\alpha}^{p} \{\partial \mu_B(T, p, x_C^\alpha)/\partial p\}_{T, x_C^\alpha} \, dp$$
$$= \int_{p^\beta}^{p} V_B(T, p, x_C^\beta) \, dp - \int_{p^\alpha}^{p} V_B(T, p, x_C^\alpha) \, dp, \tag{10.3.3}$$

where we have used equation (6.9.3) in the third step and where $V_B(T, p, x_C)$ is the partial molar volume of the substance B in a mixture with mole fractions x_C at temperature T and pressure p.

We now have a recipe for the measurement of the difference:

$$\{\mu_B(T, p, x_C^\beta) - \mu_B(T, p, x_C^\alpha)\}, \tag{10.3.4}$$

between the chemical potentials at given temperature and pressure of any substance B in any two phases having any compositions: we use our chemical potentiometer to measure the pressures p^α and p^β needed to ensure diffusive equilibrium with respect to the substance B when the two mixtures are separated by a semi-permeable membrane, and the method described in §§ 2.8 and 2.9 to measure the partial molar volume of the substance B at the two compositions over the appropriate ranges of pressure.

In the special case that the phase α is *pure* substance B the equilibrium condition (2) can be written in the form:

$$\mu_B(T, p + \Pi, x_C^\beta) = \mu_B^*(T, p), \tag{10.3.5}$$

where we have made the substitutions:

$$p^\beta = p + \Pi, \quad \text{and} \quad p^\alpha = p. \tag{10.3.6}$$

The excess pressure Π which must be applied to a mixture to ensure osmotic equilibrium between a substance B in the mixture and the pure substance B is called the *osmotic pressure* of the mixture with respect to the substance B.

In practice we might of course fail to find a semi-permeable membrane for a particular mixture. Before learning how to cope with such situations we shall discuss some kinds of system for which suitable membranes are known.

For gaseous mixtures containing hydrogen a palladium membrane is permeable only to hydrogen and so can be used to measure the pressures p^α and p^β corresponding to the equilibrium condition:

$$\mu(H_2, T, p^\beta, x_C^\beta) = \mu(H_2, T, p^\alpha, x_C^\alpha), \tag{10.3.7}$$

and so to obtain the difference:

$$\{\mu(H_2, T, p, x_C^\beta) - \mu(H_2, T, p, x_C^\alpha)\}, \tag{10.3.8}$$

between the chemical potentials of hydrogen in two mixtures having different compositions at the same temperature and pressure.

For a liquid solution of a solute having a much greater molar mass than that of the solvent a membrane can usually be found which is permeable to the solvent but not to the solute. Such membranes include animal membranes, copper ferrocyanide precipitated in the interstices of a porous pot, and regenerated cellulose (Cellophane). When a liquid solution containing mole fraction x of a solute B in a solvent A is separated from pure solvent A by a membrane permeable only to the solvent A then the condition for osmotic equilibrium is again of the form:

$$\mu_A(T, p + \Pi, x) = \mu_A^*(T, p), \tag{10.3.9}$$

where Π is the osmotic pressure of the solution (the phrase 'with respect to the solvent' being customarily omitted). The difference:

$$\{\mu_A(T, p, x) - \mu_A^*(T, p)\},\tag{10.3.10}$$

between the chemical potentials of solvent A in the solution and in the pure state both at the same pressure and temperature is then given by

$$\mu_A(T, p, x) - \mu_A^*(T, p) = -\int_p^{p+\Pi} V_A(T, p, x)\,dp,\tag{10.3.11}$$

where $V_A(T, p, x)$ is the partial molar volume of the solvent A in a solution containing mole fraction x of the solute B at temperature T and pressure p.

Various more or less satisfactory kinds of apparatus have been used to measure osmotic pressure. All are identical in principle with the chemical potentiometer shown in figure 10.1. Sometimes the manometer in contact with the solvent is omitted so that the solvent is always at atmospheric pressure. Sometimes the manometer in contact with the solution is replaced by a piston of known area to which weights can be added or by a Bourdon gauge, and sometimes the solution itself is used as the manometer.

Presumably none of the membranes mentioned above is absolutely impermeable to any substance. Presumably any of our systems of two phases separated by a membrane would eventually reach a state of *complete* equilibrium in which the two phases would have become identical, instead of the state of partial or osmotic equilibrium that we have been discussing. A sufficient condition for the successful study of an osmotic equilibrium is that the *rate* of diffusion through the membrane should be negligibly slow for all substances to which the membrane is supposed to be 'impermeable'. This being understood it is easy to see that a vapour phase will behave as a membrane permeable only to the solvent when it separates two solutions of *involatile* solutes. An apparatus with such a 'vapour membrane' is shown in figure 10.2.

A column of solvent *under tension* hangs from a porous disk D and is separated by a vapour phase from the solution. At osmotic equilibrium the heights h_1 and h_2 remain stationary and the osmotic pressure Π is given by

$$\Pi = (h_1\rho_A^* + h_2\rho_{Hg})g,\tag{10.3.12}$$

where ρ_A^* and ρ_{Hg} are the densities of pure solvent and of mercury, and g is the acceleration of free fall. This apparatus has not often been used† but might yet prove useful especially for the study of dilute solutions of electrolytes. There is little difficulty in 'stretching' a column of liquid provided that it has been freed from dissolved air. The main difficulty is that of thermal equilibration. Temperature differences between solvent and solution, caused by evaporation from one and condensation into the other, cause the direction of osmosis to be reversed so that osmotic equilibrium is reached only after a long succession of slowly attenuated oscillations.

† Williamson, A. T. *Proc. Roy. Soc.* (*London*) **1948**, A195, 97.

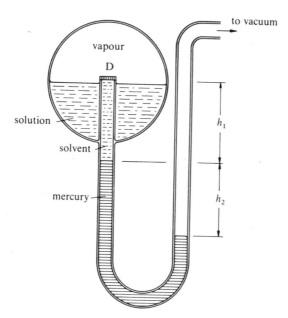

Figure 10.2. A 'vapour-membrane' apparatus for the measurement of the osmotic pressure of a solution of an involatile solute.

We now return to the point made on page 143 that in practice we might fail to find a semi-permeable membrane for a particular mixture. Indeed we must admit that usually no suitable membrane is known. Consider for example the problem of measuring the difference between the chemical potentials of nitrogen in two different mixtures of nitrogen + oxygen. No membrane is known which is permeable to nitrogen but impermeable to oxygen (or through which the rates of diffusion differ by a factor of not less than say 10^3). As another example consider the problem of measuring the difference between the chemical potentials of water in two liquid mixtures of water + ethanol.

This difficulty is resolved by our knowledge, from sources which we shall discuss in the following section, of the results of experiments with a chemical potentiometer in the special case that the two phases are both *perfect gas mixtures*.

§ 10.4   Perfect-gas-mixture chemical potentiometer

A perfect gas mixture (pgm) is defined by the equation:

$$pV = \left(\sum_{B} n_B \right) RT. \tag{10.4.1}$$

It is known by experiment that at sufficiently low pressures any gas mixture behaves as a perfect gas mixture.

When the phases α and β are both perfect gas mixtures and are separated in a chemical potentiometer like that shown in figure 10.1 by a membrane permeable only to the substance B then at osmotic equilibrium it is known that the pressures p^α and p^β of equation (10.3.2), which we now write in the form:

$$\mu_B^{\text{pgm}}(T, p^\beta, x_C^\beta) = \mu_B^{\text{pgm}}(T, p^\alpha, x_C^\alpha),\qquad(10.4.2)$$

are related by the equation:

$$x_B^\beta p^\beta = x_B^\alpha p^\alpha,\qquad(10.4.3)$$

where x_B^α and x_B^β are the mole fractions of the substance B in the two phases α and β.

Equation (3) has been verified directly[†] for hydrogen by experiments with a palladium membrane on various hydrogen-containing gas mixtures at pressures around atmospheric. Equation (3) has not been verified *directly* for other gases. There are, however, good reasons for our belief that if only suitable membranes could be found then the results of such experiments would be in accord with equation (3) for any substance in any gaseous mixture at sufficiently low pressures. One of the reasons is the agreement with experiment, for a wide variety of chemical reactions in gaseous mixtures at low pressures, of the *equilibrium law for chemical reactions in perfect gas mixtures* which depends, as we shall see, on the assumption of equation (3). Another reason is that equation (3) is, as we shall also see, one of the consequences of the statistical-mechanical treatment of perfect gas mixtures on the plausible assumption that a perfect gas mixture behaves as if it were an assembly of non-interacting particles. We shall henceforth accept and make use of equation (3) and may expect to gain increasing confidence in it as we encounter thermodynamic equations which depend on its truth and which agree accurately with experiment.

We shall now show how equation (3) can be used to enable us to measure the difference:

$$\{\mu_B^{\text{pgm}}(T, p, x_C^\beta) - \mu_B^{\text{pgm}}(T, p, x_C^\alpha)\},\qquad(10.4.4)$$

between the chemical potentials at given temperature and presssure of any substance B in any two perfect gas mixtures. We evaluate the difference (4) by rewriting it in the form:

$$\mu_B^{\text{pgm}}(T, p, x_C^\beta) - \mu_B^{\text{pgm}}(T, p, x_C^\alpha) = \{\mu_B^{\text{pgm}}(T, p, x_C^\beta) - \mu_B^{\text{pgm}}(T, p^\beta, x_C^\beta)\}$$

$$+ \{\mu_B^{\text{pgm}}(T, p^\beta, x_C^\beta) - \mu_B^{\text{pgm}}(T, p^\alpha, x_C^\alpha)\}$$

$$+ \{\mu_B^{\text{pgm}}(T, p^\alpha, x_C^\alpha) - \mu_B^{\text{pgm}}(T, p, x_C^\alpha)\}$$

[†] Ramsay, W. *Phil. Mag.* **1894**, 38, 206.

$$= \int_{p^\beta}^{p} V_B^{pgm}(T, p, x_C^\beta)\, dp$$

$$+ \{\mu_B^{pgm}(T, p^\beta, x_C^\beta) - \mu_B^{pgm}(T, p^\alpha, x_C^\alpha)\}$$

$$- \int_{p^\alpha}^{p} V_B^{pgm}(T, p, x_C^\alpha)\, dp$$

$$= RT \ln(p^\alpha/p^\beta)$$

$$+ \{\mu_B^{pgm}(T, p^\beta, x_C^\beta) - \mu_B^{pgm}(T, p^\alpha, x_C^\alpha)\}, \qquad (10.4.5)$$

where we have used the relation:

$$V_B^{pgm}(T, p, x_C) = RT/p, \qquad (10.4.6)$$

which follows from the definition (1) of a perfect gas mixture. When we now interpret p^α and p^β as the pressures corresponding to osmotic equilibrium with respect to the substance B then by use of equations (2) and (3) equation (5) becomes

$$\mu_B^{pgm}(T, p, x_C^\beta) - \mu_B^{pgm}(T, p, x_C^\alpha) = RT \ln(x_B^\beta/x_B^\alpha). \qquad (10.4.7)$$

We note that the difference (4) or the left-hand side of equation (7) for perfect gas mixtures depends only on the mole fractions of B in the two mixtures and not at all on the nature or relative amounts of other substances present and not at all on the pressure.

In view of the experimental fact that any gaseous mixture becomes more and more nearly perfect as the pressure is decreased towards zero, we may write instead of equation (7)

$$\mu_B^g(T, p \to 0, x_C^\beta) - \mu_B^g(T, p \to 0, x_C^\alpha) = RT \ln(x_B^\beta/x_B^\alpha). \qquad (10.4.8)$$

§ 10.5   Determination of the dependence of chemical potential on composition without a chemical potentiometer

Our next step is to show how equation (10.4.8) can be used to enable us to measure the difference:

$$\{\mu_B^g(T, p, x_C^\beta) - \mu_B^g(T, p, x_C^\alpha)\}, \qquad (10.5.1)$$

between the chemical potentials at given temperature and pressure of any substance B in any two gaseous mixtures without the help of a chemical potentiometer. We write the difference (1) in the form:

$$\mu_B^g(T, p, x_C^\beta) - \mu_B^g(T, p, x_C^\alpha) = \int_0^p \{V_B^g(T, p, x_C^\beta) - V_B^g(T, p, x_C^\alpha)\}\, dp$$

$$+ \{\mu_B^g(T, p \to 0, x_C^\beta) - \mu_B^g(T, p \to 0, x_C^\alpha)\}, \qquad (10.5.2)$$

where $V_B^g(T, p, x_C)$ is the partial molar volume of B in the gaseous mixture having mole fractions x_C at the temperature T and pressure p. We may replace the second term in $\{\cdots\}$ on the right of equation (2) by the right-hand side of equation (10.4.8) and so obtain

$$\mu_B^g(T, p, x_C^\beta) - \mu_B^g(T, p, x_C^\alpha) = \int_0^p \{V_B^g(T, p, x_C^\beta) - V_B^g(T, p, x_C^\alpha)\}\, dp$$

$$+ RT \ln(x_B^\beta / x_B^\alpha). \tag{10.5.3}$$

Since the partial molar volume V_B^g can be measured for any substance B in any gaseous mixture at any temperature and pressure it follows from (3) that the difference (1) can be measured for any two gaseous mixtures, without recourse to a chemical potentiometer but on the assumption that equation (10.4.8) is correct.

Our final step in this section is to show how equation (10.4.8) can be used to enable us to measure the difference:

$$\{\mu_B^l(T, p, x_C^{l\beta}) - \mu_B^l(T, p, x_C^{l\alpha})\}, \tag{10.5.4}$$

between the chemical potentials at given temperature and pressure of any substance B in any two *liquid* (or by simple extension *solid*) mixtures. We begin by writing the identity:

$$\mu_B^l(T, p, x_C^l) = \{\mu_B^l(T, p, x_C^l) - \mu_B^l(T, p^{l+g}, x_C^l)\} + \{\mu_B^l(T, p^{l+g}, x_C^l) - \mu_B^g(T, p^{l+g}, x_C^g)\}$$

$$+ \{\mu_B^g(T, p^{l+g}, x_C^g) - \mu_B^g(T, p \to 0, x_C^g)\} + \mu_B^g(T, p \to 0, x_C^g), \tag{10.5.5}$$

where we choose to let p^{l+g} be the particular pressure, called the *equilibrium vapour pressure*, at which the liquid mixture having mole fractions x_C^l can coexist at equilibrium with a gaseous mixture having mole fractions x_C^g at temperature T. We then have

$$\mu_B^l(T, p^{l+g}, x_C^l) = \mu_B^g(T, p^{l+g}, x_C^g), \tag{10.5.6}$$

so that equation (5) becomes

$$\mu_B^l(T, p, x_C^l) = \int_{p^{l+g}}^p V_B^l(T, p, x_C^l)\, dp + 0 + \int_0^{p^{l+g}} V_B^g(T, p, x_C^g)\, dp + \mu_B^g(T, p \to 0, x_C^g). \tag{10.5.7}$$

Substituting (7) into (4) and using equation (10.4.8) we then obtain

$$\mu_B^l(T, p, x_C^{l\beta}) - \mu_B^l(T, p, x_C^{l\alpha}) = \int_{p^{l+g,\beta}}^p V_B^l(T, p, x_C^{l\beta})\, dp - \int_{p^{l+g,\alpha}}^p V_B^l(T, p, x_C^{l\alpha})\, dp$$

$$+ \left\{ \int_0^{p^{l+g,\beta}} V_B^g(T, p, x_C^{g\beta})\, dp - \int_0^{p^{l+g,\alpha}} V_B^g(T, p, x_C^{g\alpha})\, dp \right\}$$

$$+ RT \ln(x_B^{g\beta} / x_B^{g\alpha}). \tag{10.5.8}$$

(Each of the integrals inside the $\{\cdots\}$ diverges, but their difference does not.) All the quantities on the right-hand side of equation (8) can be measured and so therefore can the difference (4), without recourse to a chemical potentiometer but on the assumption that equation (10.4.8) is correct. In particular we can measure the equilibrium vapour pressures $p^{l+g,\alpha}$ and $p^{l+g,\beta}$, and the sets of mole fractions $x_C^{g\alpha}$ and $x_C^{g\beta}$ in the gaseous phases gα and gβ which coexist at equilibrium with the liquid phases lα and lβ having compositions determined by the sets of mole fractions $x_C^{l\alpha}$ and $x_C^{l\beta}$; and we can measure the partial molar volumes of the substance B, V_B^l in the liquid phases and V_B^g in the gaseous phases, at the two compositions over the appropriate ranges of pressure. We shall make good use of equation (8) in chapter 16.

§ 10.6   Another kind of chemical potentiometry

In § 10.3 our chemical potentiometry was based on measurements of differences of pressure and the use of equation (6.9.3), namely $(\partial \mu_B/\partial p)_T = V_B$. For our present purpose we shall regard it as having been based on the use of the equivalent equation (6.9.6), namely

$$(\partial \ln \lambda_B/\partial p)_T = V_B/RT, \qquad (10.6.1)$$

and shall now ask whether we can instead base a method of chemical potentiometry on measurements of differences of temperature and the use of the 'corresponding' equation (6.8.11), namely

$$(\partial \ln \lambda_B/\partial T)_p = -H_B/RT^2. \qquad (10.6.2)$$

We shall find that an important method of chemical potentiometry is based on the use of equation (2), but that the method has severe practical limitations.

There are at least two important differences between equations (1) and (2) when they are regarded as bases for chemical potentiometry. First, whereas V_B can be measured, H_B cannot be measured; only differences ΔH_B can be measured. Second, whereas osmotic equilibrium can exist, no comparable partial equilibrium can exist in which two phases separated by a semi-permeable membrane have different temperatures; a semi-permeable membrane is necessarily diathermic.

We now consider the two-phase equilibria shown diagrammatically in figure 10.3, where we have used the symbols T_B^* and T_B for the equilibrium temperatures. Such equilibria can be realized. For example pure ice can coexist with pure liquid water, and pure ice can coexist with a liquid solution of say sucrose in water; pure gaseous water can coexist with pure liquid water, and pure gaseous water can coexist with a liquid solution of an involatile solute dissolved in water. We shall now show how measurements of the temperatures T_B^* and T_B defined in the legend to figure 10.3 can be used to obtain the ratio of absolute activities:

$$\lambda_B^{*\alpha}(T_B, p)/\lambda_B^\alpha(T_B, p, x_C^\alpha), \qquad (10.6.3)$$

corresponding to the chemical-potential difference:

$$\{\mu_B^{*\alpha}(T_B, p) - \mu_B^\alpha(T_B, p, x_C^\alpha)\}, \qquad (10.6.4)$$

where x_C^α denotes the set of mole fractions in the mixture α.

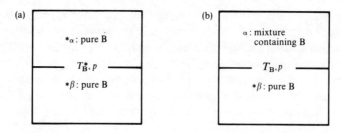

Figure 10.3. Equilibria at pressure p: (a), between two phases $*\alpha$ and $*\beta$ of the pure substance B at the temperature T_B^*, and (b), between two phases α and $*\beta$, the first a mixture containing the substance B and the second the pure substance B, at the temperature T_B.

For diffusive equilibrium of the substance B between the phases $*\alpha$ and $*\beta$ we have, for (a),

$$\lambda_B^{*\alpha}(T_B^*, p) = \lambda_B^{*\beta}(T_B^*, p),\qquad(10.6.5)$$

and between the phases α and $*\beta$, for (b),

$$\lambda_B^{\alpha}(T_B, p, x_C^{\alpha}) = \lambda_B^{*\beta}(T_B, p).\qquad(10.6.6)$$

By use of equations (5) and (6) we write the logarithm of the ratio (3) in the form:

$$\ln\{\lambda_B^{*\alpha}(T_B, p)/\lambda_B^{\alpha}(T_B, p, x_C^{\alpha})\} = -\ln\{\lambda_B^{*\alpha}(T_B^*, p)/\lambda_B^{*\alpha}(T_B, p)\}$$

$$+\ln\{\lambda_B^{*\beta}(T_B^*, p)/\lambda_B^{*\beta}(T_B, p)\}$$

$$=\int_{T_B}^{T_B^*} \{-(\partial \ln \lambda_B^{*\alpha}/\partial T)_p + (\partial \ln \lambda_B^{*\beta}/\partial T)_p\}\, dT$$

$$=\int_{T_B}^{T_B^*} \{(H_B^{*\alpha} - H_B^{*\beta})/RT^2\}\, dT$$

$$=\int_{T_B}^{T_B^*} (\Delta_\beta^\alpha H_B^*/RT^2)\, dT,\qquad(10.6.7)$$

where we have used equation (2) to obtain the third equality, and where $\Delta_\beta^\alpha H_B^*$ denotes the molar enthalpy of transition of the pure substance B from phase β to phase α at the temperature T and pressure p.

A molar enthalpy difference $\Delta_\beta^\alpha H_B^*$ (for example the molar enthalpy of melting $\Delta_s^l H_B^*$ or the molar enthalpy of evaporation $\Delta_l^g H_B^*$) of a pure substance B can be measured calorimetrically at the equilibrium transition temperature T_B^* (for example the melting temperature or the boiling temperature) which is fixed by the chosen pressure p. In principle the enthalpy of transition can also be measured calorimetrically at other temperatures sufficiently close to T_B^* (see figure 10.4) and so the integral in equation (7) can in principle be evaluated.

In practice it is always difficult and often impossible to measure $\Delta_\beta^\alpha H_B^*$ at any temperature other than T_B^*. We accordingly expand the integrand $\Delta_\beta^\alpha H_B^*/RT^2$ of

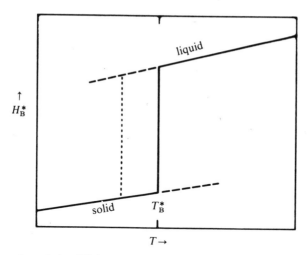

Figure 10.4. The molar enthalpy H_B^* (relative to an arbitrary zero) of a pure substance B plotted against temperature T at temperatures above and below the melting temperature T_B^* at given pressure. The unbroken vertical line determines the value of $\Delta_s^l H_B^*$ at the equilibrium melting temperature T_B^*. The broken continuations of the curves for solid and liquid at temperatures greater than and less than T_B^*, respectively, relate to metastable states which are realizable, always in principle and sometimes in practice. The dotted vertical line determines the value of $\Delta_s^l H_B^*$ at a temperature less than T_B^*.

equation (7) as a Taylor series around $T = T_B^*$ and obtain

$$\Delta_\beta^\alpha H_B^*(T, p)/RT^2 = \Delta_\beta^\alpha H_B^*(T_B^*, p)/RT_B^{*2} + \{2\Delta_\beta^\alpha H_B^*(T_B^*, p)/RT_B^{*2}$$
$$- \Delta_\beta^\alpha C_{p,B}^*/RT_B^*\}(1 - T/T_B^*) + O(1 - T/T_B^*)^2 . \quad (10.6.8)$$

Substitution of (8) into (7) and integration then leads to the equation:

$$\ln\{\lambda_B^{*\alpha}(T_B, p)/\lambda_B^\alpha(T_B, p, x_C^\alpha)\} = \{\Delta_\beta^\alpha H_B^*(T_B^*, p)/RT_B^*\}(1 - T_B/T_B^*)$$
$$+ \{\Delta_\beta^\alpha H_B^*(T_B^*, p)/RT_B^*$$
$$- \Delta_\beta^\alpha C_{p,B}^*(T_B^*, p)/2R\}(1 - T_B/T_B^*)^2$$
$$+ O(1 - T_B/T_B^*)^3 . \quad (10.6.9)$$

The ratio (3) is always greater than unity, and the difference (4) is always positive; the absolute activity or the chemical potential of a pure substance is always greater than that of the same substance in a mixture at the same temperature and pressure.† The enthalpy of transition $\Delta_\beta^\alpha H_B^*$ is positive if α is the high-temperature phase and β is the low-temperature phase, and conversely. It follows from equation (7) that $T_B < T_B^*$ when α is the high-temperature phase and that $T_B > T_B^*$ when α is the low-temperature phase.

---

† This is a consequence of one of the thermodynamic inequalities dealt with in chapter 7: $(\partial \mu_B/\partial n_B)_{T,p,nc\neq B} > 0$.

Two special cases are of overwhelming importance. If we make the identification $\alpha = l$ and $\beta = g$ then α is the low-temperature phase and $T_B > T_B^*$; at given pressure the boiling temperature of a solvent is raised by the addition of an involatile solute. If we make the identification $\alpha = l$ and $\beta = s$ then α is the high-temperature phase and $T_B < T_B^*$; at given pressure the freezing temperature of a solvent is lowered by the addition of a solute. We shall return to measurements of the elevation of the boiling temperature and of the depression of the freezing temperature in chapter 18.

The right-hand side of equation (9) can be evaluated only when T_B is close enough to T_B^* to ensure convergence of the Taylor's series. That is a severe practical limitation on this method of chemical potentiometry. In practice T_B should differ from T_B^* by at most a few kelvins and preferably by less than 1 K, depending on the accuracy required.

The method can nevertheless be used as follows, with the help of further calorimetric measurements, to determine the ratio:

$$\lambda_B^{*\alpha}(T, p)/\lambda_B^\alpha(T, p, x_C^\alpha), \tag{10.6.10}$$

at *any* temperature T, and not only at the equilibrium temperature T_B (which depends of course on the composition of the phase α of figure 10.3). We use equation (2) to obtain the relation:

$$\ln\{\lambda_B^{*\alpha}(T, p)/\lambda_B^\alpha(T, p, x_C^\alpha)\} = \ln\{\lambda_B^{*\alpha}(T_B, p)/\lambda_B^\alpha(T_B, p, x_C^\alpha)\}$$
$$- \int_{T_B}^{T} [\{H_B^{*\alpha}(T, p) - H_B^\alpha(T, p, x_C^\alpha)\}/RT^2]\, dT. \tag{10.6.11}$$

The quantity in $\{\cdots\}$ in the integrand can be obtained from measurements of the molar enthalpy of mixing $\Delta_{mix}H_m$, or from the molar enthalpy of dissolution $\Delta_{soln}H_m$, made over a range of compositions at temperatures between T_B and T according to the methods explained in chapter 4. By substitution from (11) into (9) we obtain

$$\ln\{\lambda_B^{*\alpha}(T, p)/\lambda_B^\alpha(T, p, x_C^\alpha)\}$$
$$= \{\Delta_\beta^\alpha H_B^*(T_B^*, p)/RT_B^*\}(1 - T_B/T_B^*) + \{\Delta_\beta^\alpha H_B^*(T_B^*, p)/RT_B^*$$
$$- \Delta_\beta^\alpha C_{p,B}(T_B^*, p)/2R\}(1 - T_B/T_B^*)^2 + O(1 - T_B/T_B^*)^3$$
$$- \int_{T_B}^{T} [\{H_B^{*\alpha}(T, p) - H_B^\alpha(T, p, x_C^\alpha)\}/RT^2]\, dT. \tag{10.6.12}$$

Another severe practical limitation on this method of chemical potentiometry is the need to be able to realize phase equilibria like those shown in figure 10.3. For example if $\alpha = l$ and $\beta = g$ we are restricted to solutions in which B is a volatile solvent and the solutes are involatile. If $\alpha = l$ and $\beta = s$ we are restricted to solutions that can coexist in equilibrium with the *pure* solid B. There are many examples in the older literature of the use of the lowering of the freezing temperature to determine the approximate molar mass of a solute having been bedevilled by a solid phase (corresponding to $*\beta$ in (b) of figure 10.3) not being pure solid solvent but a (single-phase) solid solution. Indeed, under these conditions the freezing temperature of a solvent can sometimes be raised by the addition of a solute; at least one then knows that the solid phase in equilibrium with the solution cannot be the pure solvent.

We shall return to the use of equation (12), and to the points made in the last paragraph, in chapter 18.

§ 10.7   Galvanic cells as chemical potentiometers

No galvanic cell can strictly be regarded as being in diffusive equilibrium, and so any discussion of galvanic cells properly belongs outside classical thermodynamics. We shall nevertheless discuss the dependence of the electromotive force (e.m.f.) of a galvanic cell on the composition in chapter 20, after introducing in chapter 19 the non-thermodynamic equation† that we shall use as the basis for that discussion.

In the meantime we describe here one example of a galvanic cell that can be used as an accurate chemical potentiometer, the non-thermodynamic terms in the expression for the e.m.f. in this case virtually vanishing.

For the cell:

$$\text{Pt}|H_2(g)|HCl(aq, m)|AgCl(s)|Ag(s)\,, \qquad (10.7.1)$$

in which a hydrogen electrode and a (silver + silver chloride) electrode are in contact with an aqueous solution containing molality m of HCl, all at given temperature and pressure, the e.m.f. E can be accurately expressed in the form:

$$-EF = \mu^*(\text{Ag, s}) + \mu(\text{HCl, aq, } m) - \tfrac{1}{2}\mu(H_2, g) - \mu^*(\text{AgCl, s})\,, \qquad (10.7.2)$$

where F denotes the Faraday constant. For two cells with $m = m_1$, $E = E_1$, and $m = m_2$, $E = E_2$, we obtain from (2) by subtraction

$$\mu(\text{HCl, aq, } m_2) - \mu(\text{HCl, aq, } m_1) = (E_1 - E_2)F\,, \qquad (10.7.3)$$

the remaining terms cancelling.‡ We note that the left-hand side of (3) is of the same form as (10.3.1), for which we have been seeking methods of measurement throughout this chapter.

§ 10.8   Summary

Let us now summarize the conclusions reached in §§ 10.2 to 10.6.

Whenever a membrane can be found which is permeable to a substance B but

---

† The general equation for the e.m.f. of a galvanic cell contains among other things the *speeds* with which ions move in a gradient of electric field strength, and therefore is *not* a thermodynamic equation.
‡ Strictly, the chemical potential of $H_2(g)$ in contact with the solution varies a little with the composition of the solution; we shall take account of this in chapter 20 but in the meantime shall assume that the corrections have been made.

impermeable to at least one of the other substances present, the difference:

$$\{\mu_B(T, p, x_C^\beta) - \mu_B(T, p, x_C^\alpha)\}, \tag{10.8.1}$$

between the chemical potentials of the substance B in any two phases can be measured by application of equation (10.3.3).

For perfect-gas mixtures the result which would be obtained for the difference (1) if a semi-permeable membrane were available is known from other sources to be given by

$$\{\mu_B^{pgm}(T, p, x_C^\beta) - \mu_B^{pgm}(T, p, x_C^\alpha)\} = RT \ln(x_B^\beta/x_B^\alpha). \tag{10.8.2}$$

By making use of equation (2) the difference (1) can be measured for *any* gas mixtures according to equation (10.5.3).

By way of the trick of studying the gaseous phase which can coexist at equilibrium with a liquid (or solid) mixture the difference (1) can be measured for any liquid (or solid) mixtures according to equation (10.5.8).

The difference (1) can sometimes be obtained for dilute solutions in the solvent B from measurements of the raising of the boiling temperature or of the lowering of the freezing temperature.

The difference (1) can sometimes be obtained for a dissolved electrolyte from measurements of the e.m.f. of a galvanic cell.

§ 10.9   Use of the Gibbs–Duhem equation

Only $(\mathscr{C} - 1)$ of the $\mathscr{C}$ differences of the form of (10.8.1) are independent. That conclusion follows at once from the Gibbs–Duhem equation (5.7.1) written in the form:

$$\sum_B x_B \, d\mu_B = 0, \quad (T, p \text{ constant}). \tag{10.9.1}$$

In particular for a binary mixture of A and B containing mole fraction x of B equation (1) becomes

$$(1 - x) \, d\mu_A + x \, d\mu_B = 0, \quad (T, p \text{ constant}), \tag{10.9.2}$$

so that

$$\mu_A(T, p, x^\beta) - \mu_A(T, p, x^\alpha) = -\int_{x=x^\alpha}^{x=x^\beta} \{x/(1-x)\} \, d\mu_B(T, p, x). \tag{10.9.3}$$

Thus if the difference:

$$\{\mu_B(T, p, x) - \mu_B(T, p, x')\}, \tag{10.9.4}$$

has been measured with constant x' (for example $x' = 1$) for values of x over the range $x = x^\alpha$ to $x = x^\beta$, then the difference:

$$\{\mu_A(T, p, x^\beta) - \mu_A(T, p, x^\alpha)\}, \tag{10.9.5}$$

can be determined according to equation (3).

§ 10.10   Dependence of Gibbs function on composition

We recall from § 10.2 that there are two kinds of change of composition for which the enthalpy difference ΔH can be measured. These are change of composition as a result of *mixing* and change of composition as a result of *chemical reaction*. In this section we shall use the results of §§ 10.3 to 10.9 to show how the difference of Gibbs function ΔG can also be measured for these two kinds of change of composition. Before doing that we remind the reader of the relation (5.4.4):

$$G = \sum_B n_B \mu_B, \quad \text{or} \quad G_m = \sum_B x_B \mu_B, \tag{10.10.1}$$

between the Gibbs function G or molar Gibbs function G_m and the chemical potentials μ_B.

The molar Gibbs function of mixing $\Delta_{mix}G_m$ is defined for a binary mixture of A and B containing mole fraction x of B by the relation, parallel to equation (10.2.2) for the corresponding enthalpy change,

$$\Delta_{mix}G_m = G_m(T, p, x) - (1-x)G_m(T, p, 0) - xG_m(T, p, 1). \tag{10.10.2}$$

By use of equation (1) we may rewrite (2) in the form:

$$\Delta_{mix}G_m = (1-x)\{\mu_A(T, p, x) - \mu_A(T, p, 0)\}$$
$$+ x\{\mu_B(T, p, x) - \mu_B(T, p, 1)\}. \tag{10.10.3}$$

We have seen in §§ 10.3 to 10.9 how to measure each of the two quantities in $\{\cdots\}$ in equation (3) and so can now see how to measure $\Delta_{mix}G_m$. It is left as an exercise for the reader to verify that ΔG can be measured for the mixing at constant temperature and pressure of any number of phases having any compositions.

The change $\Delta_r G$ in Gibbs function which accompanies a change in the extent ξ of a chemical reaction at given temperature and pressure and for given initial amounts of substance is

$$\Delta_r G = G(T, p, \xi'') - G(T, p, \xi')$$
$$= \int_{\xi'}^{\xi''} (\partial G / \partial \xi)_{T,p} \, d\xi. \tag{10.10.4}$$

In view of the relation:

$$(\partial G / \partial \xi)_{T,p} = \sum_B \nu_B \mu_B, \tag{10.10.5}$$

which follows from equation (5.9.5), equation (4) becomes

$$\Delta_r G = \int_{\xi'}^{\xi''} \left\{ \sum_B \nu_B \mu_B(T, p, \xi) \right\} d\xi. \tag{10.10.6}$$

It is not yet obvious that $\Delta_r G$ given by equation (6) is measurable. When, however, we use the fundamental equation (§ 7.7) for chemical equilibrium:

$$\sum_B \nu_B \mu_B(T, p, \xi^{eq}) = 0 , \qquad (10.10.7)$$

where ξ^{eq} is the extent of reaction at chemical equilibrium, we can recast equation (6) into the form:

$$\Delta_r G = \int_{\xi'}^{\xi''} \left[\sum_B \nu_B \{ \mu_B(T, p, \xi) - \mu_B(T, p, \xi^{eq}) \} \right] d\xi . \qquad (10.10.8)$$

We have seen in §§ 10.3 to 10.9 how to measure the quantity in $\{ \cdot \cdot \}$ in equation (8) and so can now see how to measure the change $\Delta_r G$ in Gibbs function which accompanies a change in the extent ξ of any chemical reaction from any value ξ' to any other value ξ'' at constant temperature and pressure.

§ 10.11   Dependence of entropy on composition

Since at constant temperature T,

$$\Delta S = (\Delta H - \Delta G)/T , \qquad (10.11.1)$$

it follows at once from the conclusions reached in §§ 10.2 and 10.10 that the entropy difference ΔS can also be measured for changes of composition as a result of mixing or as a result of chemical reaction. Alternatively this statement follows from the conclusions of § 10.10 in view of the relation:

$$\Delta S = -(\partial \Delta G / \partial T)_p . \qquad (10.11.2)$$

§ 10.12   Conclusions

We have now completed the three tasks which we set ourselves towards the end of § 5.2, and have learnt not only how to measure thermodynamic temperature T, and the dependence of energy U, enthalpy H, and entropy S on temperature, pressure, and composition, and of Gibbs function G and chemical potential μ_B (or absolute activity λ_B) on pressure and composition, but also that each of the quantities occurring in the fundamental axiom (5.2.1) is completely and independently defined.

We have now also completed the thermodynamic framework needed to tackle any equilibrium problem for any system that can be described by the independent variables temperature T, pressure p, and amounts of substance n_B, n_C, $\cdot \cdot \cdot$.

Problems for chapter 10

Problem 10.1

Briefly describe an experimental method for the measurement of the molar entropy of mixing $\Delta_{mix} S_m$ of two gases, and formulate the equation needed to calculate the result from the measured quantities.

Problem 10.2

Formulate an equation for the molar entropy of mixing $\Delta_{mix}S_m$ of two perfect gases. Calculate the value of $\Delta_{mix}S_m$ for $x = 0.5$, where x denotes the mole fraction of either substance in the perfect-gas mixture.

Problem 10.3

Formulate equation (10.5.8) for a binary mixture $\{(1-x)A + xB\}$ when phase β is the mixture and phase α is pure A, assuming that V_A^l and V_A^{*l} are independent of pressure and that the molar volume of a gaseous mixture is given by

$$V_m^g = RT/p + (1-x^g)^2 B_{AA} + 2x^g(1-x^g)B_{AB} + (x^g)^2 B_{BB}.$$

Write the corresponding equation for the substance B, and derive the formula for the molar Gibbs function of mixing $\Delta_{mix}G_m$.

Problem 10.4

The osmotic pressure of a solution containing mole fraction 0.001 of a solute B is 0.139 MPa at 300 K when the solvent A is at a pressure of 0.100 MPa. Assuming as approximations (i) that $V_A^l \approx V_A^{*l} = 18 \text{ cm}^3 \text{ mol}^{-1}$, and (ii) that V_A^l is independent of pressure over the range 0.1 to 0.25 MPa, calculate the value of

$$\mu(A, l, 300 \text{ K}, 0.1 \text{ MPa}, x = 0.001) - \mu(A, l, 300 \text{ K}, 0.1 \text{ MPa}, x = 0),$$

where x denotes the mole fraction of the solute B.

Problem 10.5

The freezing temperature of sea water is 271.240 K. Given that the freezing temperature of pure water is 273.150 K, the molar volume of water is $18 \text{ cm}^3 \text{ mol}^{-1}$, and the molar enthalpy of melting of ice is 6000 J mol^{-1}, calculate the osmotic pressure of sea water at about 273 K.

Problem 10.6

Calculate $\ln\{\lambda(H_2O, l, m = 0)/\lambda(H_2O, l, m = 0.10459 \text{ mol kg}^{-1})\}$ for a solution of HCl of molality $m = 0.10459 \text{ mol kg}^{-1}$ in water at 273.15 K and atmospheric pressure, given that the depression of the freezing temperature is 0.3684 K, and for pure water at its normal freezing temperature 273.15 K that $\Delta_s^l H_m = 6008 \text{ J mol}^{-1}$ and $\Delta_s^l C_{p,m} = 38.09 \text{ J K}^{-1} \text{ mol}^{-1}$. How much difference would it make if $\Delta_s^l C_{p,m}$ were neglected?

Problem 10.7

For the galvanic cell:

$$Pt|H_2(g)|HCl(aq, m)|HgCl(s)|Hg(l) ,$$

at 298.15 K the e.m.f. E is 0.38948 V when the molality $m = 0.119304$ mol kg^{-1} and is 0.50532 V when $m = 0.0109474$ mol kg^{-1}. Calculate the difference between the chemical potentials of HCl in the two solutions.

Problem 10.8

For a liquid mixture of A and B containing mole fraction x' of B show that if

$$\mu(B, l, T, p, x') - \mu(B, l, T, p, 0) = RT \ln x' + A(1-x')^2 ,$$

then

$$\mu(A, l, T, p, x') - \mu(A, l, T, p, 1) = RT \ln(1-x') + A(x')^2 .$$

Are the corresponding formulae with $A(1-x')$ in place of $A(1-x')^2$ and Ax' in place of $A(x')^2$ possible formulae?

Chapter 11

Standard thermodynamic functions

§ 11.1 Introduction

Chapter 10 was devoted to experimental methods for the measurement of the chemical potential difference:

$$\{\mu_B(T, p, x_C^\beta) - \mu_B(T, p, x_C^\alpha)\}, \tag{11.1.1}$$

or of course the corresponding ratio of absolute activities:

$$\lambda_B(T, p, x_C^\beta)/\lambda_B(T, p, x_C^\alpha), \tag{11.1.2}$$

of a substance B in two different mixtures at the same temperature and pressure. The results of § 6.9 can be used to extend the experimental methods to the measurement of the chemical potential difference:

$$\{\mu_B(T, p^\beta, x_C^\beta) - \mu_B(T, p^\alpha, x_C^\alpha)\}, \tag{11.1.3}$$

or the corresponding ratio of absolute activities:

$$\lambda_B(T, p^\beta, x_C^\beta)/\lambda_B(T, p^\alpha, x_C^\alpha), \tag{11.1.4}$$

of a substance B in two different mixtures when the pressures (but not the temperatures) are also different.

It is expedient to rewrite the difference (3) in the form:

$$\{\mu_B(T, p^\beta, x_C^\beta) - \mu_B^\ominus(T)\} - \{\mu_B(T, p^\alpha, x_C^\alpha) - \mu_B^\ominus(T)\}, \tag{11.1.5}$$

where $\mu_B^\ominus$ is called the *standard chemical potential* of the substance B,[†] and the ratio (4) in the form:

$$\{\lambda_B(T, p^\beta, x_C^\beta)/\lambda_B^\ominus(T)\}/\{\lambda_B(T, p^\alpha, x_C^\alpha)/\lambda_B^\ominus(T)\}, \tag{11.1.6}$$

where $\lambda_B^\ominus$ is called the *standard absolute activity* of the substance B and is related to the standard chemical potential $\mu_B^\ominus$ by the definition:

$$\lambda_B^\ominus \overset{\text{def}}{=} \exp(\mu_B^\ominus/RT). \tag{11.1.7}$$

We have not yet defined $\mu_B^\ominus$ or $\lambda_B^\ominus$ but we note that each will be defined in such a way that it depends *only* on the temperature T and not at all on pressure or composition.

[†] We shall consistently use the superscript $^\ominus$ (which may be pronounced 'standard') with the meaning 'standard'. The 'degree sign' $^\circ$ has been widely used for 'standard'; unfortunately it has also been used with other meanings such as 'pure'.

Considerable economy can now be effected by tabulation of values of

$$\{\mu_B(T, p, x_C) - \mu_B^\ominus(T)\}, \quad \text{or} \quad \lambda_B(T, p, x_C)/\lambda_B^\ominus(T), \quad (11.1.8)$$

rather than values of (3) or (4).†

The accepted definitions of $\mu_B^\ominus$ or $\lambda_B^\ominus$ are different from one another (i) for gases, (ii) for liquids and solids, and (iii) for the solvent and for the solute(s) in a solution. We shall give those particular definitions: for gases in chapter 12, for liquids and solids in chapter 16, and for solutions in chapter 18. Here we shall introduce some general standard thermodynamic functions derived from $\mu_B^\ominus$ or $\lambda_B^\ominus$, and especially the standard equilibrium constant $K^\ominus$ of a chemical reaction.

§ 11.2   Derived standard thermodynamic functions

Once the standard chemical potential $\mu_B^\ominus$ or standard absolute activity $\lambda_B^\ominus$ has been defined, formulae follow for all the other standard thermodynamic functions obtained by differentiation of $\mu_B^\ominus$ or $\lambda_B^\ominus$ with respect to temperature T. In particular

$$S_B^\ominus = -d\mu_B^\ominus/dT = -R \ln \lambda_B^\ominus - RT \, d \ln \lambda_B^\ominus/dT, \quad (11.2.1)$$

$$H_B^\ominus = \mu_B^\ominus - T \, d\mu_B^\ominus/dT = -RT^2 \, d \ln \lambda_B^\ominus/dT, \quad (11.2.2)$$

$$G_B^\ominus = \mu_B^\ominus = RT \ln \lambda_B^\ominus, \quad (11.2.3)$$

$$C_{p,B}^\ominus = -T \, d^2\mu_B^\ominus/dT^2 = -2RT \, d \ln \lambda_B^\ominus/dT - RT^2 \, d^2 \ln \lambda_B^\ominus/dT^2, \quad (11.2.4)$$

where $S_B^\ominus$, $H_B^\ominus$, $G_B^\ominus$, and $C_{p,B}^\ominus$ are called the *standard molar entropy*, the *standard molar enthalpy*, the *standard molar Gibbs function*, and the *standard molar heat capacity*, each of the substance B; and where each depends only on the temperature T.‡

§ 11.3   Standard equilibrium constant

For the general chemical reaction:

$$0 = \sum_B \nu_B B, \quad (11.3.1)$$

the criterion of chemical equilibrium is according to § 7.7

$$\sum_B \nu_B \mu_B(T, p, \xi^{eq}) = 0, \quad (11.3.2)$$

where ξ^{eq} denotes the extent of reaction at chemical equilibrium. We rewrite

---

† In just the same way considerable economy in the recording of differences in height of various points on the Earth's surface can be effected by tabulation of the difference between the height of each point and a standard height (such as 'mean sea level'). For N points a table of $(h_j - h_i)$ for each pair of points i and j has $N(N-1)/2$ entries while a table of $(h_j - h^\ominus)$ has only N entries.

‡ We use d/dT rather than $\partial/\partial T$ because the standard thermodynamic functions depend only on T.

equation (2) in the form:

$$\sum_B \nu_B\{\mu_B(T, p, \xi^{eq}) - \mu_B^\ominus(T)\} = -\sum_B \nu_B \mu_B^\ominus(T) \overset{\text{def}}{=} RT \ln\{K^\ominus(T)\}, \quad (11.3.3)$$

where $K^\ominus$, defined by

$$K^\ominus(T) \overset{\text{def}}{=} \exp\left\{-\sum_B \nu_B \mu_B^\ominus(T)/RT\right\}, \quad (11.3.4)$$

or by the equivalent relation:

$$K^\ominus(T) \overset{\text{def}}{=} \prod_B \{\lambda_B^\ominus(T)\}^{-\nu_B}, \quad (11.3.5)\dagger$$

is called the *standard equilibrium constant* of the chemical reaction, and for a given reaction depends only on temperature and not at all on pressure or composition.
By use of equations (11.2.1) to (11.2.3) we obtain

$$\sum_B \nu_B S_B^\ominus = R \ln K^\ominus + RT \, d \ln K^\ominus/dT, \quad (11.3.6)$$

$$\sum_B \nu_B H_B^\ominus = RT^2 \, d \ln K^\ominus/dT, \quad (11.3.7)$$

$$\sum_B \nu_B \mu_B^\ominus = \sum_B \nu_B G_B^\ominus = -RT \ln K^\ominus. \quad (11.3.8)$$

The algebraic sums $\sum_B \nu_B G_B^\ominus$, $\sum_B \nu_B H_B^\ominus$, and $\sum_B \nu_B S_B^\ominus$ are commonly denoted by the symbols $\Delta G_m^\ominus$, $\Delta H_m^\ominus$, and $\Delta S_m^\ominus$, and called the standard molar (change of) Gibbs function for the reaction, the standard molar (change of) enthalpy for the reaction, and the standard molar (change of) entropy for the reaction. These symbols and names can be misleading because it is not obvious that the operator Δ is being used with its usual meaning: $\Delta X = X_2 - X_1$, and indeed it is not being so used for any realizable initial and final states 1 and 2 of any real reaction. It is true that $\sum_B \mu_B G_B^\ominus$ or $\Delta G_m^\ominus$ can be interpreted as the excess of the sum of the standard molar Gibbs functions of the products, each multiplied by its stoichiometric number, over the sum of the standard molar Gibbs functions of the reactants, each multiplied by the modulus of its stoichiometric number:

$$\Delta G_m^\ominus = \sum_B \tfrac{1}{2}(|\nu_B| + \nu_B)G_B^\ominus - \sum_B \tfrac{1}{2}(|\nu_B| - \nu_B)G_B^\ominus, \quad (11.3.9)$$

and similarly for $\Delta H_m^\ominus$ and $\Delta S_m^\ominus$. In general, however, $\Delta G_m^\ominus$ (or $\Delta S_m^\ominus$) is not even approximately equal to the actual molar change of Gibbs function ΔG_m (or the actual

---

† The operator Π is used to denote a continued product, and corresponds to the use of the operator Σ to denote a continued sum. Thus for example $\Sigma_i (\ln x_i) = \ln(\Pi_i x_i)$.

molar change of entropy ΔS_m) which accompanies a real chemical reaction.† An example of the difference between $\Delta G_m^\ominus$ and ΔG_m for a reaction of perfect gases will be discussed in § 12.9. However, as we shall see, $\Delta H_m^\ominus \approx \Delta H_m$ where ΔH_m is the actual molar enthalpy change, which can be measured calorimetrically for a reaction that goes to completion. Whether it is ever pardonable to fail to distinguish by name and by symbol between the molar enthalpy change ΔH_m of a reaction and its standard molar enthalpy change $\Delta H_m^\ominus$, it is certainly never pardonable to fail to distinguish between the molar change of Gibbs function ΔG_m and the standard molar change of Gibbs function $\Delta G_m^\ominus$, or between the molar entropy change ΔS_m and the standard molar entropy change $\Delta S_m^\ominus$.

Equation (8), whether written in that form or in the more usual form: $\Delta G_m^\ominus = -RT \ln K^\ominus$, is merely a definition. It acquires its importance only from the custom of tabulating values not of $K^\ominus$ itself, or of $\ln K^\ominus$, or of $-\ln K^\ominus$ (though tables of these are not uncommon), but of $-RT \ln K^\ominus$, and of using the symbol $\Delta G_m^\ominus$ as shorthand for $-RT \ln K^\ominus$. A table of values of $\ln K^\ominus$ is regarded as more convenient than a table of values of $K^\ominus$ by those who prefer to add and subtract rather than to multiply and divide; in just the same way values of the acidity constant $K_A^\ominus$ (see chapter 20) are commonly tabulated for weak acids in the form of values of $pK_A^\ominus \stackrel{\text{def}}{=} -\log_{10} K_A^\ominus$. The advantages of multiplication by RT before tabulation are far from obvious.‡

§ 11.4   Van't Hoff's equation

Equation (11.3.7), which is usually written in the form:

$$\mathrm{d} \ln K^\ominus / \mathrm{d}T = \Delta H_m^\ominus / RT^2 , \qquad (11.4.1)$$

is experimentally testable and most useful. It may be called *van't Hoff's equation*.§ The left-hand side can be obtained from measurements of the dependence on temperature of the extent of reaction at equilibrium (see for example § 12.12), and $\Delta H_m^\ominus$ can be obtained calorimetrically (see § 12.10 for an example of the calculation

---

† Those who like to think in terms of 'physical pictures' can think of $\Delta G^\ominus$ for a chemical reaction *of perfect gases* as $(G_{\text{final}} - G_{\text{initial}})$ for the process in which the initial state consists of amounts of substance $(-\nu_B)n$ of the pure (separate) reactants each at the standard pressure $p^\ominus$ (see § 12.9) and the final state consists of amounts of substance $\nu_B n$ of the pure (separate) products each at the standard pressure $p^\ominus$; the process must be thought of as one in which the pure separate reactants each at the pressure $p^\ominus$ are first mixed and then react completely to form products which are then separated and each brought to the pressure $p^\ominus$. Then $\Delta G_m^\ominus = (G_{\text{final}} - G_{\text{initial}})/n$. The author prefers to think about real experiments rather than imaginary ones and prefers algebra to word-pictures and so has little use for such descriptions.

‡ The advantages of tabulating values of the standard equilibrium constant $K^\ominus$ for electron-transfer reactions in solution (see chapter 20) in the form of values of $E^\ominus \stackrel{\text{def}}{=} (RT/zF) \ln K^\ominus$, where F denotes the Faraday constant and z the stoichiometric number of e^- transferred in the reaction for which $K^\ominus$ is the standard equilibrium constant, and where $E^\ominus$ has the dimension of an electric potential difference, are also far from obvious.

§ Equation (11.4.1) is sometimes called van't Hoff's 'isochore', a curiously thoughtless name since the equation has nothing to do with any volume, constant or otherwise.

of $\Delta H_m^\ominus$ from a measured ΔH). More importantly if the value of $K^\ominus$ for a reaction is known at one temperature T_1 its value at another temperature T_2 can be obtained by integration of equation (1):

$$\ln\{K^\ominus(T_2)\} = \ln\{K^\ominus(T_1)\} + \int_{T_1}^{T_2} \{\Delta H_m^\ominus(T)/RT^2\}\, dT, \qquad (11.4.2)$$

provided that $\Delta H_m^\ominus$ is known over the temperature range T_1 to T_2. It is often sufficient to assume that $\Delta H_m^\ominus$ is independent of temperature or to use an average value $\langle \Delta H_m^\ominus \rangle$, in which case

$$\ln\{K^\ominus(T_2)\} = \ln\{K^\ominus(T_1)\} + \langle\Delta H_m^\ominus\rangle(T_2 - T_1)/RT_1 T_2. \qquad (11.4.3)$$

When equation (3) is not sufficiently accurate, $\Delta H_m^\ominus$ can be expressed as a function of T, and the integration in equation (2) can be carried out accurately. If, as is often the case in practice, the standard molar heat capacities at constant pressure, $C_{p,B}^\ominus$, have been measured and tabulated (either as such or as coefficients of a series in powers of T) at temperatures over the required range, we use the relation:

$$\Delta H_m^\ominus(T) = \Delta H_m^\ominus(T_0) + \int_{T_0}^{T} \sum_B \nu_B C_{p,B}^\ominus(T)\, dT, \qquad (11.4.4)$$

where T_0 is some temperature (which may be outside the range T_1 to T_2) at which $\Delta H_m^\ominus$ is accurately known, to rewrite equation (2) in the form:

$$\ln\{K^\ominus(T_2)\} = \ln\{K^\ominus(T_1)\} + \Delta H_m^\ominus(T_0)(T_2 - T_1)/RT_1 T_2$$

$$+ \int_{T_1}^{T_2} \left\{\int_{T_0}^{T} \sum_B \nu_B C_{p,B}^\ominus(T)\, dT\right\}(1/RT^2)\, dT. \qquad (11.4.5)$$

§ 11.5   Determination of standard equilibrium constants by calorimetry

By eliminating $d \ln K^\ominus/dT$ from equations (11.3.6) and (11.3.7) we obtain

$$-RT \ln\{K^\ominus(T)\} = \sum_B \nu_B H_B^\ominus(T) - T \sum_B \nu_B S_B^\ominus(T). \qquad (11.5.1)$$

Equation (1) can be rewritten in the form:

$$-RT \ln\{K^\ominus(T)\} = \Delta H_m^\ominus(T) - T \sum_B \nu_B\{S_B^\ominus(T) - S_B(s, T \to 0)\}$$

$$- T \sum_B \nu_B S_B(s, T \to 0), \qquad (11.5.2)$$

where $S_B(s, T \to 0)$ denotes the molar entropy of pure solid B in the limit as the temperature T tends to zero.

Equation (2) is the key to the solution of a famous old problem: 'Can the standard equilibrium constant of a chemical reaction be determined from calorimetric measurements alone?'. The standard molar enthalpy change $\Delta H_m^\ominus$ can be determined, apart from corrections that are usually small and easily (though rather tediously) made (see for example § 12.10), from calorimetric measurements of energies or enthalpies of combustion or of other reactions that can be carried out quantitatively and quickly enough in calorimeters (see chapter 4). Each of the molar entropy increments $\{S_B^\ominus(T) - S_B(s, T \to 0)\}$ can be determined, apart from corrections that are usually small and easily made (for example see § 12.11), from calorimetric measurements of heat capacity and of enthalpy of phase transition for the pure substance B. The problem is therefore solved except for the last term of equation (2).

According to Nernst's heat theorem

$$\sum_B \nu_B S_B(s, T \to 0) = 0, \tag{11.5.3}$$

so that, provided that Nernst's heat theorem is obeyed,† equation (2) can be rewritten in the form:

$$-RT \ln\{K^\ominus(T)\} = \Delta H_m^\ominus(T) - T \sum_B \nu_B \{S_B^\ominus(T) - S_B(s, T \to 0)\}, \tag{11.5.4}$$

which can be used to calculate the standard equilibrium constant $K^\ominus(T)$ from the results of calorimetric measurements alone. (Even the small corrections referred to above can in principle all be measured calorimetrically, although in practice they are often determined from (p, V_m, T) or other non-calorimetric measurements.)

The standard molar Giauque function $\Phi_m^\ominus(T, T^\dagger)$ is defined‡ by the equation:

$$\Phi_m^\ominus(T, T^\dagger) \overset{\text{def}}{=} -T^{-1} \int_{T^\dagger}^T C_{p,m}^\ominus(T)\,\mathrm{d}T + \int_0^T C_{p,m}^\ominus(T)\,\mathrm{d}\ln T$$

$$\overset{\text{def}}{=} -\{H_m^\ominus(T) - H_m^\ominus(T^\dagger)\}/T$$

$$+\{S_m^\ominus(T) - S_m(s, T \to 0)\}. \tag{11.5.5}$$

It may be used to rewrite equation (4) in the form:

$$R \ln\{K^\ominus(T)\} = \Delta\Phi_m^\ominus(T, T^\dagger) - \Delta H_m^\ominus(T^\dagger)/T, \tag{11.5.6}$$

which is convenient when values of the standard molar Giauque function and of the standard molar enthalpy change have been tabulated, usually for $T^\dagger = 298.15$ K. In equation (6) we have assumed Nernst's heat theorem (equation 3) as we did in equation (4).

---

† It is a curious kind of theorem, especially when it is called 'the third law of thermodynamics', that is apparently sometimes not obeyed. How do we know when $\sum_B \nu_B S_B(s, T \to 0)$ is *not* zero? Strictly, the only answer we can give to that question is: 'When equation (11.5.4) is found to be incorrect', so that we have apparently failed in our aim of obtaining $K^\ominus(T)$ from calorimetric measurements alone by implying the need for a 'direct' non-calorimetric measurement of $K^\ominus(T)$ before we can be confident that equation (11.5.3), and so equation (11.5.4), will hold for any particular reaction. Happily, however, we know experimentally that apparent exceptions to Nernst's heat theorem are confined to relatively few substances. Moreover, we know, when $\sum_B \nu_B S_B(s, T \to 0)$ is not zero, that it is never greater than about $R \ln 3$, so that the value of $K^\ominus(T)$ calculated from equation (11.5.4) is never wrong by a factor of more than about 4, an inaccuracy that is often not as serious as it might appear to be at first sight.

‡ The molar Giauque function was introduced in a footnote on p. 59. It has the advantage for most substances of varying quite slowly with temperature and so of allowing easy interpolation from a table giving its value only for widely spaced temperatures.

We shall deal with Nernst's heat theorem and the experimental testing of it in chapter 15, where we shall also mention the exceptions to it. In the meantime we shall assume either that Nernst's heat theorem is obeyed or, for reactions involving a few exceptional substances like CO, N_2O, NO, and H_2O, that we know how to calculate the necessary small corrections and have included them in $\Phi_m^{\ominus}(T, T^+)$.

§ 11.6 Primary thermodynamic tables

Chemical-thermodynamic tables may be divided into two kinds: primary and secondary. Primary tables, with which we shall deal here, contain values of standard molar thermodynamic functions related by the equations of § 11.5 to the standard equilibrium constant $K^{\ominus}(T)$. Secondary tables deal with quantities like fugacities or virial coefficients, activity coefficients, and osmotic coefficients which are needed, except for reactions in mixtures of perfect gases or in ideal-dilute solutions, to calculate the extent of reaction ξ under given conditions of pressure and initial composition from the value of $K^{\ominus}(T)$. For primary tables it is necessary to make measurements only on each *pure* substance: for example the enthalpy of combustion at one temperature and the heat capacity as a function of temperature. We may look forward to the time when primary tables are essentially complete and are kept up to date as new compounds are prepared. For secondary tables it is necessary to make measurements on each *mixture*, often of more than two substances. The increase of work needed is prodigious and such that we cannot look forward to the completion of the task. That is why theories of mixtures, tested on carefully chosen mixtures, are so important.

An example of a useful form of primary table is given in table 11.1. Values of $\ln\{K_f^{\ominus}(T)\}$, the natural logarithm of the standard equilibrium constant for the formation of the substance, are given for several temperatures, where (see § 4.7) formation $_f$ denotes the reaction in which the substance is formed from its elements each in the form most stable at the temperature in question and atmospheric

Table 11.1. Values for a few substances of $\ln\{K_f^{\ominus}(T)\}$ over a range of temperatures and of $\Delta_f H_m^{\ominus}$ (298.15 K). ($R = 8.31441$ J K^{-1} mol^{-1}.)

(T/K):	298.15	400	500	600	800	1000	$\Delta_f H_m^{\ominus}$ (298.15 K)
			$\ln\{K_f^{\ominus}(T)\}$				$\overline{R K}$
CO(g)	55.38	44.04	37.43	33.01	27.48	24.13	−13295
CO$_2$(g)	159.09	118.66	94.99	79.21	59.47	47.61	−47328
H$_2$O(g)	92.22	67.33	52.70	42.91	30.60	23.17	−29086
H$_2$O(l)	95.68	—	—	—	30.60	23.17	−34378
HI(g)	−0.64	1.92	2.43	2.20	1.88	1.69	3170
AgCl(s)	44.30	—	—	—	—	—	−15283
(H$^+$ + Cl$^-$)(aq)	52.95	—	—	—	—	—	−20105
(Ag$^+$ − H$^+$)(aq)	−31.11	—	—	—	—	—	12698
C$_6$H$_5$·C$_2$H$_5$(g)	−52.67	−50.00	−48.87	−48.34	−47.99	−47.95	3583
C$_6$H$_5$·CH:CH$_2$(g)	−86.24	−71.39	−63.01	−57.59	−51.05	−47.26	17724
C$_6$H$_5$·C:CH(g)	−145.93	−112.50	−93.12	−80.31	−64.45	−55.02	39362

pressure.† Such a table is sufficient to allow $K^{\ominus}(T)$ to be calculated for any reaction among the substances listed (and the elements in their reference states) at any temperature within the given range.

For two reasons we also give in table 11.1 values of $\Delta_f H_m^{\ominus}(T^{\dagger})$. First, the values of $\ln\{K_f^{\ominus}(T)\}$ are usually obtained by subtracting the best experimental value of $\Delta_f H_m^{\ominus}(T^{\dagger})/RT$, measured by reaction calorimetry at the temperature $T^{\dagger}$, from the best experimental values of $\Delta_f \Phi_m^{\ominus}(T, T^{\dagger})/R$, measured by heat-capacity calorimetry over the temperature range from 0 to T; if either of the terms of equation (6) needs revision in the light of new measurements then, provided that $\Delta_f H_m^{\ominus}(T^{\dagger})$ is included, it is easy to revise the values of $\ln\{K_f^{\ominus}(T)\}$. Second, for some gases the values of $\ln\{K_f^{\ominus}(T)\}$ are those evaluated according to the methods of statistical mechanics (see chapter 14) from spectroscopic quantities together with the value of $\Delta_f H_m^{\ominus}(T^{\dagger})$ at any one temperature $T^{\dagger}$.

In table 11.1 the values of $\ln\{K_f^{\ominus}(T)\}$ for $H_2O(g)$ and $H_2O(l)$ coincide at temperatures greater than the critical temperature (647.14 K). For ions, either one may do as we have done and tabulate values for electrically neutral pairs each composed of an ion and the ion H^+, or (as is more commonly done) one may omit the H^+ and import the further convention that H^+ contributes nothing to the reaction called 'formation'. For gases like CO, CO_2, H_2O, and HI the values of $\Phi_m^{\ominus}(T, T^{\dagger})$ used to obtain the values of $\ln\{K_f^{\ominus}(T)\}$ are those calculated by statistical mechanics; the values for CO and for H_2O accordingly do not contain mistakes arising from the asumption that those substances obey Nernst's heat theorem as they would do if uncorrected calorimetric values of $\Phi_m^{\ominus}(T, T^{\dagger})$ had been used.

The only other standard thermodynamic quantity that the industrial chemist is likely to need is values of $\Delta_f H_m^{\ominus}(T)$, so as to allow him to balance the heat flows in the plant he is designing. Those can be obtained from the values in table 11.1 by use of the relation:

$$\Delta_f H_m^{\ominus}(T) = -R\,d\ln\{K_f^{\ominus}(T)\}/d(1/T),\qquad(11.6.1)$$

if they are not listed in a separate primary table.

Selected references to primary thermodynamic tables are given below.‡ References to older and more specialized primary tables can be found in an article by

---

† Thus C(s, graphite), $O_2(g)$, $Cl_2(g)$, $Br_2(l)$, $I_2(s)$, and so on, are normally implied. For P, however, the form implied is P(s, white) rather than the more stable but thermochemically indefinite P(s, red).

‡ *Selected Values of Chemical Thermodynamic Properties*, Technical Note 270, National Bureau of Standards (U.S.). Part 1, **1965**; part 2, **1966**; part 3, **1968**; part 4, **1969**; part 5, **1971**; part 6, **1971**.
JANAF Thermochemical Tables, second edition. Nat. Stand. Ref. Data Ser., Nat. Bur. Stand. (U.S.) **1971**, 37, 1141; with supplement in *J. Phys. Chem. Ref. Data* **1975**, 4(1), 1.
Thermodynamic Properties of Pure Substances. Glushko, V. P.; Editor in Chief. Nauka: Moscow. Volume 1, Parts 1 and 2, **1978**. (The remaining volumes are promised by 1980.)
Cox, J. D.; Pilcher, G. *Thermochemistry of Organic and Organometallic Compounds*. Academic Press: London. **1970**.
Stull, D. R.; Westrum, E. F., Jr.; Sinke, G. C. *The Chemical Thermodynamics of Organic Compounds*. Wiley: New York. **1969**.
Ashcroft, S. J.; Mortimer, C. T. *Thermochemistry of Transition Metal Complexes*. Academic Press: London. **1970**.
Hultgren, R.; Orr, R. L.; Anderson, P. D.; Kelley, K. K. *Selected Values of the Thermodynamic Properties of Metals and Alloys*. Wiley: New York. **1963**.

Herington.[†] A set of 'key values' of standard thermodynamic functions is being produced by an expert international task group for CODATA.[‡]

One might be forgiven for supposing that some of these published primary thermodynamic tables had been compiled more with ease of revision by the compiler in mind than with ease of use by the reader (and perhaps also to provide arithmetical exercises for undergraduates). "Technical Note 270", for example, lists for each substance $\Delta_f H_m^\ominus (T \to 0)$, $\Delta_f H_m^\ominus (298.15 \text{ K})$, $\Delta_f G_m^\ominus (298.15 \text{ K})$ $[= -RT \ln\{K_f^\ominus (298.15 \text{ K})\}]$, $\{H_B^\ominus (298.15 \text{ K}) - H_B^\ominus (0)\}$, $S_B^\ominus (298.15 \text{ K})$ $[= \{S_B^\ominus (298.15 \text{ K}) - S_B(s, T \to 0)\}]$, and $C_{p,B}^\ominus (298.15 \text{ K})$. From such a table the value of $K^\ominus (298.15 \text{ K})$ can be calculated for any reaction involving the substances listed, and values of $K^\ominus (T)$ at other temperatures can be estimated by use of equation (11.4.3), or better by use of (11.4.5) with $\Sigma_B \nu_B C_{p,B}^\ominus$ ($\Delta C_p^\ominus$) taken as independent of temperature. The "JANAF" tables contain for a wide range of temperatures the quantities: $C_{p,B}^\ominus (T)$, $S_B^\ominus (T)$ $[= \{S_B^\ominus (T) - S_B(s, T \to 0)\}]$, $\Phi_B^\ominus (T, 298.15 \text{ K})$, $\{H_B^\ominus (T) - H_B^\ominus (298.15 \text{ K})\}$, $\Delta_f H_m^\ominus (T)$, $\Delta_f G_m^\ominus (T)$ $[= -RT \ln\{K_f^\ominus (T)\}]$, and $\log_{10}\{K_f^\ominus (T)\}$.

Problems for chapter 11

Problem 11.1

Use table 11.1 to calculate the standard equilibrium constant $K^\ominus$ for the reaction: $C(s) + H_2O(g) = CO(g) + H_2(g)$ at 400 K and at 1000 K.

Problem 11.2

Use the following values from other tables to calculate the standard equilibrium constant $K^\ominus$ for the reaction: $C(s) + H_2O(g) = CO(g) + H_2(g)$ at 1000 K.

	C(s)	H$_2$O(g)	CO(g)	H$_2$(g)
$\Delta_f H_m^\ominus (298.15 \text{ K})/R$ K	0	−29086	−13295	0
$\Phi_m^\ominus (1000 \text{ K}, 298.15 \text{ K})/R$	1.521	24.854	25.634	17.493

Problem 11.3

Use the following values from yet other tables to estimate the standard equilibrium constant $K^\ominus$ for the reaction: $C(s) + H_2O(g) = CO(g) + H_2(g)$ at 1000 K (a), by use of equation (11.4.3) and (b), by use of equation (11.4.5).

	C(s)	H$_2$O(g)	CO(g)	H$_2$(g)
$\Delta_f H_m^\ominus (298.15 \text{ K})/R$ K	0	−29086	−13295	0
$\{S_B^\ominus (298.15 \text{ K})$ $- S_B(s, T \to 0)\}/R$	0.685	22.700	23.802	15.706
$C_{p,B}^\ominus (298.15 \text{ K})/R$	1.040	4.041	3.507	3.468

[†] Herington, E. F. G. *SPRCT(I)*, chapter 2.
[‡] *CODATA Recommended Key Values for Thermodynamics*, 1977. J. Chem. Thermodynamics **1978**, 10, 903.

Problem 11.4

Use table 11.1 to estimate the standard molar enthalpy change $\Delta H_m^{\ominus}(T)$ for the reaction: $H_2(g) + I_2(g) = 2HI(g)$ at 400 K and at 800 K.

Problem 11.5

Use table 11.1 to calculate the standard equilibrium constant $K^{\ominus}$ for the reaction: $AgCl(s) = (Ag^+ + Cl^-)(aq)$ at 298.15 K. (This $K^{\ominus}$ is also called—see chapter 20—the standard solubility product of AgCl.)

Problem 11.6†

Ethynylbenzene (phenylacetylene) is an undesirable by-product of the industrial pyrolysis of ethylbenzene to form ethenylbenzene (styrene); ethynylbenzene leads to unwanted branching and cross-linking during the subsequent polymerization of ethenylbenzene. Use table 11.1 to calculate the mole fractions x of $C_6H_5 \cdot CH:CH_2$ and y of $C_6H_5 \cdot C:CH$ produced by the pyrolysis of $C_6H_5 \cdot C_2H_5$ at 1000 K and atmospheric pressure.

Problem 11.7†

Use table 11.1 to calculate for H_2O at 298.15 K (a), its vapour pressure and (b), its molar enthalpy of evaporation.

---

† Problems 11.6 and 11.7 anticipate the elementary conclusions (see equation 12.13.3 and § 12.15) for a perfect gas B that $\lambda_B/\lambda_B^{\ominus} = x_B p/p^{\ominus}$ where $p^{\ominus}$ is the standard pressure 101.325 kPa, and for a pure liquid B that $\lambda_B/\lambda_B^{\ominus} \approx 1$.

Chapter 12

Gases and gaseous mixtures

§ 12.1 Experimental methods for the study of pure gases

The apparatus used to study the form of the equation of state:

$$f(p, V_m, T) = 0 , \qquad (12.1.1)$$

of a gas is always a more or less sophisticated 'Boyle's tube'. Usually the pressures p are measured at known volumes V at a known temperature T. The amount of substance n is usually inferred from the value of pV/RT as $p \to 0$. For example the piezometer might consist of a set of bulbs separated by narrow-bore tubes, mercury being used to define the volume of the gas and to transmit the pressure to a pressure balance. A wire sealed through each tube can be arranged to complete an electrical circuit when it makes contact with mercury; the piezometer can then be used unseen inside a furnace and the operator is protected against the consequences of the piezometer bursting at high pressures. The most demanding step is the measurement of the volumes above each wire. They can be measured by weighing mercury run from the evacuated apparatus down to each wire, due allowance being made for the dependence of the volumes on the temperature and pressure. Alternatively, the volumes can be measured with a gas having a known equation of state, but the method then becomes a relative one.

At temperatures at which its vapour pressure is appreciable mercury should be avoided; the gas under study would be a mixture containing gaseous mercury.

Burnett described a method that can be used without mercury. It has the further advantage that no volume or amount of substance need be measured, though that advantage is won at the expense of a greater demand for accuracy in the measurements of pressure. Two bulbs A of volume V_A and B of volume V_B are separated by a valve I as shown in figure 12.1. The bulb A is connected to a pressure transducer and thence to a manometer, and the bulb B is connected through another valve II to a vacuum pump. A sample of gas is introduced into bulbs A and B, valve I is closed, and the pressure $p_1 = p(n_1, V_A)$ is measured. The bulb B is evacuated, valve II is closed, and valve I is opened. When the gas has had time to equilibrate thermally, valve I is closed, and the pressure $p_2 = p(n_1, V_A + V_B) = p(n_2, V_A)$ is measured. The procedure is repeated until the pressure can no longer be measured with sufficient accuracy. To analyse the results we write the equation of state in the form:

$$p = (n/V)(RT + Bp + C'p^2 + \cdots) . \qquad (12.1.2)$$

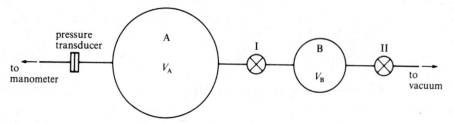

Figure 12.1. A schematic diagram of a Burnett's apparatus.

By eliminating n_1 between the results of the first and second measurements we obtain

$$\frac{p_1}{p_2} = \frac{(V_A + V_B)}{V_A} \frac{(RT + Bp_1 + C'p_1^2 + \cdots)}{(RT + Bp_2 + C'p_2^2 + \cdots)}, \tag{12.1.3}$$

and by eliminating n_2 between the results of the second and third measurements,

$$\frac{p_2}{p_3} = \frac{(V_A + V_B)}{V_A} \frac{(RT + Bp_2 + C'p_2^2 + \cdots)}{(RT + Bp_3 + C'p_3^2 + \cdots)}. \tag{12.1.4}$$

Division of (3) by (4) then eliminates the ratio $(V_A + V_B)/V_A$ leaving

$$\frac{p_1 p_3}{p_2^2} = \frac{(RT + Bp_1 + C'p_1^2 + \cdots)(RT + Bp_3 + C'p_3^2 + \cdots)}{(RT + Bp_2 + C'p_2^2 + \cdots)^2}. \tag{12.1.5}$$

Thus, if the number of measured pressures is large enough, the equation of state of the gas can be determined. The principal limitation of Burnett's method is finding sufficiently reproducible vacuum-tight valves that will operate at the temperature in question. The principal uncertainties arise from the assumption that the ratio $(V_A + V_B)/V_A$ is independent of the pressure and of the opening and closing of the valves, and as in all (p, V, n, T) measurements that adsorption of the gas on the walls of the apparatus is negligible. Although the series (2) has been used in presenting this analysis, the coefficient B and especially the higher coefficients $C', D', \cdots$ will be the coefficients of an arbitrarily truncated series of the form of (2) and might differ appreciably from the coefficients of the infinite series (2).

Full accounts of (p, V, n, T) measurements can be found in the literature.[†]

The behaviour of $p(V_m, T)$ for CO_2 at each of three temperatures has already been shown in figure 9.2(b). In figure 12.2 the compression factor:

[†] Mason, E. A.; Spurling, T. H. *The Virial Equation of State*. Pergamon: Oxford. **1969**.
  Cox, J. D.; Lawrenson, I. J. *SPRCT(I)*, chapter 5.
  Saville, G. *ETd(II)*, chapter 6.
  Brielles, J.; Dédit, A.; Lallemand, M.; Le Neindre, B.; Leroux, Y.; Vermesse, J.; Vidal, D. *ETd(II)*. chapter 7.
  Malbrunot, P. *ETd(II)*, chapter 8.

$$Z \overset{\text{def}}{=} p V_m / RT, \tag{12.1.6}$$

is plotted against pressure p for CO_2 at each of six temperatures T. At sufficiently low pressures $Z \to 1$; the fluid becomes a perfect gas with $pV_m = RT$. At low pressures the imperfections increase as the temperature is lowered; no such generalization is possible at high pressures.

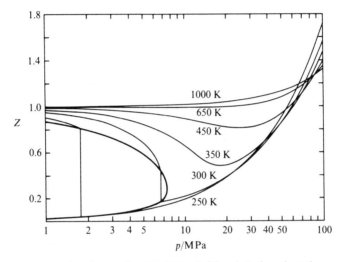

Figure 12.2. The compression factor $Z = pV_m/RT$ of CO_2 plotted against the pressure p (on a logarithmic scale) at each of six temperatures. The coexistence curve is shown as a heavy line.

The equation of state in the form:

$$f(H_m, p, T) = 0, \tag{12.1.7}$$

can be studied by adiabatic flow calorimetry (see § 1.23). In particular the quantity:

$$H_m(T, p_2) - H_m(T, p_1) = \int_{p_1}^{p_2} \{V_m - T(\partial V_m/\partial T)_p\} \, dp, \tag{12.1.8}$$

can be studied by isothermal Joule–Thomson throttling; the quantity:

$$T(H_m, p_2) - T(H_m, p_1) = -\int_{p_1}^{p_2} [\{V_m - T(\partial V_m/\partial T)_p\}/C_{p,m}] \, dp, \tag{12.1.9}$$

by isenthalpic Joule–Thomson throttling; and the quantity:

$$H_m(T_2, p) - H_m(T_1, p) = \int_{T_1}^{T_2} C_{p,m} \, dT$$

$$= \int_{T_1}^{T_2} \left\{ C_{p,m}(p \to 0) - T \int_0^p (\partial^2 V_m/\partial T^2)_p \, dp \right\} dT, \tag{12.1.10}$$

by flow calorimetry in the absence of any throttle. All these methods are free from errors due to adsorption. In the first two methods it has proved difficult experiment-ally to overcome with the highest accuracy the problems associated with heat leaks arising from temperature gradients around the throttle. All three methods are experimentally exacting; in any steady-state experiment it is necessary to 'walk round and round' the measuring instruments (in isothermal Joule–Thomson calori-metry for example: the instruments measuring flow rate, electric power, temperature difference, and pressures) until all are independent of time. Full accounts of the flow calorimetry of gases can be found in the literature.†

Measurements of the speed of sound u (in the absence of dispersion) can be used to study the equation of state of a gas through the relation (see § 6.12)

$$RT/u^2M = -(RT/V_m^2)(\partial V_m/\partial p)_S$$

$$= (RT/V_m^2)\left[-(\partial V_m/\partial p)_T - T\{(\partial V_m/\partial T)_p\}^2 \Big/ \left\{C_{p,m}(p \to 0)\right.\right.$$

$$\left.\left. - T\int_0^p (\partial^2 V_m/\partial T^2)_p \, dp\right\}\right],$$
(12.1.11)

where M is the molar mass of the gas. Such measurements of u can nowadays be made with high precision, and computers have banished any fear of analysing the results of measurements of the right-hand side of equation (11) in terms for example of any proposed form of the equation of state $V_m = V_m(p, T)$. In the limit as $p \to 0$ the quantity RT/u^2M becomes equal to $\{1 - R/C_{p,m}(p \to 0)\}$.‡ An account of methods of measurement of the speed of sound in a gas and of the analysis of the results can be found in the literature.§

§ 12.2   The virial equation of state of a gas

The fundamental form of the equation of state of a gas is known to be

$$pV_m = RT(1 + B/V_m + C/V_m^2 + D/V_m^3 + \cdots),$$
(12.2.1)

where $B, C, D, \cdots$ are called the *second, third, fourth, $\cdots$ virial coefficients*, and for a given substance depend only on the temperature.

Equation (1) converges more rapidly than the equation:

$$pV_m = RT + B'p + C'p^2 + D'p^3 + \cdots,$$
(12.2.2)

from which we obtain by inversion of the series:

$$B' = B, \quad C' = (C - B^2)/RT, \quad D' = (D - 3BC + 2B^3)/(RT)^2, \quad \cdots.$$
(12.2.3)

† Counsell, J. F. *SPRCT(I)*, chapter 6.
 McCullough, J. P.; Waddington, G. *ETd(I)*, chapter 10.
  ‡ Unless the period of the sound waves is much greater than the time of adjustment of translational or rotational energy and vibrational energy, the measured heat capacity will be lower than the true value; the measured heat capacity will be that of the gas behaving as if it had no vibrational degrees of freedom (see chapter 14).
 § Van Dael, W. *ETd(II)*, chapter 11.

The coefficients C', D', $\cdots$ are *not* called virial coefficients; those names are reserved for B, C, D, $\cdots$.

The virial coefficients B, C, D, $\cdots$ are known from statistical-mechanical arguments to be proportional to 'irreducible-cluster' integrals for two, three, four, $\cdots$ molecules at a time.[†]

In particular, for a pair-interaction energy $u(r)$ that depends only on the separation r of the centres of mass of two molecules the second virial coefficient B is given by

$$B(T) = 2\pi L \int_0^\infty [1 - \exp\{-u(r)/kT\}]r^2 \, dr, \qquad (12.2.4)$$

where L denotes the Avogadro constant and k ($=R/L$) the Boltzmann constant. Such a pair-interaction energy is shown in figure 12.3(a). Much progress has been

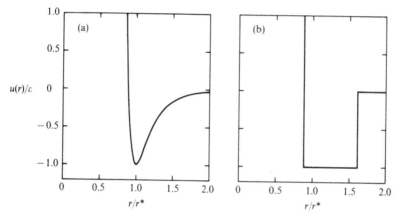

Figure 12.3. The pair-interaction energy $u(r)$ plotted against the separation r. (a), A realistic example; (b), a crude approximation: the 'square-well potential' with $g = 1.8$.

made in inferring the detailed shape of $u(r)$ from experimental values of $B(T)$ measured over wide ranges of temperature. Lennard–Jones's form:

$$u(r)/\varepsilon = \{m(r^*/r)^n - n(r^*/r)^m\}/(n - m), \qquad (12.2.5)$$

where ε and r^* are the coordinates of the minimum, has been widely used, usually with $n = 12$ and $m = 6$, but has been shown to differ significantly from the

---

[†] Mayer, J. E.; Mayer, M. G. *Statistical Mechanics*. Wiley: New York. **1940**.
  Rushbrooke, G. S. *Introduction to Statistical Mechanics*. Clarendon Press: Oxford. **1949**.
  Münster, A. *Statistical Thermodynamics*, Vol. I. Springer-Verlag: Berlin; Academic Press: New York **1969**. Chapter IX, pp. 537–614.

experimentally derived form of $u(r)$ for the simplest gases. The apparently crude approximation:

$$\left.\begin{aligned} u(r) &= \infty, \quad (r < \sigma), \\ u(r) &= -\varepsilon, \quad (\sigma \leqslant r \leqslant g\sigma), \\ u(r) &= 0, \quad (r > g\sigma), \end{aligned}\right\} \qquad (12.2.6)$$

shown in figure 12.3(b) and known as the 'square-well potential', leads by integration in equation (4) to

$$B(T) = (2\pi L\sigma^3/3)\{g^3 - (g^3 - 1)\exp(\varepsilon/kT)\}, \qquad (12.2.7)$$

an equation with three adjustable parameters which fits measured second virial coefficients almost within experimental error.

For the higher virial coefficients the formulae analogous to equation (4) for B are multiple integrals over the distances between molecular centres. These formulae† are not given here, if only because of uncertainty about the extent to which for example the interaction energy $u(r_{ij}, r_{jk}, r_{ki})$ of a triplet differs from the sum $\{u(r_{ij}) + u(r_{jk}) + u(r_{ki})\}$ of three pair-interaction energies.

In general the interaction of two molecules depends on angles as well as on the distance between the centres of mass, and the formulae are correspondingly more complicated.

§ 12.3   The second virial coefficient

Although second virial coefficients can be extracted from (p, V_m, T) measurements made over large ranges of pressure, the most reliable values are usually obtained from measurements made at low pressures in an apparatus designed for the purpose. This may be a more or less sophisticated 'Boyle's tube' or a Burnett apparatus. Alternatively, if the properties of say nitrogen are known then relative measurements in a differential-compression apparatus have some advantages. All these methods suffer from adsorption of the gas on to the walls of the apparatus, an effect that can give rise to large errors at temperatures below the critical. The buoyancy-balance method is relatively free from errors due to adsorption. A quartz or gold-plated-glass bulb balanced on a quartz fibre against a flattened bulb with the same surface finish and the same surface area but of very different volume is contained in a vessel connected to a manometer. The balance is used to identify the pressures at which the densities of the gas in question and of nitrogen are the same. Adsorption should be the same on each side of the balance. Incidentally, the 'method of limiting densities', usually with a buoyancy balance, was used primarily to determine relative molar masses and only secondarily to determine second virial coefficients, until mass spectrometry proved a more accurate way of determining relative molar masses.

† Hirschfelder, J. O.; Curtiss, C. F.; Bird, R. B. *Molecular Theory of Gases and Liquids.* Wiley: New York. **1964**.

Flow-calorimetric measurements at low pressures may be used to determine $(B - T\,dB/dT)$, and speed-of-sound measurements at low pressures to extract $B(T)$ from equation (12.1.11). Both of these methods are free of errors due to adsorption.

In figure 12.4 we show the second virial coefficient B plotted against temperature T for N_2. A 'secondary table' of second virial coefficients (and of some third virial coefficients) is available.† The values of B plotted against temperature in figure 12.4 are the results of six sets of measurements taken from that source.

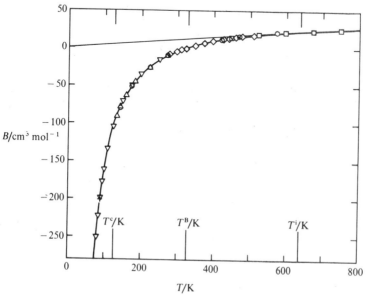

Figure 12.4. Plot of the second virial coefficient B against temperature T for six sets of measurements on N_2. The curve is that calculated from $B/cm^3\,mol^{-1} = 195.4 - 150.8\,\exp(85.40\,K/T)$. The critical temperature T^c, the Boyle temperature T^B at which $B = 0$, and the inversion temperature T^i at which $(B - T\,dB/dT) = 0$, are marked on the abscissa.

The temperature at which $B = 0$ is called the Boyle temperature T^B. For N_2, $T^B = 329\,K$. The temperature at which $(B - T\,dB/dT) = 0$ is called the low-pressure inversion temperature $T^i(p \to 0)$; it is the temperature at which the Joule–Thomson coefficient changes sign. It can be determined by the temperature of contact of the tangent to $B(T)$ drawn from the origin. For N_2, $T^i(p \to 0) = 638\,K$.

In view of the scatter among the third virial coefficients C that have been measured for N_2, we shall postpone consideration of the temperature dependence of C until the next chapter. There we shall combine the values of $C(p^c/RT^c)^2$ for several gases that conform with the principle of corresponding states and consider them together as a function of T/T^c.

† Dymond, J. H.; Smith, E. B. *The Virial Coefficients of Gases. A Critical Compilation.* Clarendon Press: Oxford. **1969**.

§ 12.4   Chemical potential of a pure gas

The chemical potential μ of a pure gas is given by the identity:

$$\mu(g, T, p) = \mu(pg, T, p^{\ominus}) + \{\mu(pg, T, p) - \mu(pg, T, p^{\ominus})\}$$

$$- \{\mu(pg, T, p) - \mu(pg, T, p \to 0)\} + \{\mu(g, T, p) - \mu(g, T, p \to 0)\}$$

$$= \mu^{\ominus}(g, T) + RT \ln(p/p^{\ominus}) + \int_0^p (V_m - RT/p) \, dp, \tag{12.4.1}$$

where pg denotes perfect gas, where we have identified $\mu(pg, T, p \to 0)$ with $\mu(g, T, p \to 0)$ (all gases become perfect at zero pressure), where $p^{\ominus}$ denotes a standard pressure, and where we have written $\mu^{\ominus}(g, T)$ for $\mu(pg, T, p^{\ominus})$, a function only of temperature.

In terms of T and V_m rather than T and p as independent variables, equation (1) can be recast† in the form:

$$\mu(g, T, V_m) = \mu^{\ominus}(g, T) + pV_m - RT - RT \ln(p^{\ominus}V_m/RT)$$

$$- \int_\infty^{V_m} (p - RT/V_m) \, dV_m. \tag{12.4.2}$$

We shall need equation (2) in the next section.

§ 12.5   Van der Waals's and some related equations of state

Van der Waals's equation of state:

$$p = RT/(V_m - b) - a/V_m^2, \tag{12.5.1}$$

has retained its advantages over the many empirical equations of state that have since been proposed. Though it never agrees more than qualitatively with experiment it never leads to physical nonsense: the underlying physics, except for the absence of an appropriate singularity at the critical point, seems to be sound. The first term represents that part of the volume excluded by the molecules themselves, that is to say the repulsive interactions. The second term represents the attractive interactions of the molecules.

By use of the criteria:

$$(\partial p/\partial V_m)_T = 0, \quad (\partial^2 p/\partial V_m^2)_T = 0, \quad (T = T^c, p = p^c, V_m = V_m^c), \tag{12.5.2}$$

$$† \int_0^p (V_m - RT/p) \, dp = \int_0^p (p - RT/V_m)(V_m/p) \, dp$$

$$= \int_{RT}^{pV_m} (1 - RT/pV_m) \, d(pV_m) - \int_\infty^{V_m} (p - RT/V_m) \, dV_m$$

$$= pV_m - RT - RT \ln(pV_m/RT) - \int_\infty^{V_m} (p - RT/V_m) \, dV_m.$$

we derive for van der Waals's equation:

$$b = V_m^c/3, \quad a = 9RT^c V_m^c/8,\tag{12.5.3}$$

and by substitution in (1) obtain the equation in *reduced* form:

$$p/p^c = 8(T/T^c)/(3V_m/V_m^c - 1) - 3(V_m^c/V_m)^2,\tag{12.5.4}$$

and the critical compression factor:

$$Z^c = p^c V_m^c/RT^c = \tfrac{3}{8}.\tag{12.5.5}$$

The chemical potential of a pure van der Waals gas is given by substitution of (1) into (12.4.2) and integration:

$$\mu(T, V_m) = \mu^\ominus(T) + RTV_m/(V_m - b) - 2a/V_m - RT - RT\ln(p^\ominus V_m/RT)$$
$$+ RT\ln\{V_m/(V_m - b)\}.\tag{12.5.6}$$

Coexisting gaseous and liquid phases of a pure substance are given by solving (preferably with the help of a computer) the simultaneous equations:

$$\mu(T, V_m^g) = \mu(T, V_m^l),\tag{12.5.7}$$

$$p(T, V_m^g) = p(T, V_m^l),\tag{12.5.8}$$

with

$$(\partial p/\partial V_m^g)_T < 0, \quad \text{and} \quad (\partial p/\partial V_m^l)_T < 0.\tag{12.5.9}$$

For van der Waals's equation we have

$$(V_m^g - V_m^l)b/(V_m^g - b)(V_m^l - b) - (2a/RT)(V_m^g - V_m^l)/V_m^g V_m^l$$
$$+ \ln\{(V_m^g - b)/(V_m^l - b)\} = 0,\tag{12.5.10}$$

$$V_m^g V_m^l/(V_m^g - b)(V_m^l - b) - (a/RT)(V_m^g + V_m^l)/V_m^g V_m^l = 0,\tag{12.5.11}$$

with

$$RT/(V_m^g - b)^2 > 2a/(V_m^g)^3, \quad \text{and} \quad RT/(V_m^l - b)^2 > 2a/(V_m^l)^3,\tag{12.5.12}$$

or

$$3(\phi^g - \phi^l)/(3\phi^g - 1)(3\phi^l - 1) - 9(\phi^g - \phi^l)/4\theta\phi^g\phi^l$$
$$+ \ln\{(3\phi^g - 1)/(3\phi^l - 1)\} = 0,\tag{12.5.13}$$

$$\phi^g\phi^l/(3\phi^g - 1)(3\phi^l - 1) - (\phi^g + \phi^l)/8\theta\phi^g\phi^l = 0,\tag{12.5.14}$$

with

$$4\theta(\phi^g)^3(3\phi^g - 1)^2 > 1, \quad \text{and} \quad 4\theta(\phi^l)^3/(3\phi^l - 1)^2 > 1,\tag{12.5.15}$$

where we have written $\phi^\alpha \overset{\text{def}}{=} V_m^\alpha/V_m^c$ and $\theta \overset{\text{def}}{=} T/T^c$. The coexistence curve drawn for van der Waals's equation in figure 9.6 was calculated by solving equations (13) and (14) with (15) for several values of θ.

We begin our discussion of related equations of state by rewriting van der Waals's equation in the form:

$$pV_m/RT = (1-4y)^{-1} - a/RTV_m$$
$$= 1 + 4y + 16y^2 + 64y^3 + 256y^4$$
$$+ 1024y^5 + 4096y^6 + \cdots - a/RTV_m, \qquad (12.5.16)$$

where we have used the substitution $y = b/4V_m$. The equation of state of non-attracting hard spheres is known to the seventh virial coefficient:

$$pV_m/RT = 1 + 4y + 10y^2 + 18.365y^3 + 28.26y^4 + 39.5y^5 + 57y^6. \qquad (12.5.17)$$

The 'hard-sphere' term in van der Waals's equation is accurate only to the second virial coefficient. Of the several forms that have been proposed for the replacement of $(1-4y)^{-1}$ in van der Waals's equation, only that of Carnahan and Starling will be discussed in this book. Their equation of state is

$$pV_m/RT = (1 + y + y^2 - y^3)/(1-y)^3 - a/RTV_m$$
$$= 1 + 4y + 10y^2 + 18y^3 + 28y^4 + 40y^5 + 54y^6 + \cdots - a/RTV_m, \qquad (12.5.18)$$

in much closer accord with (17). We shall make use of equation (18) in our discussions of liquid mixtures in chapter 16 and of fluid mixtures in chapter 17.

§ 12.6 Gaseous mixtures

In the homogeneous state gaseous mixtures are studied in the same ways as pure gases. They can also be studied in mixing experiments. The molar Gibbs function of mixing $\Delta_{mix}G_m$ (and thence by differentiation with respect to temperature the molar entropy of mixing $\Delta_{mix}S_m$) can in principle be obtained from osmotic-pressure measurements as explained in §§ 10.3, 10.9, and 10.10. Alternatively, equation (10.4.8) may be assumed for the mixing of perfect gases, and equation (10.5.3) can then be used with the results of §§ 10.9 and 10.10 to evaluate $\Delta_{mix}G_m$. For a binary mixture $\{(1-x)A + xB\}$, $\Delta_{mix}G_m$ takes the form (see equation 12.8.2):

$$\Delta_{mix}G_m(g, T, p, x) = G_m(g, T, p, x) - (1-x)G_m(g, T, p, 0) - xG_m(g, T, p, 1)$$
$$= (1-x)(\mu_A - \mu_A^*) + x(\mu_B - \mu_B^*)$$
$$= RT\{(1-x)\ln(1-x) + x\ln x\} + \int_0^p \Delta_{mix}V_m \, dp. \qquad (12.6.1)$$

The molar volume of mixing $\Delta_{mix}V_m$ can be measured at constant temperature and pressure by opening a tap between two bulbs each containing one of the pure gases in an apparatus designed for the measurement of the change of volume after mixing. (Alternatively, the change of pressure resulting from mixing at constant temperature and volume can be measured.) The molar enthalpy of mixing $\Delta_{mix}H_m$ can be measured in an adiabatic flow calorimeter having twin inlet tubes and a single outlet tube.

For perfect gaseous mixtures we have

$$\Delta_{mix}G_m(\text{pgm},\, T, p, x) = RT\{(1-x)\ln(1-x) + x\ln x\} < 0\,, \qquad (12.6.2)$$

$$\Delta_{mix}S_m(\text{pgm},\, T, p, x) = -R\{(1-x)\ln(1-x) + x\ln x\} > 0\,, \qquad (12.6.3)$$

$$\Delta_{mix}H_m(\text{pgm},\, T, p, x) = 0\,, \qquad (12.6.4)$$

$$\Delta_{mix}V_m(\text{pgm},\, T, p, x) = 0\,. \qquad (12.6.5)$$

It is again emphasised that equations (2) and (3) (and the first term in the last expression of equation 1) are based on the well founded but *extrathermodynamic* assumption (10.4.3) or (10.4.7).

The extension of the virial equation of state to a binary gaseous mixture $\{(1-x)A + xB\}$ has the form:

$$pV_m = RT\{1 + B(T, x)/V_m + C(T, x)/V_m^2 + \cdots\}\,, \qquad (12.6.6)$$

where

$$B(T, x) = (1-x)^2 B_{AA} + 2(1-x)xB_{AB} + x^2 B_{BB}\,, \qquad (12.6.7)$$

$$C(T, x) = (1-x)^3 C_{AAA} + 3(1-x)^2 xC_{AAB} + 3(1-x)x^2 C_{ABB} + x^3 C_{BBB}\,, \qquad (12.6.8)$$

and so on. In equation (7), B_{AA} and B_{BB} are the second virial coefficients of pure A and of pure B, and B_{AB} is a new second virial coefficient that takes account of the pairwise molecular interactions of A with B. Similarly the C's in equation (8) relate to the four kinds of interactions of three molecules at a time. The generalization of equations (6) to (8) to mixtures of more than two components is straightforward.

The new second virial coefficient B_{AB}, and *a fortiori* the two new third virial coefficients C_{AAB} and C_{ABB}, have been much less often studied than those for pure substances. Values of B_{AB} can be obtained from measurements of $B(T, x)$, B_{AA}, and B_{BB}. Alternatively, the molar volume of mixing $\Delta_{mix}V_m$ leads to

$$\Delta_{mix}V_m = 2(1-x)x\delta_{AB}\,, \qquad (12.6.9)$$

where

$$\delta_{AB} \overset{\text{def}}{=} B_{AB} - \tfrac{1}{2}(B_{AA} + B_{BB})\,. \qquad (12.6.10)$$

Similarly, the molar enthalpy of mixing $\Delta_{mix}H_m$ leads to

$$\Delta_{mix}H_m = 2(1-x)x(\delta_{AB} - T\, d\delta_{AB}/dT)p\,. \qquad (12.6.11)$$

Much effort has been devoted to the expression of B_{AB} in terms of B_{AA} and B_{BB}. The simplest rule is that of Lewis and Randall:

$$B_{AB} = \tfrac{1}{2}(B_{AA} + B_{BB}),\qquad (12.6.12)$$

or

$$\delta_{AB} = 0,\qquad (12.6.13)$$

so that

$$B(T, x) = (1 - x)B_{AA} + xB_{BB}.\qquad (12.6.14)$$

Unfortunately, such experimental evidence as there is suggests that Lewis and Randall's rule is seldom accurately obeyed, though in the absence of experimental results for B_{AB} it is often used. For molecules with central interactions a method based on the principle of corresponding states will be introduced in § 13.5.

The chemical potential μ_B of a substance B in a gaseous mixture is given by the identity:

$$\mu_B(g, T, p, x_B, x_C, \cdots) = \mu_B^*(\mathrm{pg}, T, p^\ominus) + \{\mu_B^*(\mathrm{pg}, T, p) - \mu_B^*(\mathrm{pg}, T, p^\ominus)\}$$

$$-\{\mu_B^*(\mathrm{pg}, T, p) - \mu_B^*(\mathrm{pg}, T, p \to 0)\}$$

$$-\{\mu_B^*(g, T, p \to 0) - \mu_B(g, T, p \to 0, x_B, x_C, \cdots)\}$$

$$+\{\mu_B(g, T, p, x_B, x_C, \cdots) - \mu_B(g, T, p \to 0, x_B, x_C, \cdots)\}$$

$$= \mu_B^\ominus(g, T) + RT \ln(x_B p/p^\ominus) + \int_0^p (V_B - RT/p)\, dp,\ (12.6.15)$$

where we have used equation (10.4.8) to evaluate the third $\{\cdots\}$ in the first expression. Equation (15) differs from equation (12.4.1) for a pure gas only by the inclusion of the mole fraction x_B of B in the second term and by the presence of the partial molar volume V_B of B in the integrand in place of the molar volume V_m of a pure gas.

For a binary gas mixture $\{(1 - x)A + xB\}$ the molar Gibbs function G_m follows from the identity $G_m = (1 - x)\mu_A + x\mu_B$:

$$G_m(g, T, p, x) = (1 - x)\mu_A^\ominus(g, T) + x\mu_B^\ominus(g, T) + RT\{(1 - x)\ln(1 - x) + x \ln x\}$$

$$+ RT \ln(p/p^\ominus) + \int_0^p (V_m - RT/p)\, dp.\qquad (12.6.16)$$

In terms of T and V_m rather than T and p as independent variables, equation (16) can be recast in the form:

$$A_m(g, T, V_m, x) = G_m(g, T, V_m, x) - p(g, T, V_m, x)V_m$$

$$= (1 - x)\mu_A^\ominus(g, T) + x\mu_B^\ominus(g, T) - RT + RT\{(1 - x)\ln(1 - x) + x \ln x\}$$

$$- RT \ln(p^\ominus V_m/RT) - \int_\infty^{V_m} (p - RT/V_m)\, dV_m,\qquad (12.6.17)$$

where A_m is the molar Helmholtz function. We shall make use of equation (17), with the integration carried out for an explicit equation of state, in chapters 16 and 17.

Although we shall postpone until chapters 16 and 17 a discussion of phase equilibria in fluid mixtures, it will be convenient to present here the thermodynamic equations for coexisting phases and for critical points in binary fluid mixtures.

The mole fractions and molar volumes of two coexisting phases in a binary mixture are determined by the simultaneous equations:

$$\mu_A(T, V_m^\alpha, x^\alpha) = \mu_A(T, V_m^\beta, x^\beta), \tag{12.6.18}$$

$$\mu_B(T, V_m^\alpha, x^\alpha) = \mu_B(T, V_m^\beta, x^\beta), \tag{12.6.19}$$

$$p(T, V_m^\alpha, x^\alpha) = p(T, V_m^\beta, x^\beta), \tag{12.6.20}$$

with the diffusional stability conditions (equivalent to equation 7.10.5):

$$(\partial\mu_A/\partial x)_{T,p}^\alpha < 0, \quad (\partial\mu_B/\partial x)_{T,p}^\alpha > 0,$$

$$(\partial\mu_A/\partial x)_{T,p}^\beta < 0, \quad (\partial\mu_B/\partial x)_{T,p}^\beta > 0. \tag{12.6.21}$$

By making use of the rule for changing the variable held constant (appendix I) from p to V_m, the conditions (21) can be written in the more complicated but, for use with an explicit equation of state, more useful forms:

$$(\partial\mu_A/\partial x)_{T,p}^\alpha = (\partial\mu_A/\partial x)_{T,V_m}^\alpha - (\partial\mu_A/\partial V_m)_{T,x}^\alpha (\partial p/\partial x)_{T,V_m}^\alpha / (\partial p/\partial V_m)_{T,x}^\alpha < 0, \tag{12.6.22}$$

and so on. It turns out that the hydrostatic stability conditions play no independent part in determining phase equilibria in mixtures.†

The criteria for a critical point in a mixture can be found by consideration of figure 12.5 where the change of molar Gibbs function on mixing, $\Delta_{mix}G_m(T, p, x)$, is plotted against the mole fraction x of substance B at the critical temperature T^c and at a temperature on each side of it. For stable phases $(\partial^2 G_m/\partial x^2)_{T,p}$ cannot be negative; curves of G_m against x must be everywhere concave upwards. That follows from $G_m = (1-x)\mu_A + x\mu_B$ and $(\partial G_m/\partial x)_{T,p} = \mu_B - \mu_A$ so that

$$(\partial^2 G_m/\partial x^2)_{T,p} = (\partial\mu_B/\partial x)_{T,p} - (\partial\mu_A/\partial x)_{T,p} > 0, \tag{12.6.23}$$

where we have used the inequalities (21) to determine the sign. Derivatives of $\Delta_{mix}G_m$ after the first are identical with corresponding derivatives of G_m. In figure 12.5, at the temperature T_1 (not necessarily less than the critical temperature T^c) phase separation occurs. The compositions of the coexisting phases are determined by the common tangent, at which (18) and (19) are obeyed. Between the coexisting phases the (broken) continuations of the curves correspond to phases that are metastable with respect to separation into the two coexisting phases. Each of the metastable curves terminates at a point of inflexion where $(\partial^2 G_m/\partial x^2)_{T,p} = 0$. Any curve joining the two metastable curves must be convex upwards and so cannot

† Prigogine, I.; Defay, R. *Chemical Thermodynamics.* Everett, D. H.: Translator. Longmans Green: London. **1954**.

Rowlinson, J. S. *Liquids and Liquid Mixtures.* Second edition. Butterworth: London. **1969**.

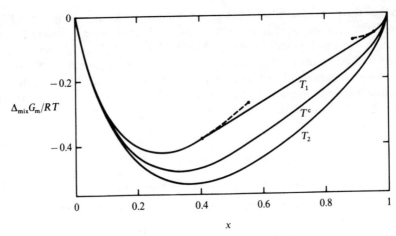

Figure 12.5. Plot of $\Delta_{mix}G_m(T, p, x)/RT$ against x for three temperatures around the critical temperature. At a temperature T_1 on one side of the critical, stable phases are shown as full curves and the two coexisting phases are joined by a straight line; metastable continuations of the two curves are shown as broken curves. At $T = T^c$ two points of inflexion coincide. At the temperature T_2 on the other side of the critical there is no phase separation.

occur. At the critical temperature T^c the two points of inflexion coincide, and so at the critical temperature

$$(\partial^2 G_m/\partial x^2)_{T,p} = 0, \quad \text{and} \quad (\partial^3 G_m/\partial x^3)_{T,p} = 0, \tag{12.6.24}$$

with

$$(\partial^4 G_m/\partial x^4)_{T,p} > 0. \tag{12.6.25}$$

At the temperature T_2 the curve is everywhere concave upwards; all points on the curve represent stable phases. Equations (24) and (25) can be recast in terms of T, x, and V_m rather than T, x, and p as independent variables in the more complicated but, for use with an explicit equation of state, more useful forms:

$$\partial^2 A_m/\partial x^2 - 2\chi\, \partial^2 A_m/\partial V_m\, \partial x + \chi^2\, \partial^2 A_m/\partial V_m^2 = 0, \tag{12.6.26}$$

$$\partial^3 A_m/\partial x^3 - 3\chi\, \partial^3 A_m/\partial V_m\, \partial x^2 + 3\chi^2\, \partial^3 A_m/\partial V_m^2\, \partial x$$
$$-\chi^3\, \partial^3 A_m/\partial V_m^3 = 0, \tag{12.6.27}$$

with

$$\partial^4 A_m/\partial x^4 - 4\chi\, \partial^4 A_m/\partial V_m\, \partial x^3 + 6\chi^2\, \partial^4 A_m/\partial V_m^2\, \partial x^2$$
$$-4\chi^3\, \partial^4 A_m/\partial V_m^3\, \partial x + \chi^4\, \partial^4 A_m/\partial V_m^4 > 0, \tag{12.6.28}$$

where χ denotes $(\partial^2 A_m/\partial V_m\, \partial x)/(\partial^2 A_m/\partial V_m^2)$, and where for simplicity the subscripts denoting the independent variables, T and V_m or T and x, held constant during partial differentiation have been omitted. By use of the definition of χ

equation (26) can be simplified to the form:

$$(\partial^2 A_m/\partial x^2)(\partial^2 A_m/\partial V_m^2) - (\partial^2 A_m/\partial V_m \, \partial x)^2 = 0 . \qquad (12.6.29)$$

In an *azeotropic mixture* the compositions of two coexisting phases are identical but the molar volumes are different. When $x^\alpha = x^\beta \; (= x^{az})$ but $V_m^\alpha \neq V_m^\beta$ equations (18) to (21) lead to the conditions for an azeotrope.

The generalization of equations (18) to (21) to three coexisting phases is straightforward, as is their generalization to mixtures of more than two substances.

Van der Waals generalized his equation of state to a binary mixture by writing

$$p = RT/\{V_m - b(x)\} - a(x)/V_m^2 , \qquad (12.6.30)$$

with

$$b(x) = (1-x)^2 b_{AA} + 2(1-x)xb_{AB} + x^2 b_{BB} , \qquad (12.6.31)$$

$$a(x) = (1-x)^2 a_{AA} + 2(1-x)xa_{AB} + x^2 a_{BB} . \qquad (12.6.32)$$

The two new quantities b_{AB} and a_{AB} are usually expressed in terms of the corresponding quantities for the pure substances A and B by the formulae:

$$b_{AB}^{1/3} = \tfrac{1}{2}(b_{AA}^{1/3} + b_{BB}^{1/3}) , \qquad (12.6.33)$$

which is correct for hard spheres and is probably an accurate approximation for real molecules, and

$$a_{AB} = \zeta(a_{AA}a_{BB}/b_{AA}b_{BB})^{1/2} b_{AB} , \qquad (12.6.34)$$

where ζ takes account of the small but significant failure of the geometric-mean rule $\varepsilon_{AB} = (\varepsilon_{AA}\varepsilon_{BB})^{1/2}$ for the characteristic interaction energies.[†]

Equations (31) to (34) will be used with the equation (12.5.18) of Carnahan and Starling in chapters 16 and 17.

Accounts of the volumetric properties of gaseous mixtures can be found in the literature.[‡]

§ 12.7 Fugacity

The *fugacity* f_B of the substance B in a gas or gaseous mixture, sometimes used as a measure of the departure of the gas or gaseous mixture from perfection, is defined by the equation:

$$f_B \stackrel{\text{def}}{=} (x_B p) \exp\left\{ \int_0^p (V_B/RT - 1/p) \, dp \right\} , \qquad (12.7.1)$$

and so has the dimension of pressure.

[†] We use ζ rather than the more usual ξ so as to avoid possible confusion with our use of ξ as the symbol for extent of reaction.

[‡] Knobler, C. M. *SPRCT*(*II*), chapter 7.

Mason, E. A.; Spurling, T. H. *The Virial Equation of State.* Pergamon: Oxford. **1969**.

For the virial equation of state with $C' = 0$, $D' = 0$, $\cdots$, equation (1) becomes

$$f_B = (x_B p) \exp\left[\left\{B_{BB} - (1 - x_B)^2 B_{BB} - \sum_{C \neq B} x_C^2 B_{CC} + \sum_{C \neq B} 2(1 - x_B) x_C B_{BC}\right.\right.$$
$$\left.\left. - \sum_{C \neq B} x_C \sum_{D \neq C, D \neq B} x_D B_{CD}\right\} p/RT\right]. \tag{12.7.2}$$

When the gas obeys the Lewis and Randall rule (12.6.12), equation (2) reduces to the strikingly simple form:

$$f_B = (x_B p) \exp(B_{BB} p/RT). \tag{12.7.3}$$

The form (3) is often used to correct 'classical' equilibrium constants $\Pi_B (x_B p/p^\ominus)^{\nu_B}$ towards accurate ('thermodynamic') equilibrium constants $\Pi_B (f_B/p^\ominus)^{\nu_B}$. This subject will be taken up again in §§ 12.13 and 12.14.

§ 12.8   Standard thermodynamic functions of a gaseous substance

The standard chemical potential $\mu_B^\ominus (g, T)$ of a gaseous substance B is defined by the relation:

$$\mu_B(g, T, p, x_C) \stackrel{\text{def}}{=} \mu_B^\ominus (g, T) + RT \ln(x_B p/p^\ominus)$$

$$+ \int_0^p \{V_B(g, T, p, x_C) - RT/p\} \, dp, \tag{12.8.1}$$

where $\mu_B(g, T, p, x_C)$ is the chemical potential, and $V_B(g, T, p, x_C)$ is the partial molar volume, of the substance B in a gaseous mixture of composition specified by the set of mole fractions x_C at the pressure p and temperature T, x_B is the mole fraction of the substance B in the mixture, and $p^\ominus$ is a standard pressure.[†]
  It might be helpful to see how such a definition is arrived at, and in the process to verify that $\mu_B^\ominus (g, T)$, the chosen standard pressure having been specified, does in fact depend only on the temperature and not at all on pressure or composition. We begin by writing the identity:

$$\mu_B(g, T, p, x_C) = \mu_B^*(g, T, p \to 0) + \{\mu_B(g, T, p \to 0, x_C) - \mu_B^*(g, T, p \to 0)\}$$

$$+ \{\mu_B(g, T, p, x_C) - \mu_B(g, T, p \to 0, x_C)\}$$

$$= \mu_B^*(g, T, p \to 0) + RT \ln x_B + \int_0^p V_B(g, T, p, x_C) \, dp, \tag{12.8.2}$$

---

† In chemical thermodynamics the standard pressure is usually, though not necessarily, chosen to be 101.325 kPa.

where we have used equation (10.4.8) to replace the first $\{\cdots\}$ by $RT \ln x_B$. Because the integral in equation (2) diverges we rewrite the equation in the form:

$$\mu_B(g, T, p, x_C) = \left\{ \mu_B^*(g, T, p \to 0) + \int_0^{p^{\ominus}} (RT/p)\, dp \right\} + \int_{p^{\ominus}}^p (RT/p)\, dp + RT \ln x_B$$

$$+ \int_0^p \{V_B(g, T, p, x_C) - RT/p\}\, dp, \qquad (12.8.3)$$

and now replace the first $\{\cdots\}$, which we observe to be independent of p and of x_C, by $\mu_B^{\ominus}(g, T)$, and so recover equation (1).

All the other standard thermodynamic functions of a gaseous substance B follow by differentiation of equation (1) with respect to T. Thus

$$\lambda_B(g, T, p, x_C) = \lambda_B^{\ominus}(g, T)(x_B p/p^{\ominus}) \exp\left\{ \int_0^p (V_B/RT - 1/p)\, dp \right\}, \qquad (12.8.4)$$

$$S_B(g, T, p, x_C) = S_B^{\ominus}(g, T) - R \ln(x_B p/p^{\ominus}) - \int_0^p \{(\partial V_B/\partial T)_p - R/p\}\, dp, \quad (12.8.5)$$

$$H_B(g, T, p, x_C) = H_B^{\ominus}(g, T) + \int_0^p \{V_B - T(\partial V_B/\partial T)_p\}\, dp, \qquad (12.8.6)$$

$$C_{p,B}(g, T, p, x_C) = C_{p,B}^{\ominus}(g, T) - \int_0^p T(\partial^2 V_B/\partial T^2)_p\, dp. \qquad (12.8.7)$$

We note that in each of equations (1) and (4) to (7) the integral vanishes when the gaseous mixture is perfect, that is to say when $pV = (n_A + n_B + \cdots)RT$ so that $V_B = RT/p$, $(\partial V_B/\partial T)_p = R/p$, and $(\partial^2 V_B/\partial T^2)_p = 0$. We also see now the reason for the warnings given on pages 114 and 161: whereas for a reaction in a real gaseous mixture at a pressure not too far above atmospheric pressure $\Delta H_m \approx \Delta H_m^{\ominus}$ because the integral in equation (6) is small, and similarly $\Delta C_{p,m} \approx \Delta C_{p,m}^{\ominus}$, no such approximation can be made for the chemical potential or molar Gibbs function or molar entropy; the term containing $\ln(x_B p/p^{\ominus})$ is in general not small.

§ 12.9   Comparison of 'changes of molar Gibbs function for a chemical reaction'

In this section we shall compare the change in molar Gibbs function ΔG_m when a chemical reaction proceeds from one extent of reaction ξ_1 to another ξ_2, the quantity $(\partial G/\partial \xi)_{T,p}$, and the *standard* molar change of Gibbs function $\Delta G_m^{\ominus}$ of the reaction. Each of these quantities is widely confused with the others. In particular the word 'standard' is often inexcusably omitted, and the quantity $(\partial G/\partial \xi)_{T,p} = -A$, where A is the affinity of the reaction, is often loosely called the 'change of molar Gibbs function'.

We shall illustrate the meanings of these quantities and the differences between them by considering a reaction $0 = \Sigma_B \nu_B B$ *in a perfect gas mixture*. For any extent of reaction ξ the amounts of substance $n_B(\xi)$ are given by

$$n_B(\xi) = n_B(0) + \nu_B \xi, \tag{12.9.1}$$

where the reader is reminded that ν_B is the stoichiometric number of the substance B and is positive if B is a 'product' and negative if B is a 'reactant'.

The Gibbs function of the mixture for extent of reaction ξ is then given by

$$G(\text{pgm}, T, p, \xi) = \sum_B n_B(\xi) \mu_B(\text{pgm}, T, p, \xi)$$

$$= \sum_B \{n_B(0) + \nu_B \xi\} \mu_B^\ominus(g, T)$$

$$+ RT \sum_B \{n_B(0) + \nu_B \xi\} \ln\left[\{n_B(0) + \nu_B \xi\} \middle/ \left\{\sum_B n_B(0) + \xi \sum_B \nu_B\right\}\right]$$

$$+ RT \sum_B \{n_B(0) + \nu_B \xi\} \ln(p/p^\ominus), \tag{12.9.2}$$

where we have used equation (12.8.1) with the integral set equal to zero for a perfect gas mixture. For a change of the extent of reaction from ξ_1 to ξ_2 we then have for the molar change of Gibbs function[†]

$$\Delta G_m \stackrel{\text{def}}{=} \Delta G/(\xi_2 - \xi_1)$$

$$= \Delta G_m^\ominus$$

$$+ RT(\xi_2 - \xi_1)^{-1} \sum_B \{n_B(0) + \nu_B \xi_2\} \ln\left[\{n_B(0) + \nu_B \xi_2\} \middle/ \left\{\sum_B n_B(0) + \xi_2 \sum_B \nu_B\right\}\right]$$

$$- RT(\xi_2 - \xi_1)^{-1} \sum_B \{n_B(0) + \nu_B \xi_1\} \ln\left[\{n_B(0) + \nu_B \xi_1\} \middle/ \left\{\sum_B n_B(0) + \xi_1 \sum_B \nu_B\right\}\right]$$

$$+ RT\left(\sum_B \nu_B\right) \ln(p/p^\ominus), \tag{12.9.3}$$

where we have replaced $\sum_B \nu_B \mu_B^\ominus(g, T)$ by the standard molar change of Gibbs function $\Delta G_m^\ominus$, and we remind the reader that it is equal by definition to $-RT \ln K^\ominus$, where $K^\ominus$ is the standard equilibrium constant.

[†] In accord with common practice we have called $\Delta G/(\xi_2 - \xi_1)$ 'the molar change of Gibbs function' and have denoted it by 'ΔG_m', in spite of the rule (§ 2.4) that 'molar' should mean 'divided by amount of substance' and not as here 'divided by the change in the extent of reaction'. The *dimension* of this 'ΔG_m' is of course the same as that of any real change of molar Gibbs function, but that does not justify the use of the word 'molar' to describe it.

For $(\partial G/\partial \xi)_{T,p}$ we have

$$(\partial G/\partial \xi)_{T,p} = \Delta G_m^\ominus + RT \sum_B \nu_B \ln\left[\{n_B(0) + \nu_B \xi\}\Big/\left\{\sum_B n_B(0) + \xi \sum_B \nu_B\right\}\right]$$

$$+ RT\left(\sum_B \nu_B\right)\ln(p/p^\ominus). \qquad (12.9.4)$$

In figure 12.6 we show $\{G(T, p, \xi) - G(T, p, 0)\}/RT$ plotted against ξ for

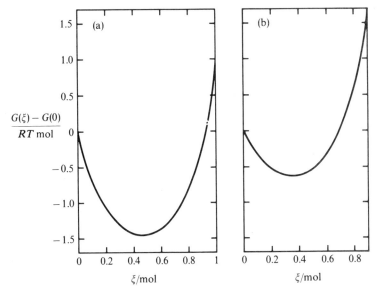

Figure 12.6. The change in Gibbs function $\{G(T, p, \xi) - G(T, p, 0)\}$ divided by RT plotted against the extent of reaction ξ for the reaction $N_2 + 3H_2 = 2NH_3$ in a perfect gas mixture for $K^\ominus = 0.15$ and $p/p^\ominus = 5$, and for initial mole fractions: (a) 0.25 of N_2, 0.75 of H_2, and 0 of NH_3, and (b) 0.25 of N_2, 0.675 of H_2, and 0.075 of NH_3.

the synthesis of ammonia:

$$N_2 + 3H_2 = 2NH_3, \qquad (12.9.5)$$

at a temperature (about 492 K) at which the standard equilibrium constant $K^\ominus = 0.15$ and at a pressure $p = 5p^\ominus$ (≈ 500 kPa). The initial mole fractions were taken: for figure 12.6(a) as 0.25 of N_2, 0.75 of H_2, and 0 of NH_3, and for figure 12.6(b) as 0.25 of N_2, 0.675 of H_2, and 0.075 of NH_3. In figure 12.6(a) the equilibrium value ξ^{eq} of ξ is 0.47 mol, corresponding to the mole fraction $2\xi^{eq}/(4 \text{ mol} - 2\xi^{eq}) = 0.31$ of NH_3. At $\xi = 0$, $\Delta G/\Delta \xi = -\infty$ and $(\partial G/\partial \xi)_{T,p} = -\infty$. At $\xi = 0.2$ mol, say, $\Delta G/\Delta \xi = -21.9$ kJ mol^{-1} while $(\partial G/\partial \xi)_{T,p} = -12.3$ kJ mol^{-1}. At $\xi = \xi^{eq}$, $\Delta G/\Delta \xi = -12.7$ kJ mol^{-1} while $(\partial G/\partial \xi)_{T,p} = 0$. The standard molar change of Gibbs function $\Delta G_m^\ominus$ is defined as $-RT \ln K^\ominus$ and so is $+7.8$ kJ mol^{-1}. We note that it does *not* follow that because $\Delta G_m^\ominus > 0$ the reaction is not 'feasible'. In figure 12.6(b) the equilibrium value ξ^{eq} of ξ is 0.35 mol, corresponding to the mole fraction $(0.075\ mol + 2\xi^{eq})/(4\ mol - 2\xi^{eq}) = 0.23$ of NH_3. At $\xi = 0$, $\Delta G/\Delta\xi = -\infty$ while $(\partial G/\partial\xi)_{T,p} = -16.1\ kJ\ mol^{-1}$. At $\xi = 0.2\ mol$, $\Delta G/\Delta\xi = -10.7\ kJ\ mol^{-1}$ while $(\partial G/\partial\xi)_{T,p} = -6.0\ kJ\ mol^{-1}$. At $\xi = \xi^{eq}$, $\Delta G/\Delta\xi = -7.4\ kJ\ mol^{-1}$ while $(\partial G/\partial\xi)_{T,p} = 0$. The standard molar change of Gibbs function $\Delta G_m^\ominus$ is still of course $+7.8\ kJ\ mol^{-1}$. Of all these laboriously calculated quantities it is only the equilibrium values ξ^{eq} of the extent of reaction ξ that are of any importance, and they can be calculated for any initial mole fractions, any pressure, and any value of the standard equilibrium constant $K^\ominus$, without reference to the Gibbs function. Figure 12.6 serves only to remind us that if anything is happening in a system of fixed temperature and pressure then the Gibbs function is decreasing.

§ 12.10 Calculation of the standard molar enthalpy change from a measured enthalpy change. The Washburn correction

The calculation of a standard molar enthalpy change $\Delta H_m^\ominus$ from a calorimetrically measured value of ΔH involves many small corrections.

For simplicity we shall consider as an example a reaction:

$$0 = \sum_B \nu_B B(g)\,, \tag{12.10.1}$$

confined to a single gaseous phase. We suppose that amounts of substance n_B^i are caused to react in a thermally insulated bomb calorimeter at the temperature T_1 and pressure p_1 so as to give final amounts of substance n_B^f at the temperature T_2 and pressure p_2, the n_B^f's and n_B^i's being related by

$$n_B^f = n_B^i + \nu_B\xi\,, \tag{12.10.2}$$

where ξ denotes the extent of reaction. We then have

$$U(T_2, p_2, \xi) - U(T_1, p_1, 0) = W^i\,, \tag{12.10.3}$$

where W^i is any work done in initiating the reaction. Next we suppose that the temperature of the calorimeter is returned to T_1 and that the pressure is then found (or calculated) to be p_3. Then we suppose that the work W^{el} that causes the temperature to change exactly to T_2 is measured (electrically), the pressure then being p_2. We then have

$$U(T_2, p_2, \xi) - U(T_1, p_3, \xi) = W^{el}\,. \tag{12.10.4}$$

Subtracting (4) from (3) we obtain

$$U(T_1, p_3, \xi) - U(T_1, p_1, 0) = -W^{el} + W^i\,, \tag{12.10.5}$$

which we rewrite in the form:

$$H(T_1, p_3, \xi) - H(T_1, p_1, 0)$$
$$= -W^{el} + W^i + \{p_3 V(T_1, p_3, \xi) - p_1 V(T_1, p_1, 0)\}\,. \tag{12.10.6}$$

In deriving equations (4) to (6) we have assumed either that the thermal insulation of the calorimeter is perfect, or more realistically that accurate corrections have been made for heat leaks (see §§ 1.19 and 4.2). We next assume either that the energy of the fabric of the calorimeter remains unchanged during each part of the experiment, or less drastically that any change is exactly the same in each part of the experiment and so has cancelled out in equations (5) and (6). We may then write

$$H(T, p, \xi) = \sum_B (n_B^i + \nu_B \xi) H_B(T, p, \xi), \qquad (12.10.7)$$

and

$$V(T, p, \xi) = \sum_B (n_B^i + \nu_B \xi) V_B(T, p, \xi), \qquad (12.10.8)$$

where H_B and V_B are the partial molar enthalpy and the partial molar volume of the substance B.

By use of equation (12.8.6) we rewrite equation (7) in the form:

$$H(T, p, \xi) = \sum_B (n_B^i + \nu_B \xi) H_B^\ominus (T)$$

$$+ \sum_B (n_B^i + \nu_B \xi) \int_0^p [V_B(T, p, \xi) - T\{\partial V_B(T, p, \xi)/\partial T\}_p] \, dp. \qquad (12.10.9)$$

Substituting (9) into (6) and rearranging we obtain finally

$$\xi \sum_B \nu_B H_B^\ominus (T_1) \stackrel{\text{def}}{=} \xi \Delta H_m^\ominus (T_1)$$

$$= -W^{\text{el}} + W^i$$

$$- \sum_B (n_B^i + \nu_B \xi) \int_0^{p_3} [V_B(T_1, p, \xi) - T_1\{\partial V_B(T, p, \xi)/\partial T\}_{p, T=T_1}] \, dp$$

$$+ \sum_B n_B^i \int_0^{p_1} [V_B(T_1, p, 0) - T_1\{\partial V_B(T, p, 0)/\partial T\}_{p, T=T_1}] \, dp$$

$$+ \sum_B (n_B^i + \nu_B \xi) p_3 V_B(T_1, p_3, \xi) - \sum_B n_B^i p_1 V_B(T_1, p_1, 0). \qquad (12.10.10)$$

Each of the last four terms in equation (10) can be evaluated if enough is known about $V_B(T, p, \xi)$ for each substance B and the given n_B^i's. These last four terms are known as the *Washburn correction*.

If the pressures were all low enough to allow the gaseous mixtures to be regarded as perfect, equation (10) would reduce to

$$\Delta H_m^\ominus (T_1) = -W^{\text{el}}/\xi + W^i/\xi + \left(\sum_B \nu_B\right) RT_1, \qquad (12.10.11)$$

a formula often given in elementary textbooks.

Had we taken as our example the complete combustion of a sample of say $C_6H_5CO_2H(s)$ in a bomb calorimeter at room temperature to form a gaseous mixture of $O_2(g)$, $CO_2(g)$, and $H_2O(g)$, and a liquid solution of O_2(solute) and CO_2(solute) in H_2O(solvent), the analysis would have been a good deal more complicated. In particular, we should then have had to know or measure also the thermal and volumetric properties of solutions of O_2 and CO_2 in H_2O(l), the solubilities, and the volumetric properties of $C_6H_5CO_2H$(s); and we should have had to use the appropriate formula for $H_B^{\ominus}$ according as the substance B was a component in a gaseous mixture, or a pure solid (see chapter 16), or a solute or the solvent in a dilute solution (see chapter 18). Nor would the analysis have been less complicated (except for extremely dilute solutions) had we taken as our example a quantitative reaction between two aqueous electrolyte solutions brought about by causing the two solutions to mix in a thermally insulated constant-pressure ('open') calorimeter, even if we had taken precautions to prevent the large errors that can result from the evaporation or condensation of small amounts of solvent into or from the air space in the calorimeter when the composition and hence the vapour pressure change as a result of the reaction.

Authoritative accounts of the Washburn correction can be found in the literature.†

§ 12.11   Calculation of the standard molar entropy of a gas from its molar entropy

The difference $\{S_m^{\ominus}(B, g, T) - S_m(B, s, T \to 0)\}$ is commonly called the standard molar entropy of the gaseous substance B at the temperature T (see chapter 15). The quantity resulting from calorimetric measurements from the solid at $T \to 0$ to the gaseous state at temperature T and pressure p is $\{S_m(B, g, T, p) - S_m(B, s, T \to 0)\}$. In the present section we shall be concerned only with the difference $\{S_m^{\ominus}(B, g, T) - S_m(B, g, T, p)\}$. According to equation (12.8.9) that difference is given by the equation:

$$S_m^{\ominus}(B, g, T) - S_m(B, g, T, p) = R \ln(p/p^{\ominus})$$

$$+ \int_0^p \{(\partial V_m/\partial T)_p - R/p\} \, dp. \quad (12.11.1)$$

If we write $V_m = RT/p + B$ then equation (1) becomes

$$S_m^{\ominus}(B, g, T) - S_m(B, g, T, p) = R \ln(p/p^{\ominus}) + p \, dB/dT. \quad (12.11.2)$$

In practice the second term of equation (2) is usually small and sufficiently accurate. For example for Xe at 100 kPa and 298.15 K we compare

$$p \, dB/dT = 0.08 \text{ J K}^{-1} \text{ mol}^{-1},$$

† Hubbard, W. N.; Scott, D. W.; Waddington, G. *ETc(I)*, chapter 5. See also other chapters in *ETc(I)*, chapters in *ETc(II)*, and especially Månsson, M.; Hubbard, W. N. *ECTd(I)*, chapter 5.

with

$$\{S_m^\ominus(\text{Xe}, \text{g}, 298.15 \text{ K}) - S_m(\text{Xe}, \text{s}, T \to 0)\} = 169.57 \text{ J K}^{-1} \text{ mol}^{-1}.$$

§ 12.12 Experimental methods for the study of a chemical equilibrium in a gaseous mixture

All that we need to know about any chemical equilibrium is the extent of reaction ξ^{eq} as a function of pressure and temperature. We must make sure that the chemical reaction is known, that the initial mole fractions of all the substances (whether reacting or not) are known, and that the reaction has reached chemical equilibrium. It is then sufficient to measure the mole fraction x_B^{eq} of any one of the reacting substances B at a measured pressure and temperature. The chemical analysis must be done without appreciably disturbing the carefully established equilibrium. That can be achieved either by rapidly cooling ('quenching') the reacting mixture so as to lower the rate of reaction considerably and then analysing the mixture at leisure, or by using a spectrophotometric or other optical method to analyse the reacting mixture with as little disturbance as possible. The best way to make sure that the reaction has reached chemical equilibrium is to obtain the same results when the equilibrium is approached first from one direction and then from the other.

Both static and dynamic (flow) methods are used for such measurements. In static methods a sample of the reacting mixture in a thermostat is connected to a manometer usually through a pressure transducer. It is allowed to reach chemical equilibrium and is then analysed. Flow methods are usually used for reactions having negligible rates of reaction at room temperature. The initial gaseous mixture flows into a tube contained in an oven at the desired temperature and remains there for long enough (often in the presence of a solid catalyst) to reach equilibrium. The equilibrium mixture then flows out, is quenched by the lower temperature (or by the absence of a catalyst), and is analysed. The pressure is measured on the inlet and on the outlet sides of the reaction tube; an appreciable gradient of pressure would lead to unwanted Joule–Thomson effects.

When the gaseous mixture is known to be perfect the extent of reaction ξ^{eq} at equilibrium can be obtained for reactions with $\Sigma_B \nu_B \neq 0$ from the relation:

$$\xi^{eq} = \left(p^{eq} V/RT - \sum_B n_B^i \right) \Big/ \sum_B \nu_B, \qquad (12.12.1)$$

if the initial amounts of substance n_B^i are known and the equilibrium pressure p^{eq} is measured for a known volume V. Alternatively, if the initial pressure p^i can be measured at a temperature T^i then

$$\xi^{eq} = \left(p^{eq} V/RT - p^i V/RT^i \right) \Big/ \sum_B \nu_B, \qquad (12.12.2)$$

where we have assumed that the volume V is independent of T.

§ 12.13   Thermodynamics of a chemical equilibrium in a gaseous mixture

For a chemical reaction $0 = \Sigma_B \nu_B B$ in any phase or phases the standard equilibrium constant $K^{\ominus}$ is defined according to equation (11.3.5) by

$$K^{\ominus}(T) \stackrel{\text{def}}{=} \prod_B \{\lambda_B^{\ominus}(T)\}^{-\nu_B} = \prod_B \{\lambda_B(T, p, \xi^{eq})/\lambda_B^{\ominus}(T)\}^{\nu_B}, \qquad (12.13.1)$$

where the superscript ^{eq} denotes the value of a quantity at chemical equilibrium.

Substituting from equation (12.8.4) we then obtain for a chemical reaction in a gaseous mixture

$$K^{\ominus}(g, T) = \prod_B (x_B^{eq} p/p^{\ominus})^{\nu_B} \exp\left[\sum_B \nu_B \int_0^p \left\{V_B(g, T, p, x_C^{eq})/RT - 1/p\right\} dp\right].$$
$$(12.13.2)$$

The standard equilibrium constant of a gas reaction at given temperature can thus be determined from values of the mole fractions x_B^{eq} at equilibrium at the pressure p obtained by chemical analysis of the equilibrium mixture, and from values of the partial molar volumes $V_B(g, T, p, x_C^{eq})$ obtained from (p, V, T) measurements on the gaseous mixture at compositions around the equilibrium composition and at pressures between 0 and p.

For a perfect gas mixture equation (2) reduces to the familiar 'classical' form:†

$$K^{\ominus}(\text{pgm}, T) = \prod_B (x_B^{eq} p/p^{\ominus})^{\nu_B}. \qquad (12.13.3)$$

For any real gaseous mixture equation (2) can be written in the form:

$$K^{\ominus}(g, T) = \lim_{p \to 0} \left\{\prod_B (x_B^{eq} p/p^{\ominus})^{\nu_B}\right\}. \qquad (12.13.4)$$

Equation (4) usually provides an easier experimental route than equation (2) from measurements of the mole fraction x_B^{eq} at equilibrium to the standard equilibrium constant. The measurements must be made at lower and lower pressures so that $K^{\ominus}$ can be obtained by extrapolation to $p \to 0$, but no measurements of partial molar volumes need be made.

When the pressure is low enough to allow us to assume that $C' = 0$, $D' = 0$, $\cdots$, and when we further assume the Lewis and Randall rule (12.6.12), we may write equation (2) in the form:

$$K^{\ominus}(g, T) \approx \prod_B (x_B^{eq} p/p^{\ominus})^{\nu_B} \exp\left(\sum_B \nu_B B_{BB} p/RT\right). \qquad (12.13.5)$$

† When $p^{\ominus}$ is omitted from equation (12.13.3) the quantity is sometimes denoted by K_p instead of $K^{\ominus}$; whereas $K^{\ominus}$ is a number, K_p has the dimension: $(\text{pressure})^{\Sigma_B \nu_B}$. Many textbooks fail to distinguish between K_p and $K^{\ominus}$. It is only when $\Sigma_B \nu_B = 0$ (as for example it is for the reaction: $H_2 + I_2 = 2HI$) that the distinction between the dimensions of K_p and $K^{\ominus}$ disappears. We shall not use the quantity K_p in this book.

§ 12.14   Example

In figure 12.7 we show as an example the mole fraction y^{eq} of NH_3 plotted against the pressure at 723 K for equilibrium of the chemical reaction: $N_2 + 3H_2 = 2NH_3$, when the initial mole fractions are 0.25 of N_2, 0.75 of H_2, and 0 of NH_3. The mole-fraction

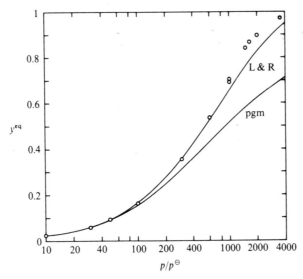

Figure 12.7. Experimental values of the mole-fraction yield y^{eq} of NH_3 plotted against $p/p^{\ominus}$ at 723 K for the chemical reaction: $N_2 + 3H_2 = 2NH_3$, when the initial mole fractions are 0.25 of N_2, 0.75 of H_2, and 0 of NH_3. The standard pressure $p^{\ominus} = 101.325$ kPa. The curve marked pgm was calculated from equation (12.11.3), and that marked L & R from equation (12.13.5) with $\{2B(NH_3) - B(N_2) - 3B(H_2)\} = -112$ cm^3 mol^{-1} obtained by extrapolation of experimental values. The standard equilibrium constant $K^{\ominus}$ (723 K) was taken to be 4.60×10^{-5}.

yield of NH_3 varies from 0.02 at 1 MPa through 0.16 at 10 MPa and 0.7 at 100 MPa to 0.97 at 350 MPa. The curves in figure 12.7 will be explained below.

We specify the stoichiometry of the reaction in table 12.1

Table 12.1. Initial amounts of substance n^i_B, equilibrium amounts of substance n^{eq}_B, and equilibrium mole fractions x^{eq}_B. In the last column the mole fractions x^{eq}_B are expressed in terms of the mole-fraction yield $y^{eq} = x^{eq}(NH_3)$ of NH_3.

B	n^i_B	n^{eq}_B	x^{eq}_B	x^{eq}_B
N_2	n	$n - \xi^{eq}$	$(n - \xi^{eq})/(n + rn - 2\xi^{eq})$	$(2 + y^{eq} - ry^{eq})/2(1 + r)$
H_2	rn	$rn - 3\xi^{eq}$	$(rn - 3\xi^{eq})/(n + rn - 2\xi^{eq})$	$(2r - 3y^{eq} - ry^{eq})/2(1 + r)$
NH_3	0	$2\xi^{eq}$	$2\xi^{eq}/(n + rn - 2\xi^{eq})$	y^{eq}
all	$n + rn$	$n + rn - 2\xi^{eq}$	1	1

For a perfect gas mixture the standard equilibrium constant has the form:

$$K^{\ominus}(\text{pgm}, T) = \{x^{\text{eq}}(\text{NH}_3)\}^2/x^{\text{eq}}(\text{N}_2)\{x^{\text{eq}}(\text{H}_2)\}^3(p/p^{\ominus})^2$$

$$= 16(y^{\text{eq}})^2(1+r)^4/(2+y^{\text{eq}}-ry^{\text{eq}})(2r-3y^{\text{eq}}-ry^{\text{eq}})^3(p/p^{\ominus})^2, \quad (12.14.1)$$

from which the yield of NH_3 can be calculated for given r, $K^{\ominus}$, and $p/p^{\ominus}$.

By differentiation of $\ln K^{\ominus}$ with respect to r we find that $\partial y^{\text{eq}}/\partial r = 0$ and $\partial^2 y^{\text{eq}}/\partial r^2 < 0$ at $r = 3$; not surprisingly the maximum yield is obtained when $n^{\text{i}}(\text{H}_2) = 3n^{\text{i}}(\text{N}_2)$. When $r = 3$ equation (1) becomes

$$K^{\ominus}(\text{pgm}, T) = 256(y^{\text{eq}})^2/27(1-y^{\text{eq}})^4(p/p^{\ominus})^2. \quad (12.14.2)$$

In figure 12.6 the curve marked pgm was calculated by solving equation (2) for y^{eq} with $K^{\ominus} = 4.60 \times 10^{-5}$. There is good agreement with experiment at pressures up to about 10 MPa. At 100 MPa, a commonly used pressure in industrial plants for the synthesis of ammonia, the calculated mole-fraction yield of NH_3 is 25 per cent less than the experimental value. The discrepancy increases at still higher pressures.

If we assume that $C' = 0$, $D' = 0, \cdots$, and that the Lewis and Randall rule is obeyed, then equation (12.13.5) takes the form:

$$K^{\ominus}(\text{g}, T) \approx \{256(y^{\text{eq}})^2/27(1-y^{\text{eq}})^4(p/p^{\ominus})^2\}$$

$$\times \exp[\{2B(\text{NH}_3) - B(\text{N}_2) - 3B(\text{H}_2)\}p/RT]. \quad (12.14.3)$$

The curve marked L & R in figure 12.6 was calculated by solving equation (3) for y^{eq} with $K^{\ominus} = 4.60 \times 10^{-5}$ and with $\{2B(\text{NH}_3) - B(\text{N}_2) - 3B(\text{H}_2)\} = -112 \text{ cm}^3 \text{ mol}^{-1}$ a value that was estimated by extrapolation of the experimental values to 723 K. There is now good agreement with experiment at pressures up to about 50 MPa.

§ 12.15 Chemical equilibrium for a reaction involving a gaseous mixture and one or more pure solid (or pure liquid) phases

For a pure solid (or pure liquid) phase we anticipate one of the conclusions of chapter 16 by writing for the standard absolute activity:

$$\lambda_B^*(\text{s}, T, p) \overset{\text{def}}{=} \lambda_B^{\ominus}(\text{s}, T) \exp\left[\int_{p^{\ominus}}^{p} \{V_B^*(\text{s}, T, p)/RT\} \, dp\right]. \quad (12.15.1)$$

Equation (1) is the exact definition of $\lambda_B^{\ominus}(\text{s}, T)$, but it is usual to make the approximation:

$$\lambda_B^*(\text{s}, T, p) \approx \lambda_B^{\ominus}(\text{s}, T). \quad (12.15.2)$$

If we accept that approximation then it follows that we may omit any factor for a pure solid (or for a pure liquid) from equation (12.13.1) for the standard equilibrium constant $K^{\ominus}$. The neglected factor is 1.004 for $(p - p^{\ominus}) = 100 \text{ kPa}$ and $V_B^* = 100 \text{ cm}^3 \text{ mol}^{-1}$ at room temperature; at higher pressures its omission might be more serious.

§ 12.16   Simultaneous equilibria

As an example, given the standard equilibrium constants $K_1^\ominus(T)$ for the reaction:

$$\tfrac{2}{3}Cr_2O_3(s) + C(s) = \tfrac{4}{3}Cr(s) + CO_2(g),\tag{12.16.1}$$

and $K_2^\ominus(T)$ for the reaction:

$$C(s) + CO_2(g) = 2CO(g),\tag{12.16.2}$$

we ask what is the lowest temperature at which carbon will continuously reduce dichromium trioxide in a furnace open to an external pressure $p^\ominus$, and what is then the composition of the effluent gas. Assuming a perfect gas mixture and approximation (12.15.2) we have for reaction (1):

$$xp/p^\ominus = K_1^\ominus(T),\tag{12.16.3}$$

and for reaction (2):

$$\{(1-x)^2/x\}(p/p^\ominus) = K_2^\ominus(T),\tag{12.16.4}$$

where x denotes the mole fraction of CO_2 in the effluent gas. When $p = p^\ominus$ equation (3) gives

$$x = K_1^\ominus(T).\tag{12.16.5}$$

Substituting for x from (5) and putting $p = p^\ominus$ equation (4) gives

$$(K_1^\ominus K_2^\ominus)^{1/2} + K_1^\ominus = 1.\tag{12.16.6}$$

From the given values of $K_1^\ominus(T)$ and $K_2^\ominus(T)$ we find the temperature T^* that satisfies equation (6). The value of $K_1^\ominus(T^*)$ is the mole fraction of CO_2 in the effluent gas.

Problems for chapter 12

Problem 12.1

The following values of the ratio $r = V(C_3H_8, 295.20\ K, p)/V(N_2, 295.20\ K, p)$ were measured in a differential-compression apparatus.

r	1.011678	1.014557	1.017274	1.018760	1.019695
p/kPa	72.619	55.439	39.022	30.246	24.625

Assuming that $pV = nRT + nBp$ and that $B(N_2, 295.20\ K) = -5\ cm^3\ mol^{-1}$, where B denotes the second virial coefficient, calculate $B(C_3H_8, 295.20\ K)$. What difference would it have made if C rather than C' had been neglected in this calculation?

Problem 12.2

The following pairs $p(N_2)$ and $p(C_6H_6)$ were pressures at which nitrogen and benzene successively balanced an adsorption-compensated buoyancy balance at 295.22 K.

| $p(N_2)$/kPa | 22.5648 | 22.5721 | 22.5711 |
| $p(C_6H_6)$/kPa | 8.0521 | 8.0548 | 8.0544 |

Assuming that $pV = nRT + nBp$ and that $B(N_2, 295.22 \text{ K}) = -5 \text{ cm}^3 \text{ mol}^{-1}$, where B denotes the second virial coefficient, calculate $B(C_6H_6, 295.22 \text{ K})$. What difference would it have made if C rather than C' had been neglected in this calculation?

Problem 12.3

Neglecting virial coefficients higher than the second calculate the speed of sound in Ar at 100 K (a) at $p \to 0$ and (b) at $p = 30$ kPa. Take $C_{p,m}(p \to 0)/R = \frac{5}{2}$ and $B/\text{cm}^3 \text{ mol}^{-1} = 138.5 - 105.4 \exp(113.0 \text{ K}/T)$.

Problem 12.4

Derive the conditions for the critical temperature T^c and for the compositions of two coexisting phases at $T = 0.7T^c$ for a binary fluid mixture for which $\Delta_{mix}G_m(T, x) = RT\{(1-x)\ln(1-x)+x \ln x\}+Ax^2(1-x)$ where A is a constant. Plot $\Delta_{mix}G_m(0.7T^c, x)$ against x, and find the values of x^α and x^β in the two coexisting phases.

Problem 12.5

Use the final equation of problem 12.3 to calculate the fugacity of pure Ar at 100 K and 0.5 MPa.

Problem 12.6

Calculate $\Delta G_m^{\ominus}$ and $\{G(T, p, \xi) - G(T, p, 0)\}$ and plot the latter against the extent of reaction ξ for the chemical reaction: $N_2 + 3H_3 = 2NH_3$, at $K^{\ominus}(492 \text{ K}) = 0.15$ and at the pressure $p = 5p^{\ominus}$ when the initial mole fractions are 0.25 of N_2 and 0.75 of H_2 and the gaseous mixture is regarded as perfect.

Problem 12.7

Use the final equation of problem 12.3 to calculate the molar entropy difference: $\{S_m^{\ominus}(\text{Ar, g, } 87.28 \text{ K}) - S_m(\text{Ar, g, } 87.28 \text{ K}, 101.325 \text{ kPa})\}$.

Problem 12.8

A mixture of 0.01 mol of H_2 and 0.02 mol of I_2 was maintained at 731 K until equilibrium was established. Assuming that the gaseous mixture is perfect calculate the extent of reaction at equilibrium and the composition of the equilibrium mixture, given that the standard equilibrium constant $K^\ominus(731 \text{ K}) = 48.7$ for the reaction: $H_2(g) + I_2(g) = 2HI(g)$.

Problem 12.9

When 0.008367 mol of PCl_5 was brought to equilibrium at 473 K in a vessel of volume 308.6 cm^3 the pressure was 150.8 kPa. Assuming that the gaseous mixture is perfect calculate the extent of the reaction: $PCl_5(g) = PCl_3(g) + Cl_2(g)$, the fraction α of PCl_5 dissociated, the mole fractions, and the standard equilibrium constant $K^\ominus$.

How would α be affected by the addition of 0.005 mol of a non-reacting gas such as Ar, (a) when the volume is maintained at 308.6 cm^3, and (b) when the pressure is maintained at 150.8 kPa?

Problem 12.10

When a known amount of substance n of I_2 was confined in a known volume V at 1274 K the pressures p and the values of pV/nRT were as follows.

p/kPa	11.37	48.05	76.31	94.31
pV/nRT	1.524	1.283	1.229	1.208

Assuming that the gaseous mixtures are perfect calculate the standard equilibrium constant $K^\ominus(1274 \text{ K})$ of the reaction: $I_2(g) = 2I(g)$.

Similar measurements led to $K^\ominus(1073 \text{ K}) = 1.09 \times 10^{-2}$. Estimate the standard molar enthalpy change $\Delta H_m^\ominus(1173 \text{ K})$.

Problem 12.11

For the reaction: $I_2(g) = 2I(g)$ the standard equilibrium constant $K^\ominus(1173 \text{ K}) = 4.80 \times 10^{-2}$ and the standard molar enthalpy change $\Delta H_m^\ominus(1173 \text{ K}) = 171 \text{ kJ mol}^{-1}$. The molar heat capacities at constant pressure may be taken as $9R/2$ for $I_2(g)$ and $5R/2$ for $I(g)$. Assuming that the gaseous mixture is perfect write an equation for the enthalpy of an equilibrium mixture of I_2 and I. Thence derive an equation for the heat capacity at constant pressure of an equilibrium mixture. Calculate the heat capacity at constant pressure of 0.2538 kg of an equilibrium mixture of I_2 and I at $T = 1173$ K and $p = p^\ominus$.

Problem 12.12

Use the quantities given in the first sentence of problem 12.11 to solve the following problem. In an adiabatic throttling experiment iodine gas flows at a rate of 0.05 g s^{-1} from a pressure of 100 kPa on the upstream side of the throttle to 10 kPa on the downstream side. If the temperature of the gas is 1173 K on the upstream side what power must be supplied on the downstream side just to restore the temperature to 1173 K?

Problem 12.13

Given the standard equilibrium constant $K^{\ominus}(673 \text{ K}) = 1.66 \times 10^{-4}$ for the reaction: $N_2(g) + 3H_2(g) = 2NH_3(g)$, calculate the mole-fraction yield of NH_3 from initial mole fractions 0.25 of N_2 and 0.75 of H_2 at each of the pressures $100p^{\ominus}$ and $1000p^{\ominus}$ (a) on the assumption that the gaseous mixtures are perfect, and (b) on the assumption that only second virial coefficients are relevant and that Lewis and Randall's rule is obeyed. At 673 K take $\{2B(NH_3) - B(N_2) - 3B(H_2)\} = -125 \text{ cm}^3 \text{ mol}^{-1}$. Compare your results with the experimental values at 673 K: 0.249 at $100p^{\ominus}$ and 0.800 at $1000p^{\ominus}$.

Chapter 13

The principle of corresponding states for fluids

§13.1 The principle and its range of validity

A set of pure fluid substances is said to conform with the principle of corresponding states when the *compression factor Z* can be written in the form:

$$Z \stackrel{\text{def}}{=} pV_m/RT = \phi(p/p^c, T/T^c), \tag{13.1.1}$$

where p^c and T^c denote the critical pressure and critical temperature, and where ϕ is the *same function* for each of the substances of the set. We shall now show that the principle leads to useful results without any knowledge of the algebraic form of the function ϕ. The ratios p/p^c and T/T^c, and V_m/V_m^c which we shall also need, are called the *reduced pressure*, the *reduced temperature*, and the *reduced volume*.

A plot of the compression factor Z against the reduced pressure p/p^c at various values of the reduced temperature T/T^c is shown in figure 13.1 for seven gaseous substances that conform with the principle of corresponding states. The curves in figure 13.1 form a family ordered with respect to the values of T/T^c. If two gases had been studied at identical values of T/T^c (rather than, as usually, at identical values of T/K) their curves would coincide. Such a diagram can be used to estimate an unmeasured Z at a given reduced pressure and reduced temperature for any substance that conforms reasonably well with the principle.

Deviations from the principle of corresponding states arise in two ways and to varying extents. The lightest molecules H_2 and He (and to a lesser extent Ne) deviate from the principle because of quantal effects at low temperatures; they can be made to conform except at the lowest temperatures by replacing the critical quantities by 'pseudocritical' quantities. The heavy monatomic molecules Ar, Kr, and Xe conform best. For these the pair-interaction energy is central, that is to say depends only on the separation of the centres of mass of the molecules. Departures from centrality cause progressive deviations from the principle; these are very roughly as follows, though the reader is warned that the sequence depends on which particular property is compared. The molecules CH_4 and N_2, O_2, and CO are nearly spherical and nearly central; they deviate at most to a very small extent from the principle. Molecules like C_2H_6 and C_3H_8 deviate rather more because of their non-linearity. Molecules like CF_4, SF_6, $C(CH_3)_4$, and CCl_4 are nearly spherical but not central (they are more nearly 'peripheral') and deviate more. Molecules like CH_3Cl and $CHCl_3$, because of

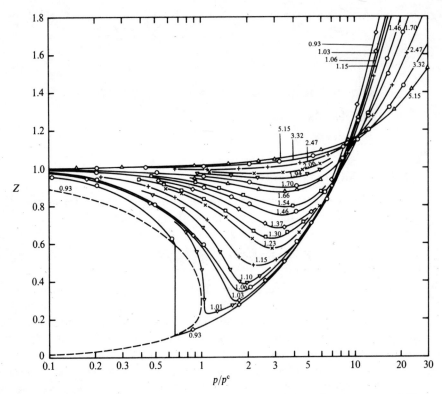

Figure 13.1. Plot of the compression factor $Z = pV_m/RT$ against the reduced pressure p/p^c for seven substances that conform with the principle of corresponding states. The numbers on the diagram are values of the reduced temperature T/T^c. $\bigcirc$, Ar; $\square$, Kr; $\diamondsuit$, Xe; $+$, CH$_4$; $\triangle$, N$_2$; $\triangledown$, O$_2$; $\times$, CO.

their large electric dipole moments, CO$_2$, because of its large electric quadrupole moment, and n-C$_6$H$_{14}$, because of its large departure from spherical symmetry, deviate still more. Very polar molecules like CH$_3$CN, CH$_3$NO$_2$, and CH$_3$COCH$_3$ deviate badly. Hydrogen-bonding molecules like NH$_3$, CH$_3$NH$_2$, and CH$_3$OH deviate very badly. For H$_2$O, and for associating molecules like NO$_2$, the deviations are so large that the principle is no longer a very useful approximation. A set of substances like C(CH$_3$)$_4$, Si(CH$_3$)$_4$, and Sn(CH$_3$)$_4$ conforms roughly with *a* principle of corresponding states but with a different one from Ar, Kr, and Xe.

When the reasons for deviations from the principle of corresponding states are systematic, for example increasing chain length l or increasing electric dipole moment p, it is possible to extend the principle to three independent variables:

$$Z = \phi_l(p/p^c, T/T^c, Ll^3/V^c), \quad \text{or} \quad Z = \phi_p(p/p^c, T/T^c, p^2/\varepsilon_0 kT^c), \quad (13.1.2)$$

where ε_0 is the electric permittivity of a vacuum. Pitzer's equation:

$$Z = \phi_\omega(p/p^c, T/T^c, \omega), \quad (13.1.3)$$

is a generalization of this kind, where ω is the 'acentric factor' defined by

$$\omega = \log_{10}\{p^c/p^{1+g}(T/T^c = 0.7)\} - 1,\qquad(13.1.4)$$

which was chosen so as to make $\omega = 0$ for Ar, Kr, and Xe.

In the rest of this chapter we shall deal mainly with simple molecules that follow equation (1) more or less closely.

§ 13.2   Some consequences of the principle for single-phase systems

By putting $p = p^c$ and $T = T^c$ in equation (13.1.1) we derive

$$Z^c = p^c V_m^c/RT^c = \phi(1, 1).\qquad(13.2.1)$$

In words, the critical compression factor Z^c has the same value for all substances in the set. Values of the critical quantities and the derived values of Z^c are shown in table 13.1 for the seven substances for which figure 13.1 was drawn. Each value of Z^c is equal to 0.290 within experimental error. (Values of V_m^c are difficult to determine accurately.)

Table 13.1. Critical quantities p^c, V_m^c, and T^c, and critical compression factors Z^c of seven substances that conform closely with the principle of corresponding states.

Substance	p^c/MPa	$V_m^c/\text{cm}^3\,\text{mol}^{-1}$	T^c/K	Z^c
Ar	4.87	75.0	150.8	0.291
Kr	5.50	91.2	209.4	0.288
Xe	5.840	118.3	289.73	0.287
CH$_4$	4.604	99.0	190.58	0.288
N$_2$	3.39	89.5	126.2	0.289
O$_2$	5.043	73.4	154.58	0.288
CO	3.499	93.1	132.91	0.295

By combining equations (1) and (13.1.1) we obtain

$$V_m/V_m^c = \phi_V(p/p^c, T/T^c),\qquad(13.2.2)$$

and a similar equation for p/p^c or T/T^c as the dependent variable. The function ϕ_V of equation (2), like all the ϕ's throughout this chapter, denotes some universal function. We could of course draw diagrams for those equations, corresponding to figure 13.1. Such diagrams are widely used in the form of large-scale wall charts (with the experimental points omitted) by industrial chemists and chemical engineers.†

Using the virial equation of state (§ 12.2):

$$Z = pV_m/RT = 1 + B/V_m + C/V_m^2 + \cdots,\qquad(13.2.3)$$

† Hougen, O. A.; Watson, K. M. *Chemical Process Principles*, Part two, *Thermodynamics*. Wiley: New York. **1947**, p. 488 *et seq.*

we see that the second virial coefficient B is defined by

$$B = \lim_{V_m \to \infty} \{(Z-1)V_m\}, \tag{13.2.4}$$

the third virial coefficient C by

$$C = \lim_{V_m \to \infty} \{(Z-1)V_m^2 - BV_m\}, \tag{13.2.5}$$

and so on, where the virial coefficients of a given substance depend only on the temperature. Substituting from (13.1.1) and (2) into (4) we obtain

$$B/V_m^c = \lim_{(p/p^c) \to 0} [\{\phi(p/p^c, T/T^c) - 1\}\phi_V(p/p^c, T/T^c)]$$

$$= \phi_B(T/T^c), \tag{13.2.6}$$

and then by successive substitution

$$C/(V_m^c)^2 = \phi_C(T/T^c), \tag{13.2.7}$$

and so on.

The reduced second virial coefficient, in the form Bp^c/RT^c (because p^c and T^c are usually known more accurately than V_m^c), is plotted against the reduced temperature

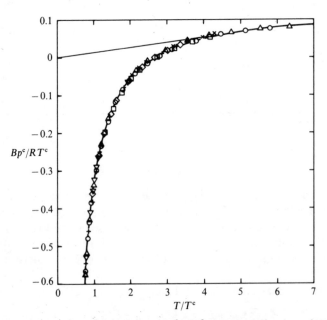

Fgure 13.2. The reduced second virial coefficient Bp^c/RT^c plotted against the reduced temperature T/T^c for seven substances that conform with the principle of corresponding states. The tangent from the origin meets the curve at the reduced inversion temperature $T^i(p \to 0)/T^c$ at zero pressure. $\bigcirc$, Ar; $\square$, Kr; $\diamond$, Xe; $+$, CH$_4$; $\triangle$, N$_2$; $\triangledown$, O$_2$; $\times$, CO.

T/T^c for the same set of substances in figure 13.2. Apart from the excellent agreement with the principle of corresponding states, we note when $T = T^c$ that $Bp^c/RT^c \approx -0.35$, when $B = 0$ at the Boyle temperature T^B that $T^B/T^c \approx 2.73$, and when $B/T = dB/dT$ at the low-pressure inversion temperature $T^i(p \to 0)$ that $T^i(p \to 0)/T^c \approx 5.28$. The curve drawn in figure 13.2 is that calculated from the square-well formula:

$$Bp^c/RT^c = 0.597 - 0.462 \exp(0.7002 T^c/T). \tag{13.2.8}$$

The reduced third virial coefficient $C(p^c/RT^c)^2$ is plotted against the reduced temperature T/T^c for the same set of substances in figure 13.3. The much greater scatter in figure 13.3 than in figure 13.2 is due to the much greater experimental

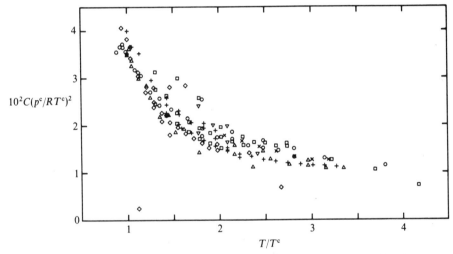

Figure 13.3. The reduced third virial coefficient $C(p^c/RT^c)^2$ plotted against the reduced temperature T/T^c for seven substances that conform with the principle of corresponding states. $\bigcirc$, Ar; $\square$, Kr; $\diamond$, Xe; $+$, CH_4; $\triangle$, N_2; $\triangledown$, O_2; $\times$, CO.

uncertainties of the results. The experimental uncertainties of the fourth and higher virial coefficients are too great to justify drawing diagrams.

§ 13.3   Some consequences of the principle for two-phase systems

For the equilibrium of (liquid + gas) of a pure substance at the temperature T and pressure $p^{l+g}(T)$ we have

$$\ln\{\lambda^l(T, p^{l+g})\} = \ln\{\lambda^g(T, p^{l+g})\}, \tag{13.3.1}$$

so that

$$\ln\{\lambda^l(T, 0)\} + \int_0^{p^{l+g}} (V_m^l/RT)\, dp = \ln\{\lambda^g(T, 0)\} + \int_0^{p^{l+g}} (V_m^g/RT)\, dp. \tag{13.3.2}$$

Since the absolute activities must be equal at $p \to 0$, equation (2) becomes

$$\int_0^{p^{l+g}} \{(V_m^g - V_m^l)/RT\}\, dp = 0 . \tag{13.3.3}$$

Substitution into (3) from (13.2.2) gives

$$\int_0^{p^{l+g}/p^c} \phi_1(p/p^c, T/T^c)\, d(p/p^c) = 0 , \tag{13.3.4}$$

which on carrying out the integration leads to

$$\phi_2(p^{l+g}/p^c, T/T^c) = 0 , \tag{13.3.5}$$

or

$$p^{l+g}/p^c = \phi_{lg}(T/T^c) . \tag{13.3.6}$$

A plot of the reduced vapour pressure on a logarithmic scale against the reciprocal of the reduced temperature is shown in figure 13.4 for the same set of substances over

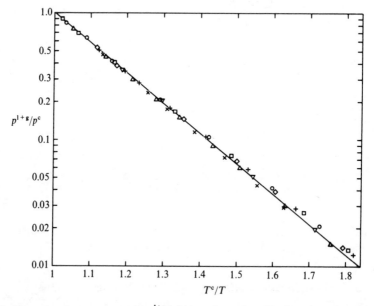

Figure 13.4. The reduced vapour pressure p^{l+g}/p^c plotted on a logarithmic scale against the reciprocal T^c/T of the reduced temperature for seven substances that conform with the principle of corresponding states. The straight line is discussed in the text. $\bigcirc$, Ar; $\square$, Kr; $\diamond$, Xe; $+$, CH_4; $\triangle$, N_2; $\triangledown$, O_2; $\times$, CO.

the whole range of existence of (liquid+gas) from the triple-point temperature $(T^{s+l+g}/T^c \approx 0.55)$ to the critical temperature. On this scale the plot is apparently nearly linear; the straight line shown was drawn according to the formula:

$$\ln(p^{l+g}/p^c) = 5.522 - 5.512(T^c/T) . \tag{13.3.7}$$

Equation (7) does not quite satisfy the relation $p^{1+g} = p^c$ when $T = T^c$, but apart from that there must be some nearly compensating effects. The curve remains close to a straight line in spite of the facts (see § 9.4) that $\Delta_l^g H_m^* \to 0$ as $T \to T^c$ and that, far from being negligible compared with V_m^g, $V_m^l \to V_m^g$ as $T \to T^c$.

By substitution of (6) into (13.2.2) we obtain the relations:

$$V_m^g / V_m^c = \phi_g(T/T^c), \qquad (13.3.8)$$

$$V_m^l / V_m^c = \phi_l(T/T^c). \qquad (13.3.9)$$

A plot of the reciprocals of the reduced orthobaric volumes, that is to say of the reduced orthobaric densities $\rho^g/\rho^c = V_m^c/V_m^g$ and $\rho^l/\rho^c = V_m^c/V_m^l$, against the reduced temperature T/T^c is shown for the same set of substances in figure 13.5. The

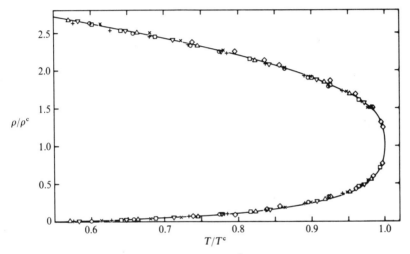

Figure 13.5. The reduced orthobaric densities ρ^g/ρ^c and ρ^l/ρ^c plotted against the reduced temperature T/T^c for seven substances that conform with the principle of corresponding states. The curve is discussed in the text. $\bigcirc$, Ar; $\square$, Kr; $\diamond$, Xe; $+$, CH$_4$; $\triangle$, N$_2$; ∇, O$_2$; $\times$, CO.

curve is that calculated according to the equations:

$$\tfrac{1}{2}(\rho^l + \rho^g)/\rho^c = 1 + 0.797(1 - T/T^c), \qquad (13.3.10)$$

$$(\rho^l - \rho^g)/\rho^c = 3.524(1 - T/T^c)^{0.326}. \qquad (13.3.11)$$

Equation (10) expresses Cailletet and Mathias's law of the rectilinear diameter. Equation (11) yields the value 0.326 for the critical exponent β discussed in § 9.8.

From Clapeyron's equation (9.3.3) we deduce

$$\Delta_l^g H_m^* / RT = \phi_H(T/T^c). \qquad (13.3.12)$$

There is little doubt that if calorimetrically determined values of $\Delta_l^g H_m^*$ were available over a range of temperatures then they would conform to equation (12).

§ 13.4 Some approximate consequences of the principle of corresponding states

All the results presented so far in this chapter follow exactly from the assumption of the principle of corresponding states in the form of equation (13.1.1). We shall now derive two less exact but useful equations from the additional very rough approximation that the critical pressure p^c is the same for all substances.

At the normal boiling temperature T^b, p^{l+g} is constant (101.325 kPa), and so equation (13.3.6) becomes

$$T^b/T^c \approx \text{`constant'} . \qquad (13.4.1)$$

The 'constant' is roughly equal to 0.6.

From equation (13.3.12) at the normal boiling temperature T^b we obtain

$$\Delta_l^g H_m^*/RT^b \approx \text{`Trouton's constant'} . \qquad (13.4.2)$$

'Trouton's constant' is roughly equal to 10 and equation (2) is known as Trouton's rule.

In table 13.2 values of p^c, T^c, Z^c, T^b, T^b/T^c, and $\Delta_l^g H_m^*(T^b)/RT^b$ are given for a range of substances with roughly increasing deviations from the principle of corresponding states. The critical compression factor Z^c is constant for molecules conforming with the principle. The values of T^b/T^c and of $\Delta_l^g H_m^*(T^b)/RT^b$ are constant for substances conforming with the principle and for which the critical pressures are equal.

Table 13.2. Values of p^c, T^c, Z^c, T^b, and $\Delta_l^g H_m^*(T^b)/RT^b$ for a range of substances with roughly increasing deviations from the principle of corresponding states.

Substance	p^c/MPa	T^c/K	Z^c	T^b/K	T^b/T^c	$\Delta_l^g H_m^*(T^b)/RT^b$
Kr	5.50	209.4	0.288	119.9	0.57	9.1
N_2	3.39	126.2	0.290	77.4	0.61	8.8
C_2H_6	4.88	305.4	0.285	184.6	0.60	9.6
CF_4	3.74	227.6	0.276	145.1	0.64	10.4
CCl_4	4.56	556.4	0.272	349.9	0.63	10.3
CH_3Cl	6.68	416.2	0.268	249.1	0.60	9.6
$CHCl_3$	5.47	536.4	0.293	334.3	0.62	10.6
CO_2	7.38	304.1	0.274	194.6	0.64	8.9
$n\text{-}C_6H_{14}$	3.01	507.5	0.264	341.9	0.67	9.9
CH_3COCH_3	4.70	508.1	0.232	329.2	0.65	11.1
NH_3	11.35	405.5	0.244	239.8	0.59	11.7
CH_3OH	8.09	512.6	0.224	337.7	0.66	12.6
H_2O	22.05	647.1	0.231	373.2	0.58	13.1

§ 13.5 The second virial coefficient of a gaseous mixture

We learnt in § 12.6 that much effort has been devoted to the expression of the second virial coefficient B_{AB} in terms of B_{AA} and B_{BB}, and that the rule of Lewis and Randall: $B_{AB} = \frac{1}{2}(B_{AA} + B_{BB})$, though simple and widely used, is seldom accurately obeyed. For substances that conform with the principle of corresponding states we may write by analogy with (13.2.4)

$$B_{AB}/V_{AB}^c = \phi_B(T/T_{AB}^c),$$ (13.5.1)

where V_{AB}^c and T_{AB}^c are the critical molar volume and the critical temperature of the hypothetical pure substance AB. By use of the combining rules (12.6.33) and (12.6.34) we have

$$V_{AB}^c = \{(V_A^c)^{1/3} + (V_B^c)^{1/3}\}^3/8,$$ (13.5.2)

$$T_{AB}^c = \zeta(T_A^c T_B^c)^{1/2}.$$ (13.5.3)

Table 13.3 shows some experimental values of B_{AB} together with values calculated from (1) with (2) and (3) for $\zeta = 1$, and for comparison values calculated from Lewis and Randall's rule. In practice experimental values of B_{AB} are nowadays more often used to determine values of ζ for use in connexion with theories of fluid mixtures (see chapter 17).

Table 13.3. Comparison of experimental values of B_{AB} with those calculated according to the principle of corresponding states from equation (13.5.1) with (13.5.2) and (13.5.3) with $\zeta = 1$, and with those calculated according to Lewis and Randall's rule: $B_{AB} = \frac{1}{2}(B_{AA} + B_{BB})$.

A	B	T/K	B_{AB}/cm^3 mol^{-1}		
			Experimental	Corresponding states with $\zeta = 1$	Lewis and Randall
Ar	Kr	116.5	−208	−213	−232
Ar	Kr	179.0	−98	−94	−101
Ar	CH$_4$	107.1	−233	−239	−255
Ar	CH$_4$	174.2	−94	−94	−100
Kr	CH$_4$	118.6	−297	−297	−297
Kr	CH$_4$	200.4	−112	−109	−109

Problem for chapter 13

Problem 13.1

For radon the critical temperature is 377 K and the critical pressure is 6.28 MPa. Use the principle of corresponding states to estimate the critical molar volume V_m^c, the compression factor Z at 471.2 K and 12.56 MPa, the second virial coefficient B at 565.5 K, the normal boiling temperature T^b, the vapour pressure p^{l+g} at 300 K, and the orthobaric molar volumes $V_m^{l,l+g}$ and $V_m^{g,l+g}$ of liquid and gas at 300 K.

Chapter 14

Digression on Boltzmann's distribution law

§ 14.1 Introduction to Boltzmann's distribution law

From the axioms of statistical mechanics we could (though we shall not do so in this book) derive not only all the formulae of thermodynamics but also relations between certain averages over the quantum states (eigenstates) to which a system has access and the observable thermodynamic properties of the system.

In the special case that a system may be regarded as being composed of *independent* molecules, statistical mechanics provides a bridge between the (microscopic) properties of one molecule and the (macroscopic) thermodynamic properties of a system of N molecules, where N is a number greater than about 10^{16}. Any thermodynamic function is then N times the appropriate average of the corresponding mechanical property of one molecule, the average being taken over all the quantum states to which the molecule has access. It is much easier to solve Schrödinger's equation for a molecule than it would be for a macroscopic system. It is much easier to measure the energy levels (energy eigenvalues) of a molecule than it would be of a macroscopic system. The differences between pairs of energy levels are measured spectroscopically; spectrometers are the experimental apparatus of the quantum theory of molecules.

Given, according to quantum theory, that the energy levels of a molecule are not continuous but discrete, it makes sense for a system of independent molecules to divide the N molecules of the system into groups, one for each quantum state. If N_i is the number of molecules in the i'th quantum state, in which the molecule has the energy ε_i, then we may write

$$N = \sum_i N_i, \tag{14.1.1}$$

$$U = \sum_i N_i \varepsilon_i, \tag{14.1.2}$$

$$p = \sum_i N_i(-d\varepsilon_i/dV). \tag{14.1.3}$$

There are now two questions that must be answered before we can calculate the energy U, the pressure p, and all the other thermodynamic functions of the system: first, 'What energies are allowed to a molecule in the system?', and second, 'How are the molecules distributed among those allowed energies?'.

The answer to the first question is obtained by solving Schrödinger's equation for the motions of and within the molecule taking due note of the boundaries implied by the volume of the system. Even for a diatomic molecule that is a formidable problem, but it becomes tractable if we may separate the modes of motion and then deal with them one at a time. The solutions of Schrödinger's equation for the allowed wave functions (eigenfunctions) are still quite complicated but for simple modes of motion the allowed energy levels have strikingly simple forms. Here we shall need only the allowed energy levels and shall simply quote the formulae we shall need for them.

The answer to the second question strictly depends on whether the molecule contains an even number or an odd number of (protons + neutrons + electrons): an even number leads to the Bose–Einstein statistics and an odd number to the Fermi–Dirac statistics. Either of these simplifies to the *Boltzmann statistics* or *Boltzmann's distribution law* for molecules of high enough mass in systems at high enough temperatures and low enough pressures. Those conditions are amply satisfied by all the molecules of interest to chemists.[†]

Boltzmann's distribution law, which we take as an axiom, is

$$N_i = \lambda \, \exp(-\varepsilon_i/kT) , \qquad (14.1.4)$$

where λ is the absolute activity of the substance and T is its temperature, and where the Boltzmann constant $k = R/L$. The quantity $\exp(-\varepsilon_i/kT)$ is called a Boltzmann factor.

Summing equation (4) over the quantum states we obtain

$$N = \lambda \sum_i \exp(-\varepsilon_i/kT) . \qquad (14.1.5)$$

The sum over states $\sum_i \exp(-\varepsilon_i/kT)$ occurs so frequently and is so important that we give it a name and a symbol; it is called the *molecular partition function*[‡] and is denoted by q. Thus

$$N = \lambda q . \qquad (14.1.6)$$

Dividing (4) by (5) we obtain for the fraction N_i/N of molecules in the i'th quantum state

$$N_i/N = \exp(-\varepsilon_i/kT) \Big/ \sum_i \exp(-\varepsilon_i/kT) = \exp(-\varepsilon_i/kT)/q . \qquad (14.1.7)$$

The fraction N_i/N becomes smaller as ε_i becomes larger with respect to kT.

---

† The Fermi–Dirac statistics are needed for the conducting electrons in a metal, and for ^{3}He at temperatures near 1 K. The Bose–Einstein statistics are needed for photons in the theory of radiation, and for ^{4}He at temperatures near 1 K.

‡ Other kinds of partition function occur in the statistical mechanics of systems. Here we shall use only the molecular partition function, and so may omit the word 'molecular'.

Substituting equation (7) into equations (2) and (3) we obtain

$$U = N \sum_i \varepsilon_i \exp(-\varepsilon_i/kT) \Big/ \sum_i \exp(-\varepsilon_i/kT)$$
$$= NkT^2(\partial \ln q/\partial T)_V, \tag{14.1.8}$$

and

$$p = N \sum_i (-d\varepsilon_i/dV) \exp(-\varepsilon_i/kT) \Big/ \sum_i \exp(-\varepsilon_i/kT)$$
$$= NkT(\partial \ln q/\partial V)_T. \tag{14.1.9}$$

From equation (6) we have for the absolute activity

$$\lambda = N/q, \tag{14.1.10}$$

or for the chemical potential

$$\mu = RT \ln N - RT \ln q. \tag{14.1.11}$$

The only problem remaining is the evaluation of the partition function q.

§ 14.2   Equilibrium constant for an isomerization

Before we attend to the evaluation of q, which will inevitably lead us into approximations, we shall in this section apply Boltzmann's distribution law exactly to the equilibrium constant of an isomerization $A = B$. It is left as an exercise for the reader to apply it to the equilibrium constants of more complex chemical reactions.

For the equilibrium $A = B$ we have $\lambda_A = \lambda_B$ so that $N_A/q_A = N_B/q_B$, or

$$K^\ominus = N_B/N_A$$

$$= q_B/q_A$$

$$= \sum_i \exp(-\varepsilon_{i,B}/kT) \Big/ \sum_i \exp(-\varepsilon_{i,A}/kT)$$

$$= \frac{\exp(-\varepsilon_{0,B}/kT) \sum_i \exp\{-(\varepsilon_{i,B}-\varepsilon_{0,B})/kT\}}{\exp(-\varepsilon_{0,A}/kT) \sum_i \exp\{(\varepsilon_{i,A}-\varepsilon_{0,A})/kT\}}$$

$$= \frac{[1+\exp\{-(\varepsilon_{1,B}-\varepsilon_{0,B})/kT\}+\exp\{-(\varepsilon_{2,B}-\varepsilon_{0,B})/kT\}+\cdots]}{[1+\exp\{-(\varepsilon_{1,A}-\varepsilon_{0,A})/kT\}+\exp\{-(\varepsilon_{2,A}-\varepsilon_{0,A})/kT\}+\cdots]} \exp(-\Delta\varepsilon_0/kT). \tag{14.2.1}$$

Two factors determine the magnitude of $K^\ominus$ at given T: the difference $\Delta\varepsilon_0 = (\varepsilon_{0,B}-\varepsilon_{0,A})$ between the lowest energies of the isomeric molecules A and B, and the relative spacings of higher levels for the two molecules. If, as is shown in figure 14.1(a), $\varepsilon_{0,B} < \varepsilon_{0,A}$ so that $\Delta\varepsilon_0 < 0$, and the energy levels are closer to one another for

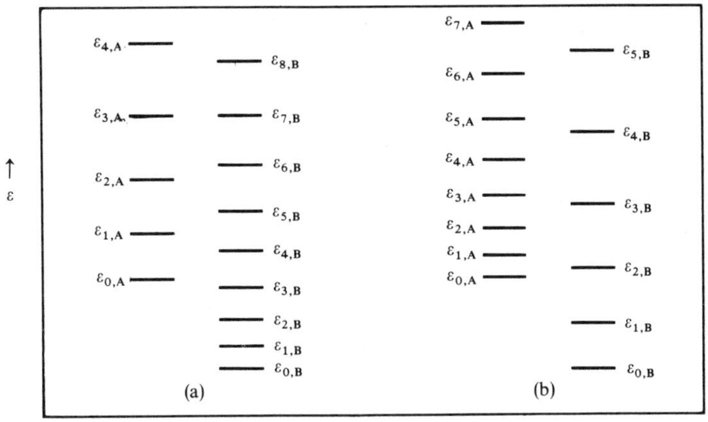

Figure 14.1. Hypothetical spacing of energy levels for an isomerization A = B. (a), $\varepsilon_{0,B} < \varepsilon_{0,A}$ and the differences of energy levels are smaller for B than for A. (b), $\varepsilon_{0,B} < \varepsilon_{0,A}$ but the differences of energy levels are greater for B than for A.

B than they are for A, then both factors favour B over A and $K^\ominus \gg 1$. If, as is shown in figure 14.1(b), $\varepsilon_{0,B} < \varepsilon_{0,A}$ so that $\Delta\varepsilon_0 < 0$, but the energy levels are further from one another for B than they are for A, then the two factors are opposed and $K^\ominus$ may be greater or less than 1.

From the relation $\mathrm{d} \ln K^\ominus/\mathrm{d}T = \Delta H_m^\ominus/RT^2$ (see § 11.3) we deduce

$$\Delta H_m^\ominus = L \sum_i (\varepsilon_{i,B} - \varepsilon_{0,B}) \exp\{-(\varepsilon_{i,B} - \varepsilon_{0,B})/kT\} \Big/ \sum_i \exp\{-(\varepsilon_{i,B} - \varepsilon_{0,B})/kT\}$$

$$-L \sum_i (\varepsilon_{i,A} - \varepsilon_{0,A}) \exp\{-(\varepsilon_{i,A} - \varepsilon_{0,A})/kT\} \Big/ \sum_i \exp\{-(\varepsilon_{i,A} - \varepsilon_{0,A})/kT\}$$

$$+ L\Delta\varepsilon_0 . \tag{14.2.2}$$

At the lowest temperatures $\Delta H_m^\ominus \to L\Delta\varepsilon_0$. In practice, unless $\Delta\varepsilon_0$ is small compared with the spacings of the energy levels, $\Delta H_m^\ominus \approx L\Delta\varepsilon_0$. In the example of figure 14.1(a) $\Delta H_m^\ominus$ is negative at the lowest temperatures and becomes less negative, and perhaps eventually positive, at higher temperatures. In the example of figure 14.1(b) $\Delta H_m^\ominus$ is negative at the lowest temperatures and becomes more negative at higher temperatures.

In principle it is always possible spectroscopically to measure all the differences of energy levels for each substance, and also to measure $\Delta\varepsilon_0$ for the chemical reaction. In principle it is therefore always possible to calculate $K^\ominus$ and $\Delta H_m^\ominus$ at any temperature from the results of spectroscopic measurements. In practice we usually simplify the task by using quantum theory with suitable approximations for the differences of energy levels of each molecule. In practice it is experimentally difficult to measure $\Delta\varepsilon_0$ spectroscopically except for a few chemical reactions like $I_2 = 2I$; for other chemical reactions we use a value of $\Delta H_m^{\ominus}(T^{\dagger})$, measured calorimetrically at a temperature $T^{\dagger}$ (usually 298.15 K), to eliminate $\Delta \varepsilon_0$ between equations (1) and (2). That is one of the reasons why access to calorimetrically derived tables of $\Delta H_m^{\ominus}(T^{\dagger})$ is so important.

§ 14.3 Degeneracy

When we write a partition function in the form of a *sum over states*:

$$q = \sum_i \exp(-\varepsilon_i/kT)$$

$$= \exp(-\varepsilon_0/kT) + \exp(-\varepsilon_1/kT) + \exp(-\varepsilon_2/kT) + \cdots, \qquad (14.3.1)$$

the subscripts of the ε's refer to independent quantum states (eigenstates). Sometimes it is more convenient to rewrite equation (1) as a *sum over discrete energy levels*:

$$q = \sum_{\varepsilon} g(\varepsilon) \exp(-\varepsilon/kT). \qquad (14.3.2)$$

The factor $g(\varepsilon)$ is called the *degeneracy* of the energy level (energy eigenvalue) ε.

As an example, in equation (1) it might be so that $\varepsilon_0 = \varepsilon_1$ and $\varepsilon_2 = \varepsilon_3 = \varepsilon_4$, that is to say that two independent quantum states each have the same energy ε_0, and that three independent quantum states each have the same energy ε_2. The lowest energy level is 'doubly degenerate' and $g(\varepsilon_0) = 2$; the next-lowest energy level is 'triply degenerate' and $g(\varepsilon_2) = 3$. It will then be convenient to write

$$q = 2 \exp(-\varepsilon_0/kT) + 3 \exp(-\varepsilon_2/kT) + \cdots. \qquad (14.3.3)$$

§ 14.4 Separability of modes of motion

The evaluation of a molecular partition function q from quantum theory depends in practice on the approximation that the modes of motion† of the molecule are *separable*, and so may be treated one at a time. If the modes of motion are separable we may write for the energy ε of a molecule in a particular eigenstate

$$\varepsilon = \varepsilon_T + \varepsilon_R + \varepsilon_V + \varepsilon_E + \varepsilon_N + \cdots + \varepsilon_0, \qquad (14.4.1)$$

where each of the ε's on the right-hand side is an energy eigenvalue: ε_T for translations of the centre of mass of the molecule, ε_R for rotations of the molecule about its centre of mass, ε_V for vibrations within the molecule, ε_E for electronic excitations, ε_N for nuclear excitations, $\cdots$ for any unknown modes of excitation, and ε_0 the lowest possible energy.

When the separability implied by equation (1) is allowed, that is to say when the modes of motion of a molecule may be regarded as independent of one another, the

---

† Or 'degrees of freedom' (compare the quite different use of this phrase in connexion with the phase rule in § 8.9).

molecular partition function q may be written in the form:

$$q = (q_T q_R q_V q_E q_N \cdots) g(\varepsilon_0) \exp(-\varepsilon_0/kT), \qquad (14.4.2)$$

where we have included the degeneracy $g(\varepsilon_0)$ of the lowest energy level ε_0 in case more than one quantum state can have that energy. The assumption of separability has allowed us to replace the problem of evaluating q by the several much more tractable problems of evaluating the factors on the right-hand side of equation (2) one at a time.

For any molecule, ε_T can be further separated into 3 independent contributions ε_x, ε_y, and ε_z for translation in each of three directions at right angles. For a molecule consisting of n atoms there are: if the molecule is linear then 2 rotational modes and $(3n-5)$ vibrational modes, and if the molecule is non-linear then 3 rotational modes and $(3n-6)$ vibrational modes. There are in all 3 modes for each atomic nucleus, corresponding to a complete specification of its position. Of these modes of nuclear motion, ε_T is unambiguously separable from ε_R and ε_V. If we assume that the vibrations are *simple harmonic* then ε_R and ε_V are also separable. Although we shall here assume with (just) sufficient accuracy that all vibrations are simple harmonic, it is easy to see that they cannot be so (if they were then no molecule could dissociate), and since they are not so, that ε_R and ε_V are not strictly separable. A rotational energy level depends on the moment of inertia of the molecule and that depends on the average nuclear separation which in turn depends on the level of vibrational excitation. The point is illustrated in figure 14.2 where the vibrational energy levels are shown for a realistic potential-energy function and for a simple harmonic oscillator. For the realistic function the average nuclear separation changes with the quantum number v of the allowed vibrational energy level. For a simple harmonic oscillator the average nuclear separation is independent of v.

Electronic modes are to a good approximation separable from nuclear modes but turn out to be relatively unimportant in chemistry. At laboratory temperatures only the lowest and occasionally the first excited energy levels are needed. For the electronic partition function q_E we then write

$$q_E = g(\varepsilon_{0,E}) \exp(-\varepsilon_{0,E}/kT) + g(\varepsilon_{1,E}) \exp(-\varepsilon_{1,E}/kT)$$

$$= g(\varepsilon_{0,E}) \exp(-\varepsilon_{0,E}/kT)[1 + \{g(\varepsilon_{1,E})/g(\varepsilon_{0,E})\} \exp\{-(\varepsilon_{1,E}-\varepsilon_{0,E})/kT\}]. \quad (14.4.3)$$

The factors $g(\varepsilon_{0,E})$ and $\exp(-\varepsilon_{0,E}/kT)$ in equation (3) may be included in the $g(\varepsilon_0)$ and $\exp(-\varepsilon_0/kT)$ of equation (2).

It is often true that $(\varepsilon_{1,E} - \varepsilon_{0,E}) \gg kT$ so that equation (3) reduces to

$$q_E = g(\varepsilon_{0,E}) \exp(-\varepsilon_{0,E}/kT). \qquad (14.4.4)$$

The electronic mode is then described as *unexcited*.

At terrestrial temperatures the nuclei are completely unexcited internally. We shall see in §14.11, however, that the two kinds of symmetry of the wave functions corresponding to the degenerate lowest energy level lead to interesting results for the lightest gases at low temperatures.

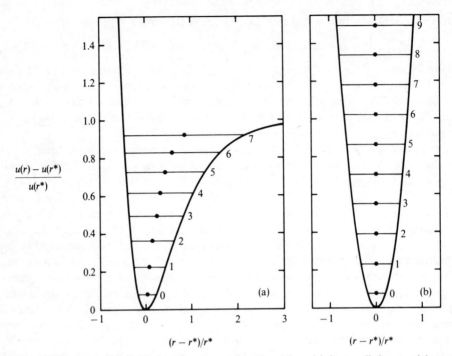

$$\frac{u(r) - u(r^*)}{u(r^*)}$$

$(r - r^*)/r^*$ $(r - r^*)/r^*$

Figure 14.2. Potential energy $u(r)$ plotted against nuclear separation r (a), for a realistic potential-energy function (actually a Morse function for the molecule H_2) and (b), for simple harmonic motion. The allowed energy levels are shown for vibrational quantum numbers v, and for each level the average nuclear separation is marked.

We may now rewrite equation (2) in the form:

$$q = q_x q_y q_z q_R q_V [1 + \{g(\varepsilon_{1,E})/g(\varepsilon_{0,E})\} \exp\{-(\varepsilon_{1,E} - \varepsilon_{0,E})/kT\}]$$
$$\times g(\varepsilon_0) \exp(-\varepsilon_0/kT), \qquad (14.4.5)$$

where the factor $[\cdots]$ may often be omitted. The product $q_R q_V$ may be further factorized into $q_{R,\text{linear}} \Pi_{i=1}^{3n-5} q_{V,i}$ for a linear molecule or into $q_{R,\text{non-linear}} \Pi_{i=1}^{3n-6} q_{V,i}$ for a non-linear molecule, each factor $q_{V,i}$ being a separately evaluated partition function for the i'th vibrational mode of the molecule. Because of the way we evaluate the rotational energy levels (using spherical polar coordinates), we do not write q_R as a product of factors, one for each rotational mode.

The following three sections are devoted to the evaluation of q_x (and hence of q_y and q_z), of q_R, and of $q_{V,i}$.

§ 14.5 Translational mode

The solutions of Schrödinger's equation for a particle moving in the x-direction in a box of length l_x are

$$|p_x| = n_x h/2l_x, \quad (n_x = 1, 2, 3, \cdots), \qquad (14.5.1)$$

where p_x denotes the momentum of the particle, h the Planck constant, and n_x a quantum number. The corresponding formula for the energy ε_x is

$$\varepsilon_x = p_x^2/2m = n_x^2 h^2/8l_x^2 m\,, \tag{14.5.2}$$

where m denotes the mass of the particle. The partition function q_x is then given by

$$q_x = \sum_{n_x=1}^{\infty} \exp(-n_x^2 h^2/8l_x^2 mkT)\,. \tag{14.5.3}$$

The sum in equation (3) cannot be evaluated in closed form, but the energy levels are so extremely closely spaced that we may replace the sum over discrete quantum states by an integral over all states. We then obtain

$$q_x = \int_0^{\infty} \exp(-n_x^2 h^2/8l_x^2 mkT)\,dn_x\,, \tag{14.5.4}$$

which on integration (see appendix III) gives

$$q_x = (2\pi mkT/h^2)^{1/2} l_x\,. \tag{14.5.5}$$

The translational partition function $q_T = q_x q_y q_z$ is then given by

$$q_T = (2\pi mkT/h^2)^{3/2} V\,, \tag{14.5.6}$$

where we have written the volume V of the gas in place of the volume $l_x l_y l_z$ of the rectangular box.

Any mode for which we may replace the sum over discrete states by the integral over all states is called a *classical* mode.

We now use the Euler–Maclaurin theorem (appendix III) to verify that translation may be regarded as a classical mode. We write

$$q_x = \sum_{n_x=0}^{\infty} \exp(-n_x^2 h^2/8l_x^2 mkT) - 1$$

$$= \int_0^{\infty} \exp(-n_x^2 h^2/8l_x^2 mkT)\,dn_x + \tfrac{1}{2} - 0 - 1$$

$$= (2\pi mkT/h^2)^{1/2} l_x - \tfrac{1}{2}\,, \tag{14.5.7}$$

and consider whether we may neglect $-\tfrac{1}{2}$ in comparison with $(2\pi mkT/h^2)^{1/2} l_x$. We choose the light molecule He ($m = 0.004$ kg mol$^{-1}/6.0 \times 10^{23}$ mol^{-1}) at the low temperature 10 K in a box of length 1 mm, and find

$$(2\pi mkT/h^2)^{1/2} l_x = 4 \times 10^6\,. \tag{14.5.8}$$

Even in this extreme case the fractional error is less than 10^{-6}. For N_2 at room temperature in a box of length 1 cm the fractional error is about 10^{-9}. We conclude that we may use equation (6) for q_T with an accuracy that is far beyond the reach of even the most precise experiments.

§ 14.6 Rotational mode

We begin by discussing linear molecules. The solutions of Schrödinger's equation for the rotation of a linear molecule have the form:

$$\varepsilon_R = j(j+1)h^2/8\pi^2 I, \quad (j=0, 1, 2, \cdots), \tag{14.6.1}$$

where I denotes the moment of inertia of the molecule and j is a quantum number, and where the j'th energy level is $(2j+1)$-fold degenerate:

$$g(\varepsilon_R) = 2j+1. \tag{14.6.2}$$

(Incidentally the method used to obtain (1) and (2), and the results, are of exactly the same form as the angular part of the solution of Schrödinger's equation for a hydrogen atom.) The partition function q_R is then given by

$$q_R = \sum_{j=0}^{\infty} (2j+1) \exp\{-j(j+1)h^2/8\pi^2 IkT\}. \tag{14.6.3}$$

The sum in (3) cannot be evaluated in closed form, but we may usually treat rotation as another classical mode and obtain

$$q_R = \int_0^{\infty} (2j+1) \exp\{-j(j+1)h^2/8\pi^2 IkT\}\, dj$$

$$= 8\pi^2 IkT/h^2. \tag{14.6.4}$$

Equation (4) is correct for a molecule like HF or OCS having no centre of symmetry. For a molecule like O_2 or OCO having a centre of symmetry it is twice as large as it should be because we have counted each distinguishable orientation twice; OCS and SCO are distinguishable orientations but OCO and OCO are not. We therefore introduce a *symmetry number s* and obtain

$$q_R = 8\pi^2 IkT/sh^2, \tag{14.6.5}$$

where $s = 1$ for a molecule having no centre of symmetry and $s = 2$ for a molecule having a centre of symmetry. This introduction of the symmetry number might appear to be somewhat arbitrary but we shall see in § 14.11 how it arises naturally from a full quantum–mechanical treatment.

For a non-linear molecule the rotational partition function q_R has the form:

$$q_R = \{8\pi^2 (I_A I_B I_C)^{1/3} kT/h^2\}^{3/2} \pi^{1/2}/s, \tag{14.6.6}$$

where I_A, I_B, and I_C are the moments of inertia about the three principal axes of the molecule. The symmetry number s is 1 for molecules like HOD or NOCl having no axis of symmetry, is 2 for HOH, 3 for NH_3, 4 for C_2H_4, 6 for BF_3 (planar), and 12 for CH_4 or C_6H_6.

The treatment of rotation as if it were classical fails for hydrogen at low temperatures, is just accurate enough for other lightish molecules at ordinary temperatures, and is a good approximation for heavy molecules at high temperatures. Applying the

Euler–Maclaurin theorem (appendix III) to the rotational partition function of a linear molecule we find

$$q_R = \sum_{j=0}^{\infty} (2j+1) \exp\{-j(j+1)h^2/8\pi^2 IkT\}$$

$$= \int_0^{\infty} (2j+1) \exp\{-j(j+1)h^2/8\pi^2 IkT\} \, dj + \tfrac{1}{2} - \tfrac{1}{12} + \tfrac{3}{720} - \cdots$$

$$= (8\pi^2 IkT/h^2) + 0.42 , \qquad (14.6.7)$$

and consider whether we may neglect 0.42 in comparison with $(8\pi^2 IkT/h^2)$. For H_2 the *rotational characteristic temperature* $\theta_R \overset{\text{def}}{=} h^2/8\pi^2 Ik$ is 85.3 K, for N_2 θ_R is 2.88 K, and for I_2 θ_R is 0.0537 K. Thus even at 300 K the fractional error for H_2 is 10^{-1}, while for N_2 at 300 K the fractional error is 4×10^{-3} and for I_2 at 300 K it is 8×10^{-5}. We conclude that we must not use equation (5) for H_2 (or HD or D_2), but that with due caution we may use equation (5) (or 6) for heavier molecules.

§ 14.7   Vibrational mode

We assume that a vibrational mode may be treated as a simple harmonic oscillator. The error caused by this assumption depends on the size of the fundamental vibration frequency ν of the oscillator and is always greater the greater the temperature, but is usually not serious in chemical problems. The solutions of Schrödinger's equation for simple harmonic motion are

$$\varepsilon_V = (v + \tfrac{1}{2})h\nu , \quad (v = 0, 1, 2, \cdots) , \qquad (14.7.1)$$

where ν is the fundamental vibration frequency of the oscillator and v is a quantum number. The partition function q_V is then given by

$$q_V = \sum_{v=0}^{\infty} \exp\{-(v + \tfrac{1}{2})h\nu/kT\} , \qquad (14.7.2)$$

which we can sum in closed form (it is a geometrical progression), obtaining

$$q_V = \{1 - \exp(-h\nu/kT)\}^{-1} \exp(-\tfrac{1}{2}h\nu/kT) . \qquad (14.7.3)$$

Molecular vibrations are usually far from classical at laboratory temperatures; indeed they are more often nearly unexcited. We must not replace the sum in equation (2) by an integral, but we are able easily to evaluate the sum.

At temperatures low enough for a vibration to be effectively unexcited equation (3) becomes

$$q_V(\text{unexcited}) = \exp(-\tfrac{1}{2}h\nu/kT) . \qquad (14.7.4)$$

At temperatures high enough for a vibration to be effectively classical equation (3) becomes

$$q_V(\text{classical}) = (kT/h\nu)\exp(-\tfrac{1}{2}h\nu/kT).$$ (14.7.5)

The factor $\exp(-\tfrac{1}{2}h\nu/kT)$ may be and often is included in the factor $\exp(-\varepsilon_0/kT)$ in equation (14.4.5).

§ 14.8 Heat capacity of a gaseous substance at constant volume

We shall now use the complete partition function q to calculate the heat capacity at constant volume, taking as an example a diatomic molecule for which the electronic mode is unexcited. The partition function then takes the form:

$$q = (2\pi mkT/h^2)^{3/2} V (8\pi^2 IkT/sh^2)\{1 - \exp(-h\nu/kT)\}^{-1}$$
$$\times g(\varepsilon_0)\exp(-\varepsilon_0/kT).$$ (14.8.1)

Using equation (14.1.8) we obtain for the energy of N molecules of a gas

$$U = \tfrac{3}{2}NkT + NkT + Nh\nu/\{\exp(h\nu/kT) - 1\} + N\varepsilon_0,$$ (14.8.2)

and for the molar heat capacity at constant volume

$$C_{V,m}/R = \tfrac{5}{2} + (h\nu/kT)^2 \exp(h\nu/kT)/\{\exp(h\nu/kT) - 1\}^2.$$ (14.8.3)

The idea that each mode of motion contributed $\tfrac{1}{2}NkT$ to the energy but that some molecules are 'rigid rotators' (the vibrational mode is unexcited) so that the energy was $\tfrac{5}{2}NkT$ rather than $\tfrac{7}{2}NkT$† was called the 'equipartition of energy' and pre-dated quantum theory. It was one of the early achievements of quantum theory to show how the energy of a diatomic gas passes smoothly from $\tfrac{5}{2}NkT$ at low temperatures to $\tfrac{7}{2}NkT$ at high temperatures. In figure 14.3 we show the heat capacity of some diatomic gases plotted against T/θ_V where $\theta_V \stackrel{\text{def}}{=} h\nu/k$ denotes the vibrational characteristic temperature and the curve is that calculated from equation (3).

For a monatomic molecule q_R and q_V are omitted from equation (1) and $C_{V,m} = \tfrac{3}{2}R$, independent of temperature and in excellent agreement with experiment. For a polyatomic molecule $(3n-5)$ factors $q_{V,i}$ must be included in equation (1) if it is linear, and $(3n-6)$ factors $q_{V,i}$ and the expression (14.6.6) for q_R if it is non-linear. If electronic excitation is important, as it is for example for NO, then the appropriate factor from (14.4.5) must also be included; the heat capacity at constant volume then passes through a maximum as the temperature is changed.

---

† For a vibration, so that the right result was obtained, kinetic energy and potential energy were each regarded as contributing $\tfrac{1}{2}NkT$ to the energy.

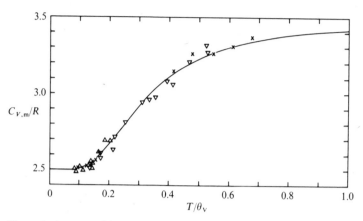

Figure 14.3. The molar heat capacities at constant volume, $C_{V,m}$, plotted against T/θ_V for some diatomic gases where θ_V denotes the characteristic vibrational temperature $h\nu/k$. Experimental values of $(C_{p,m} - R)$ at low pressures: △, N_2; ▽, O_2; ×, CO. (Adapted from Fowler, R. H.; Guggenheim, E. A. *Statistical Thermodynamics*. Cambridge University Press: Cambridge. **1949**, p. 102.)

§ 14.9 Chemical equilibrium in a gaseous reaction

The thermodynamic criterion for chemical equilibrium of the reaction $0 = \Sigma_B \nu_B B$ is

$$\prod_B (\lambda_B)^{\nu_B} = 1 . \tag{14.9.1}$$

According to Boltzmann's distribution law (equation 14.1.10)

$$\lambda_B = N_B/q_B = (x_B p/p^{\ominus})(p^{\ominus}/kT)(q_B/V)^{-1} , \tag{14.9.2}$$

so that the standard equilibrium constant $K^{\ominus}$ is given by

$$K^{\ominus} = \prod_B (x_B p/p^{\ominus})^{\nu_B} = \prod_B (kT q_B/p^{\ominus} V)^{\nu_B} . \tag{14.9.3}$$

The standard equilibrium constant $K^{\ominus}(T)$ of any gaseous reaction at any temperature T can thus be obtained by substitution of the appropriate expressions for q_B/V into equation (3).

We take as an example a reaction among diatomic molecules in which electronic modes are unexcited. Substituting for q_B/V from (14.8.1) into (3) we obtain

$$\ln\{K^{\ominus}(T)\} = \sum_B \nu_B \ln\{(kT/p^{\ominus})(2\pi M_B RT/L^2 h^2)^{3/2}\} + \sum_B \nu_B \ln(T/s_B \theta_{R,B})$$

$$- \sum_B \nu_B \ln\{1 - \exp(-\theta_{V,B}/T)\} + \sum_B \nu_B \ln\{g(\varepsilon_{0,B})\}$$

$$- \sum_B \nu_B \varepsilon_{0,B}/kT. \tag{14.9.4}$$

Equation (4) can be used to calculate $K^{\ominus}$ at any temperature T if M_B, $\theta_{R,B}$, $\theta_{V,B}$, and $g(\varepsilon_{0,B})$ are known for each of the reacting diatomic molecules B, and if $\Sigma_B \nu_B \varepsilon_{0,B}$ is also known. All these quantities can be determined spectroscopically and all except the last are usually well known; the last quantity $\Sigma_B \nu_B \varepsilon_{0,B}$ is difficult to obtain spectroscopically except for a few dissociations like $I_2 = 2I$.

If the term $\Sigma_B \nu_B \varepsilon_{0,B}$ in equation (4) is not available from spectroscopic measurements then we eliminate it by use of the exact thermodynamic equation (11.4.1):

$$\Delta H_m^{\ominus} = RT^2 \, d \ln K^{\ominus}/dT, \tag{14.9.5}$$

which leads by differentiation of equation (4) to

$$\Delta H_m^{\ominus}(T) = \tfrac{7}{2} \left(\sum_B \nu_B \right) RT + R \sum_B \nu_B [\theta_{V,B}/\{\exp(\theta_{V,B}/T) - 1\}]$$

$$+ L \sum_B \nu_B \varepsilon_{0,B}. \tag{14.9.6}$$

From equation (6) we obtain by first replacing T by $T^{\dagger}$ where $T^{\dagger}$ is some one temperature at which a value of $\Delta H_m^{\ominus}$ is available (most commonly $T^{\dagger} = 298.15$ K) and then dividing by RT

$$\Delta H_m^{\ominus}(T^{\dagger})/RT = \tfrac{7}{2} \left(\sum_B \nu_B \right) T^{\dagger}/T + (T^{\dagger}/T) \sum_B \nu_B [(\theta_{V,B}/T^{\dagger})/\{\exp(\theta_{V,B}/T^{\dagger}) - 1\}]$$

$$+ \sum_B \nu_B \varepsilon_{0,B}/kT. \tag{14.9.7}$$

When we add equations (4) and (7) we obtain

$$\ln\{K^{\ominus}(T)\} = \sum_B \nu_B \ln\{(kT/p^{\ominus})(2\pi M_B RT/L^2 h^2)^{3/2}\}$$

$$+ \sum_B \nu_B \ln(T/s_B \theta_{R,B}) - \sum_B \nu_B \ln\{1 - \exp(-\theta_{V,B}/T)\}$$

$$+ \sum_B \nu_B \ln\{g(\varepsilon_{0,B})\} - \Delta H_m^{\ominus}(T^{\dagger})/RT + \tfrac{7}{2} \left(\sum_B \nu_B \right) T^{\dagger}/T$$

$$+ (T^{\dagger}/T) \sum_B \nu_B [(\theta_{V,B}/T^{\dagger})/\{\exp(\theta_{V,B}/T^{\dagger}) - 1\}]. \tag{14.9.8}$$

Equation (8) can be used to calculate $K^{\ominus}$ at any temperature T if M_B, $\theta_{R,B}$, $\theta_{V,B}$, and $g(\varepsilon_{0,B})$ are known for each of the reacting diatomic molecules B, and if a calorimetric value of $\Delta H_m^{\ominus}(T^{\dagger})$ is known at any one temperature $T^{\dagger}$. One of the reasons why it is so important to have accurately known values of $\Delta H_m^{\ominus}$ at one temperature is to make it possible to use equation (8), or extensions of equation (8) to other than diatomic molecules, to calculate the standard equilibrium constants of gaseous reactions from spectroscopically measured quantities.

For a monatomic reactant B, q_B/V in equation (2) becomes simply $(q_{T,B}/V)g(\varepsilon_{0,B})\exp(-\varepsilon_{0,B}/kT)$, the factors q_R and q_V being omitted. For a polyatomic reactant B, $(3n - 5)$ factors $q_{V,i,B}$ must be included in q_B if it is linear, and $(3n - 6)$ factors $q_{V,i,B}$ and the expression (14.6.6) for $q_{R,B}$ if it is non-linear. If electronic excitation is important for the reactant B then the appropriate factor from (14.4.5) must be included in q_B. Finally, if the rotation of B may not be treated with sufficient accuracy as classical (for example if B is H_2 and the temperature is low) then the sum (14.6.3) (divided by the appropriate symmetry number) must be used for $q_{R,B}$ in place of the integrated form (14.6.5).

In figure 14.4 we show as a curve the results of such calculated values of $\ln K^\ominus$ plotted against $1/T$ for the reaction:

$$N_2(g) + 3H_2(g) = 2NH_3(g),\tag{14.9.9}$$

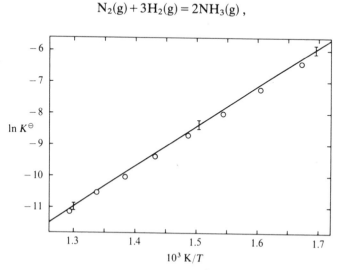

Figure 14.4. Values of the standard equilibrium constant $K^\ominus(T)$ calculated from equation (14.9.8), with appropriate substitutions as described in the text, for the reaction: $N_2(g) + 3H_2(g) = 2NH_3(g)$. $\bigcirc$, Directly measured experimental results at $p = 1$ MPa.

and for comparison the directly measured values of $K^\ominus$. The calculations were based on $M(N_2) = 0.0280134 \text{ kg mol}^{-1}$, $M(H_2) = 0.0020158 \text{ kg mol}^{-1}$, and $M(NH_3) = 0.0170304 \text{ kg mol}^{-1}$; $\theta_R(N_2) = 2.88 \text{ K}$, $\theta_R(H_2) = 85.3 \text{ K}$, and $\theta_R(NH_3) = (\theta_{R,A}\theta_{R,B}\theta_{R,C})^{1/3} = 12.3 \text{ K}$; $s(N_2) = 2$, $s(H_2) = 2$, and $s(NH_3) = 3$; $\theta_V(N_2) = 3350 \text{ K}$, $\theta_V(H_2) = 5980 \text{ K}$, and $\theta_{V,i}(NH_3)/K = 1370, 2340, 2340, 4800, 4910$, and 4910; and $\Delta H_m^\ominus(298.15 \text{ K})/R = -(11051 \pm 84) \text{ K}$. All degeneracies were taken as 1.

It would have been nice to have compared in figure 14.4 values of $K^\ominus$ calculated, on the assumption that Nernst's heat theorem is obeyed, from purely calorimetric values of $\Delta H_m^\ominus(T)$ and of $\{S_B^\ominus(g, T) - S_B(s, T \to 0)\}$; the tabulated values of $\{S_B^\ominus(g, T) - S_B(s, T \to 0)\}$ are, however, based at least partly on calculations from spectroscopic quantities (for which see the following section) and so such a comparison would have been false.

The agreement between the curve in figure 14.4 and the directly measured experimental values is probably within experimental error though the discrepancies are a little surprising. It is a challenging task for a scientist to learn to analyse these discrepancies. First he would have to examine all the approximations made in the derivation of his equation, such as the use of Boltzmann statistics, the separability of the several modes of motion, the treatment of the translational and especially the rotational modes (H_2?) as effectively classical, the assumption that the vibrations are simple-harmonic, and the omission of contributions from electronic excitation and of the degeneracies of the lowest electronic and nuclear levels. He would also have to examine both the formula he used and the calculations he made for small errors. He would then have to examine the difference to be expected between values of '$K^{\ominus}$' measured on the real gaseous mixture at a pressure of 1 MPa and those for a perfect gaseous mixture. Finally he would have to examine critically the accuracy of all the spectroscopic quantities, of the calorimetric standard molar enthalpy change, and of the direct measurements of the equilibrium constant. Only if he has a good grasp of both the theoretical and the experimental aspects of the subject will he be able to make a good job of this.

§ 14.10 Entropy of a gaseous substance

By use of the formulae: $S_B^{\ominus} = -d\mu_B^{\ominus}/dT$ and $pV = NkT$ in equation (14.1.11) we obtain for the standard molar entropy $S_B^{\ominus}$ of a pure gaseous substance B the equation:

$$S_B^{\ominus}(T)/R = \ln\{(kT/p^{\ominus})(q_B/V)\} + T(\partial \ln q_B/\partial T)_V. \quad (14.10.1)$$

By appropriate substitution for q_B the standard molar entropy of any gaseous substance can thus be evaluated.

For a monatomic molecule in which there is no electronic excitation equation (1) becomes

$$S_B^{\ominus}(T)/R - \ln\{g(\varepsilon_{0,B})\} = \ln\{(2\pi M_B)^{3/2}(RT)^{5/2}/p^{\ominus}L^4h^3\} + 2.5, \quad (14.10.2)$$

which is sometimes called the Sackur–Tetrode equation.

For a diatomic molecule in which there is no electronic excitation equation (1) becomes

$$S_B^{\ominus}(T)/R - \ln\{g(\varepsilon_{0,B})\} = \ln\{(2\pi M_B)^{3/2}(RT)^{5/2}/p^{\ominus}L^4h^3\} + \ln(8\pi^2 I_B kT/s_B h^2)$$
$$- \ln\{1 - \exp(-\theta_{V,B}/T)\} + 3.5$$
$$+ (\theta_{V,B}/T)/\{\exp(\theta_{V,B}/T) - 1\}. \quad (14.10.3)$$

Similar equations can be derived for polyatomic molecules or for molecules in which electronic excitation is important by substituting the appropriate partition function in equation (1).

For Ar, equation (2) leads to $[S_m^{\ominus}(\text{Ar, g, } 87.28 \text{ K})/R - \ln\{g(\varepsilon_{0,\text{Ar}})\}] = 15.54$. The calorimetric value of $\{S_m^{\ominus}(\text{Ar, g, } 87.28 \text{ K}) - S_m(\text{Ar, s, } T \to 0)\}/R = 15.6$. We conclude that within experimental error $S_m(\text{Ar, s, } T \to 0)/R = \ln\{g(\varepsilon_{0,\text{Ar}})\}$, or if $g = 1$ that $S_m(\text{Ar, s, } T \to 0) = 0$.

For N_2, equation (3) leads to $[S_m^{\ominus}(N_2, \text{g, } 77.35 \text{ K})/R - \ln\{g(\varepsilon_{0,N_2})\}] = 18.30$. The calorimetric value of $\{S_m^{\ominus}(N_2, \text{g, } 77.35 \text{ K}) - S_m(N_2, \text{s, } T \to 0)\}/R = 18.3$. We conclude again that within experimental error $S_m(N_2, s, T \to 0)/R = \ln\{g(\varepsilon_{0,N_2})\}$, or if $g = 1$ that $S_m(N_2, s, T \to 0) = 0$.

Similar conclusions would be obtained for almost all other gases. The outstanding exceptions are CO, NNO, NO, and HOH. For CO, for example, equation (3) leads to $[S_m^\ominus(CO, g, 81.61 K)/R - \ln\{g(\varepsilon_{0,CO})\}] = 19.22$ while the calorimetric value of $\{S_m^\ominus(CO, g, 81.61 K) - S_m(CO, s, T \to 0)\}/R = 18.6$. We conclude that $S_m(CO, s, 81.61 K)/R = \ln\{g(\varepsilon_{0,CO})\} - 0.6 \approx \ln\{g(\varepsilon_{0,CO})\} - \ln 2$. Similar discrepancies of about $\ln 2$, $\ln 2^{1/2}$, and $\ln(3/2)$ exist for NNO, NO, and HOH. We shall return to the relation between $S_m(B, s, T \to 0)/R$ and $\ln\{g(\varepsilon_{0,B})\}$ in the next chapter which will be devoted to a discussion of Nernst's heat theorem.

§ 14.11   The influence of nuclear spins: ortho and para molecules

For *homonuclear* molecules like H_2 and D_2, but not HD, the symmetries of the nuclear wave functions corresponding to the lowest energy level, called nuclear-spin states, play an important part in determining the properties of the gas if the molecule is light enough and if the temperature is low enough. If there are ρ spin states of each nucleus then there are ρ^2 nuclear-spin states of the molecule. Of these, $\frac{1}{2}\rho(\rho + 1)$ are symmetrical and $\frac{1}{2}\rho(\rho - 1)$ are antisymmetrical with respect to interchange of the nuclei.†

It is one of the axioms of quantum theory that any wave function of a molecule must be antisymmetrical with respect to the interchange of a proton or of a neutron (or of an electron—the Pauli exclusion principle—but electrons are not relevant here). The translational wave functions ψ_T are all symmetrical. The vibrational wave functions ψ_V are symmetrical for $v = 0, 2, 4, \cdots$, and antisymmetrical for $v = 1, 3, 5, \cdots$; we shall confine our discussion to vibrationally unexcited molecules for which $v = 0$ and ψ_V is symmetrical. The rotational wave functions ψ_R are symmetrical for $j = 0, 2, 4, \cdots$, and antisymmetrical for $j = 1, 3, 5, \cdots$. We must accordingly combine the nuclear and rotational wave functions so as to obtain the correct symmetry for the total wave function of the molecule. For nuclei like H of odd nucleon-number we must combine antisymmetrical nuclear-spin states with symmetrical rotational states and symmetrical nuclear-spin states with antisymmetrical rotational states so as to obtain antisymmetrical wave functions for the molecule as a whole. The rotational partition function of the molecule must then be given by

$$q_R = \tfrac{1}{2}\rho(\rho - 1) \sum_{j=0,2,4,\cdots} f(j) + \tfrac{1}{2}\rho(\rho + 1) \sum_{j=1,3,5,\cdots} f(j), \qquad (14.11.1)$$

---

† Let the nuclear wave functions be $\psi_1(a), \psi_2(a), \cdots, \psi_\rho(a)$ for nucleus 'a' and $\psi_1(b), \psi_2(b), \cdots, \psi_\rho(b)$ for nucleus 'b', the nuclei 'a' and 'b' being indistinguishable. A wave function is said to be antisymmetrical with respect to the nuclei if it changes its sign on interchange of the nuclei, and symmetrical if it does not. Nuclear wave functions of the molecule can be formed in the ρ symmetrical ways $\psi_i(a)\psi_i(b)$, and in the $\frac{1}{2}\rho(\rho - 1)$ symmetrical ways $\{\psi_i(a)\psi_j(b) + \psi_j(a)\psi_i(b)\}$ and the $\frac{1}{2}\rho(\rho - 1)$ antisymmetrical ways $\{\psi_i(a)\psi_j(b) - \psi_j(a)\psi_i(b)\}$.

where

$$f(j) = (2j+1) \exp\{-j(j+1)h^2/8\pi^2 IkT\}. \tag{14.11.2}$$

For nuclei like D of even nucleon-number we must combine nuclear-spin states with rotational states so as to obtain symmetrical wave functions of the molecule as a whole. The rotational partition function of the molecule must then be given by

$$q_R = \tfrac{1}{2}\rho(\rho+1) \sum_{j=0,2,4,\cdots} f(j) + \tfrac{1}{2}\rho(\rho-1) \sum_{j=1,3,5,\cdots} f(j). \tag{14.11.3}$$

For a heteronuclear molecule like HD no question of symmetry arises and the corresponding rotational partition function is given by

$$q_R = \rho_a\rho_b \sum_{j=0,1,2,\cdots} f(j), \tag{14.11.4}$$

where the factor $\rho_a\rho_b$ may simply be included in the factor $g(\varepsilon_0)$ in the complete partition function of the molecule.

At high temperatures, where we may replace the sums by integrals,

$$\sum_{j=0,2,4,\cdots} f(j) = \sum_{j=1,3,5,\cdots} f(j) = \tfrac{1}{2} \sum_{j=0,1,2,\cdots} f(j), \tag{14.11.5}$$

so that equations (1) and (3) both tend towards

$$q_R = \tfrac{1}{2}\rho^2 \sum_{j=0,1,2,\cdots} f(j), \tag{14.11.6}$$

which differs from (4) for a heteronuclear molecule by division by the symmetry number 2.

Incidentally, when the proton number and the neutron number are both equal and even, the nucleus has no resultant spin and $\rho = 1$. Then the rotational states corresponding to $j = 1, 3, 5, \cdots$, should be absent. That has been verified spectroscopically for molecules like $^4He_2$ and $^{16}O_2$.

In figure 14.5 we compare experimental and calculated rotational contributions to the heat capacity for H_2 ($\rho = 2$) and for D_2 ($\rho = 3$), and in figure 14.6 for HD. For each gas we have plotted the curve 2 corresponding to equation (4) ignoring nuclear spins but summing the series instead of treating the rotation as classical. Had we treated the rotation as classical then the line 1 would have resulted in each case. For HD curve 2 in figure 14.6 agrees with experiment and no more need be said. For H_2 and for D_2 the curves 2 in figure 14.5 disagree badly with experiment; we must recognize the effect of nuclear spins by turning to equations (1) and (3) respectively and using them to plot curves 3. Now the disagreement with experiment is spectacular, and at first rather puzzling.

The resolution of this paradox lies in the extreme slowness (in the absence of a catalyst) of the interconversion of *ortho* molecules (with symmetrical spin states) and *para* molecules (with antisymmetrical spin states) of H_2 or of D_2. The gas, n-H_2 or n-D_2 (n for 'normal'), distinguished from e-H_2 or e-D_2 (e for 'equilibrium') which in

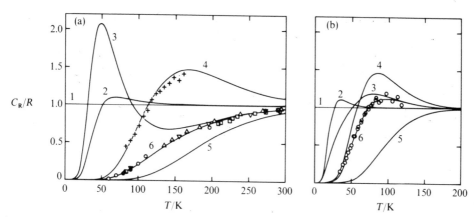

Figure 14.5. Rotational contributions C_R to the molar heat capacity plotted against temperature T (a), for H_2 and (b), for D_2. Curves: 1, classical: $q_R = 8\pi^2 IkT/2h^2$; 2, ignoring nuclear spins but summing rather than integrating equation (14.11.6); 3, e-H_2 and e-D_2 according to equation (14.11.1) for e-H_2 and equation (14.11.3) for e-D_2; 4, p-H_2 (first sum of equation 14.11.1) and p-D_2 (second sum of equation 14.11.3); 5, o-H_2 (second sum of 14.11.1) and o-D_2 (first sum of 14.11.3); 6, n-H_2 (equation 14.11.9) and n-D_2 (equation 14.11.10). (a), $\bigcirc, \square, \diamond, \triangle, \triangledown$, Experimental results for n-H_2; +, experimental results for nearly pure p-H_2. (b), $\bigcirc$, Experimental results for n-D_2.

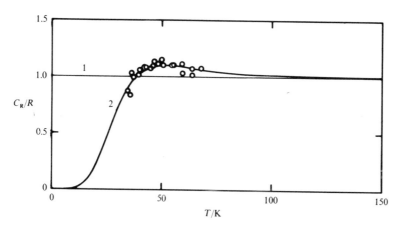

Figure 14.6. Rotational contribution C_R to the molar heat capacity plotted against temperature T for HD. Curves: 1, classical: $q_R = 8\pi^2 IkT/h^2$; 2, ignoring nuclear spins but summing rather than integrating equation (14.11.4). $\bigcirc$, Experimental results.

the presence of an efficient catalyst would presumably have resulted in the agreement of experiment with curves 3, behaves like a mixture of non-reacting o-H_2 and p-H_2 or of o-D_2 and p-D_2 (o for ortho and p for para) at the composition corresponding to equilibrium at high temperatures.

In figure 14.7 we have plotted the equilibrium mole fractions x of o-H_2 and of p-D_2 as functions of temperature. The values of x were calculated from the equations for

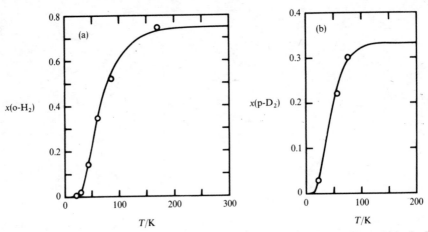

Figure 14.7. Mole fractions x at equilibrium (a), of o-H_2 in a mixture of o-H_2 and p-H_2, and (b), of p-D_2 in a mixture of o-D_2 and p-D_2, plotted aginst temperature T. $\bigcirc$, Experimental results.

the number fractions of odd rotational states for H_2 and of even rotational states for D_2:

$$x(\text{o-}H_2) = 3 \sum_{j=1,3,5,\cdots} f(j) \Big/ \Big\{ \sum_{j=0,2,4,\cdots} f(j) + 3 \sum_{j=1,3,5,\cdots} f(j) \Big\}, \qquad (14.11.7)$$

and

$$x(\text{p-}D_2) = 3 \sum_{j=1,3,5,\cdots} f(j) \Big/ \Big\{ 6 \sum_{j=0,2,4,\cdots} f(j) + 3 \sum_{j=1,3,5,\cdots} f(j) \Big\}. \qquad (14.11.8)$$

The limiting values of x at high temperatures are $\frac{3}{4}$ for H_2 and $\frac{1}{3}$ for D_2; it is these values that persist at all temperatures in n-H_2 and n-D_2. It is possible to prepare pure p-H_2 or pure o-D_2 by cooling the gas towards $T \to 0$ in the presence of a catalyst such as charcoal. When the gas is removed from the catalyst it will remain as pure p-H_2 or o-D_2 at any temperature. The other gas, o-H_2 or p-D_2, can in principle be prepared pure by differential adsorption on a suitable solid. We have plotted as curves 4 in figure 14.5 the rotational contributions to the heat capacities for pure o-H_2 and for pure o-D_2, and as curves 5 those for pure p-H_2 and for pure p-D_2. For hydrogen some measurements on nearly pure p-H_2 are also shown; the agreement is excellent.

It is now a simple matter to calculate the heat capacity of n-H_2 from the equation:

$$C_V(\text{n-}H_2) = \tfrac{1}{4}C_V(\text{p-}H_2) + \tfrac{3}{4}C_V(\text{o-}H_2), \qquad (14.11.9)$$

and of n-D_2 from the equation:

$$C_V(\text{n-}D_2) = \tfrac{2}{3}C_V(\text{o-}D_2) + \tfrac{1}{3}C_V(\text{p-}D_2). \qquad (14.11.10)$$

The curves so calculated are those numbered 6 in figure 14.5. The agreement with experiment is now excellent.

Similar effects are observable, though only at very low temperatures, for CH_4 ($s = 12$), and might be observable for other light molecules having $s > 1$. For CH_4 there are three forms with high-temperature mole fractions 0.3125, 0.5625, and 0.1250.

§ 14.12   Application to gaseous mixtures

By straightforward extension of equation (14.1.10) to a mixture of perfect gases B, C, $\cdots$, we obtain

$$\lambda_B = N_B/q_B, \quad \lambda_C = N_C/q_C, \quad \cdots, \quad \cdots, \quad \cdots, \tag{14.12.1}$$

and for the Gibbs function

$$G = kT \sum_B N_B \ln \lambda_B = kT \sum_B N_B \ln(N_B/q_B). \tag{14.12.2}$$

Each q_B contains a factor V and otherwise depends for a particular substance only on the temperature T. We accordingly use the equation: $pV = (\Sigma_B N_B)kT$ to rewrite (2) in the form:

$$G/kT = \sum_B N_B \ln(N_B V/q_B) - \left(\sum_B N_B\right) \ln\left(\sum_B N_B\right) - \left(\sum_B N_B\right) \ln(kT/p). \tag{14.12.3}$$

The change of Gibbs function on mixing perfect gases at constant pressure is thus given by

$$\Delta_{mix}G/kT = -\left(\sum_B N_B\right) \ln\left(\sum_B N_B\right) + \sum_B N_B \ln N_B, \tag{14.12.4}$$

and the molar change of Gibbs function by

$$\Delta_{mix}G_m/RT = \sum_B x_B \ln x_B, \tag{14.12.5}$$

where x_B denotes the mole fraction of B in the mixture. The corresponding equation for the molar entropy of mixing is

$$\Delta_{mix}S_m/R = -\sum_B x_B \ln x_B. \tag{14.12.6}$$

Equations (5) and (6) agree with those that can be derived from equation (10.4.8), which as the reader will recall was obtained with the help of an extra-thermodynamic assumption. They also agree with (12.6.2) and (12.6.3) which were based on the same assumption. It is pleasing to find that Boltzmann's distribution law leads to the same conclusions as those we based on the results of measurements of the osmotic pressure of a gas mixture at low pressures.

The Helmholtz function $A = G - pV = G - \Sigma_B N_B kT$ is given by

$$A/kT = -\ln \left\{ \prod_B (q_B^{N_B}/N_B^{N_B} \, e^{-N_B}) \right\}. \tag{14.12.7}$$

When we write $A/kT = -\ln Q$, the quantity Q is called the canonical partition function *of the system*. We shall say no more in this book about partition functions of systems except that they are powerful tools for dealing with systems of interacting (non-independent) particles, that they are the subject matter of statistical mechanics,[†] and that for a system of independent particles $Q = \Pi_B(q_B^{N_B}/N_B^{N_B} \, e^{-N_B})$, or as it is more usually written, $Q = \Pi_B \, (q_B^{N_B}/N_B!).$[‡]

§ 14.13 Application to crystals

A crystal composed of N identical atoms each occupying a site on the crystal lattice may be regarded as having no modes of motion except the N three-dimensional vibrations of the atoms about their lattice sites. If we regard those vibrations as simple harmonic then the partition function q of an atom in the crystal is given by

$$q^N = \prod_{i=1}^{3N} [\{1 - \exp(-h\nu_i/kT)\}^{-1} \exp(-\tfrac{1}{2}h\nu_i/kT)]. \qquad (14.13.1)$$

The problem is thus reduced to one of finding the distribution of the $3N$ lattice frequencies ν_i.

The simplest assumption, first made by Einstein, is that all the frequencies have the same value ν_E. Equation (1) then becomes

$$q = \{1 - \exp(-h\nu_E/kT)\}^{-3} \exp(-\tfrac{3}{2}h\nu_E/kT). \qquad (14.13.2)$$

The molar heat capacity at constant volume of such a crystal is given by

$$C_{V,m}/R = 3(\Theta_E/T)^2 \exp(\Theta_E/T)/\{\exp(\Theta_E/T) - 1\}^2, \qquad (14.13.3)$$

where we have written Θ_E for $h\nu_E/k$.

In figure 14.8 we have plotted against temperature experimental values of the molar heat capacity at constant volume of Ar(s), calculated from measured values of $C_{p,m}$ by use of equation (6.11.5) with experimental values of α, V_m, and κ_T, and the values of $C_{V,m}$ calculated from equation (3) with $\Theta_E = 64.0$ K. If Einstein's theory were exact, Θ_E must depend on the molar volume of the crystal. The theory should be tested against measurements of the molar heat capacity, not only at constant volume, but at the *same* constant volume at all temperatures. Such heat capacities would not be easy to obtain. To maintain the volume of Ar(s) constant from $T \to 0$ to the triple-point temperature (83.8 K) would need the application of pressures up to

---

[†] References are given to some textbooks of statistical mechanics on p. 173. Others include
Fowler, R. H.; Guggenheim, E. A. *Statistical Thermodynamics*. C.U.P.: Cambridge. **1949**.
Hill, T. L. *An Introduction to Statistical Thermodynamics*. Addison-Wesley: Reading, Mass., U.S.A. **1960**.
Davidson, N. *Statistical Mechanics*. McGraw-Hill: New York. **1962**.
Münster, A. *Statistical Thermodynamics*, Vol. II. Springer-Verlag: Berlin; Academic Press: New York. **1974**.
[‡] According to Stirling's approximation $N! \approx N^N \, e^{-N}$ and the approximation becomes nearly exact for large values of N. Even for $N = 1000$, an extremely small number compared with the number of molecules in any macroscopic system, $\ln N! = 5912$ while $(N \ln N - N) = 5908$.

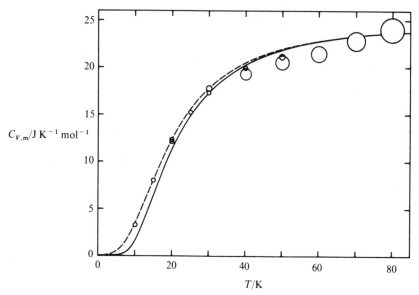

Figure 14.8. The molar heat capacity at constant volume calculated as a function of temperature: ——, according to Einstein's theory with $\Theta_E = 64.0$ K; and ----, according to Debye's theory with $\Theta_D = 81.0$ K. $\bigcirc$, Values of $C_{V,m}$, calculated from directly measured values of $C_{p,m}$, α, V_m, and κ_T, for Ar(s).

about 100 MPa. In view of these considerations the fit in figure 14.8 is surprisingly good.

At high temperatures Einstein's theory leads to

$$C_{V,m} \to 3R, \tag{14.13.4}$$

in agreement with the law of Dulong and Petit.

A better fit, especially at low temperatures, can be obtained by using Debye's theory instead of Einstein's. Debye assumed that the number of frequencies ν_i is proportional to ν_i^2 (as it is for electromagnetic radiation) up to an upper bound ν_D chosen so as to ensure that there are $3N$ frequencies in all. Debye's formula for the molar heat capacity at constant volume is

$$C_{V,m}/R = 9(T/\Theta_D)^3 \int_0^{\Theta_D/T} y^4 \, e^y (e^y - 1)^{-2} \, dy. \tag{14.13.5}$$

The curve calculated from (5) with $\Theta_D = 81.0$ K is shown as a broken line in figure 14.8. At low temperatures, where the fit is better than for Einstein's theory, Debye's theory leads to

$$C_{V,m}/R \to (12/5)\pi^4 (T/\Theta_D)^3. \tag{14.13.6}$$

The form: $C_{p,m} \propto T^3$ has been much used to obtain the enthalpy or the entropy of a crystal between the lowest temperature at which a measurement of heat capacity was made and $T \to 0$.

Except at low temperatures there is little to choose between Einstein's and Debye's theories, and if there were it would be fallacious to decide between them on the basis of experimental heat capacities at constant volume when the volume held constant in each measurement varies with temperature.

Problems for chapter 14

Problem 14.1

The first excited electronic energy level of NO exceeds the lowest level by $(\varepsilon_1 - \varepsilon_0)/k = 180$ K. Both levels are doubly degenerate. All higher energy levels have much greater energy differences. Calculate the fraction of molecules in the excited state (a) at 90 K and (b) at 360 K.

Problem 14.2

Assuming that at temperatures up to 180 K NO behaves as a rigid rotator and using the information in problem 14.1 calculate and plot against the temperature T the molar heat capacity at constant pressure of NO for T between 120 and 180 K. Compare your results with the measured values:

T/K	127	146	174	178
$C_{p,m}/R$	3.76	3.79	3.72	3.67

Problem 14.3

Calculate and plot against T from 0 to 5000 K the vibrational contribution to the molar heat capacity (a) for O_2 (for which $h\nu/k = 2240$ K) and (b) for Br_2 (for which $h\nu/k = 462$ K), where ν denotes the fundamental vibration frequency. In each case state approximate temperatures (i) below which the vibrational mode is effectively unexcited and (ii) above which the vibrational mode is effectively classical.

Problem 14.4

The planar molecule CH_2O (methanal or formaldehyde) has vibrational characteristic temperatures 1680, 1840, 2160, 2510, 4000, and 4130 K. Calculate the molar heat capacity at constant pressure (a) at 300 K and (b) at 600 K.

Problem 14.5

Write equations for the average energy $\langle \varepsilon_x \rangle$ for translation of a molecule in the x-direction, the average value $\langle n_x \rangle$ of the quantum number, and the relative spacing between successive energy levels for a molecule with the average quantum number $\langle n_x \rangle$. Calculate each of these quantities for He at 10 K in a box of length 1 mm.

Problem 14.6

Given the internuclear separation 0.07512 nm calculate the moment of inertia of the molecule H_2.

Using your moment of inertia investigate the extent to which H_2 fails to behave as a classical rotator (a) at 100 K and (b) at 300 K.

Problem 14.7

Given the following quantities calculate the standard equilibrium constant $K^{\ominus}$ of the reaction $I_2(g) = 2I(g)$. The molar mass $M(I) = 0.1269$ kg mol^{-1}. The lowest electronic energy level of I_2 is non-degenerate while that of I is four-fold degenerate; neither has an accessible excited electronic energy level. For I_2 the rotational characteristic temperature is 0.0537 K and the vibrational characteristic temperature is 306.6 K. The spectroscopically determined value of $\Delta\varepsilon_0/k$ is 17910 K.

Compare your result with the directly measured value: $K^{\ominus}(1274$ K$) = 0.167$.

Problem 14.8

Given the following quantities calculate the standard equilibrium constant $K^{\ominus}(T)$ at several temperatures T between 650 and 780 K for the reaction: $H_2(g) + I_2(g) = 2HI(g)$.

	H_2	I_2	HI
M/kg mol^{-1}	0.002016	0.2538	0.1279
θ_R/K	85.4	0.0537	9.29
θ_V/K	5980	306.6	3210

The calorimetrically determined standard molar enthalpy change is given by $\Delta H_m^{\ominus}(298.15$ K$) = -9.47$ kJ mol^{-1}.

Compare your values with those directly measured:

T/K	666.8	698.6	730.8	763.8
$K^{\ominus}$	60.9	55.1	49.6	45.8

Chapter 15

Nernst's heat theorem

§ 15.1 Statement of Nernst's heat theorem

We have already mentioned Nernst's heat theorem twice, in a footnote to § 4.6 and in § 11.5; we must now state it as unambiguously as we can and discuss its testing and consequences. We must also discuss its relation to the third law of thermodynamics, however that might be formulated.

The most useful form of Nernst's heat theorem concerns the value of the entropy change for a reaction as the temperature is lowered towards zero. Such an entropy change can be determined experimentally from measurements (a) of the standard molar entropy difference: $\{S_B^\ominus(\text{state}, T) - S_B(\text{s}, T \to 0)\}$ for each reacting substance, and (b) of the standard molar entropy change $\Sigma_B\, \nu_B S_B^\ominus(\text{state}, T)$ for the reaction at some convenient temperature T. In symbols:

$$\sum_B \nu_B S_B(\text{s}, T \to 0) = -\sum_B \nu_B\{S_B^\ominus(\text{state}, T) - S_B(\text{s}, T \to 0)\}$$

$$+\sum_B \nu_B S_B^\ominus(\text{state}, T)\,. \tag{15.1.1}$$

The measurement of the molar entropy difference $\{\cdots\}$ for a pure substance has been dealt with in §§ 6.3, 6.5, and 6.6, with corrections to standard molar entropy difference in §§ 12.11 and 12.15. The experimental realization of '$T \to 0$' has not yet been explored. Measurements of the molar heat capacity are made to the lowest experimentally feasible temperature T_0 and then $C_{p,\mathrm{m}}/T$ is smoothly extrapolated to $T = 0$, often with the help of Debye's low-temperature rule $C_{p,\mathrm{m}} \propto T^3$ leading to $\{S_B^\ominus(\text{s}, T_0) - S_B(\text{s}, T \to 0)\} = \frac{1}{3}C_{p,\mathrm{m}}(T_0)$. With liquid hydrogen in the cryostat of the heat-capacity calorimeter, T_0 can be made as low as about 10 K; with liquid helium T_0 can be made as low as 1 K or even less. Does it make any difference whether T_0 is 10 K or is 1 K? Would it make any difference if T_0 were 0.1 K?

The standard molar entropy change $\Sigma_B\, \nu_B S_B^\ominus(\text{state}, T)$ can be measured unambiguously at some convenient temperature T in a variety of ways that we shall briefly review in § 15.2.

As originally formulated, Nernst's heat theorem asserts that as the temperature tends to zero the entropy change for a reaction vanishes, or

$$\sum_B \nu_B S_B(T \to 0) = 0\,. \tag{15.1.2}$$

In that form the theorem is usually, but not always, obeyed. In the form:

$$\sum_{B} \nu_B S_B(\text{p.o.s.}, T \to 0) = 0, \tag{15.1.3}$$

where 'p.o.s.' denotes a perfectly ordered solid, it is always obeyed. Since, however, it is not possible to know in advance whether any substance is a perfectly ordered solid at the temperature T_0, we are no further forward. It is reasonable to conjecture (and ought to be explored experimentally, for example for CO) that the earlier formulation (2) would sometimes become identical with the formulation (3) if T_0 were made lower. So it turned out for H_2, and so it might turn out for CO and some other 'exceptional' substances.† For other 'exceptional' substances, especially for those that are glasses (supercooled liquids), it is unlikely that (2) and (3) would become identical however low the value of T_0.

In the formulation given we have implied, by omitting any mention of pressure or any standard sign $^\ominus$, that the entropy of a solid at very low temperatures is independent of pressure. That implication will be discussed in § 15.3.

In its original form (2), Nernst's heat theorem is obeyed within experimental error by nearly all of the vast range of reactions for which it has been tested. The exceptions are reactions involving CO, N_2O, NO, H_2O (and D_2O), CH_3D (and other partly deuterated methanes), a few hydrates like $Na_2SO_4 \cdot 10H_2O$, and all glasses or amorphous 'solids' like most polymers.

The name 'standard molar entropy of the substance B' and the custom of abbreviating $\{S_m^\ominus(\text{B, state, } T) - S_m(\text{B, s, } T \to 0)\}$ to $S_m^\ominus(\text{B, state, } T)$ seem to imply that the molar entropy of the solid at zero temperature may be taken as zero for all substances. That is a *convention*, consistent with Nernst's heat theorem; we have no means of knowing whether the entropy of a substance is or is not zero at zero temperature. To call '$S_m^\ominus(\text{B, state, } T)$' the 'absolute entropy' is indefensible.

In trying to devise a less restrictive convention some authors have said that the molar entropy of a solid tends to a '*constant*' as $T \to 0$. Unless the 'constant' is always zero that convention is inconsistent with Nernst's heat theorem; the entropy change as $T \to 0$ for say A + B = C would be '$-constant$' rather than zero. If one wishes to avoid the notion of 'absolute entropies' by avoiding ever talking of any entropy being zero then the appropriate convention is that the entropy of any substance $A_n B_m \cdots$ is the sum $(nk_A + mk_B + \cdots)$ where k_A, k_B, $\cdots$, are constants characteristic of the atoms A, B, $\cdots$. It nevertheless remains true that a convention, however useful, can reveal no scientific truth.

The values of '$S_m^\ominus(\text{B, state, } T)$' found in primary thermodynamic tables have usually been corrected, for the few exceptional substances, to the values which it is believed would have been obtained if the solid has been a perfectly ordered one at the lowest temperature reached, that is to say if it had obeyed Nernst's heat theorem.

---

† See for example Gill, E. K.; Morrison, J. A. *J. Chem. Phys.* **1966**, 45, 1585.

§ 15.2 Methods of testing Nernst's heat theorem for reactions

We have dealt in § 15.1 with the measurement of the molar entropy differences: $\{S_B^\ominus(\text{state}, T) - S_B(s, T \to 0)\}$, and shall now briefly review the measurement of the standard molar entropy change $\Sigma_B \nu_B S_B^\ominus(\text{state}, T)$ at some temperature T, often 298.15 K.

The most widely applicable method is based on equation (11.5.2) which we rewrite in the form:

$$\sum_B \nu_B S_B^\ominus(\text{state}, T) = R \ln\{K^\ominus(T)\} + \Delta H_m^\ominus(T)/T. \tag{15.2.1}$$

A direct measurement of the standard equilibrium constant $K^\ominus(T)$ and a calorimetric measurement of $\Delta H_m^\ominus(T)$ are sufficient. Alternatively we may determine $K^\ominus$ at each of several temperatures and use the equation:

$$\sum_B \nu_B S_B^\ominus(\text{state}, T) = d(RT \ln K^\ominus)/dT$$

$$= R \ln K^\ominus + RT \, d \ln K^\ominus/dT. \tag{15.2.2}$$

Another method is available for substances like sulphur, tin, cyclohexanol, and phosphine which exist in two solid forms α and β, one stable above a transition temperature $T^{\alpha+\beta}$ and the other stable below $T^{\alpha+\beta}$. Each form can be cooled below $T^{\alpha+\beta}$, though one form is then metastable with respect to the other. The two quantities $\{S_B^\ominus(s, \alpha, T^{\alpha+\beta}) - S_B(s, \alpha, T \to 0)\}$ and $\{S_B^\ominus(s, \beta, T^{\alpha+\beta}) - S_B(s, \beta, T \to 0)\}$ can each be measured calorimetrically, as can the standard molar entropy of transition $\{S_B^\ominus(s, \beta, T^{\alpha+\beta}) - S_B^\ominus(s, \alpha, T^{\alpha+\beta})\}$. The quantity $\{S_B(s, \beta, T \to 0) - S_B(s, \alpha, T \to 0)\}$ can therefore be determined.

In a galvanic cell of the kind:

$$\text{Ag(s)}|\text{AgCl(s)}|\text{solution containing Cl}^-|\text{HgCl(s)}|\text{Hg(l)}, \tag{15.2.3}$$

the e.m.f. $E = E^\ominus$ depends with high accuracy (see chapter 19) only on the chemical potentials of pure substances:

$$-E^\ominus F = \mu^\ominus(\text{Hg}, 1) + \mu^\ominus(\text{AgCl}, s) - \mu^\ominus(\text{Ag}, s) - \mu^\ominus(\text{HgCl}, s), \tag{15.2.4}$$

and is independent of the nature or molality of the electrolyte solution containing chloride ion. For such a cell we have

$$\sum_B \nu_B S_B^\ominus(s \text{ or } 1, T) = F \, dE^\ominus/dT. \tag{15.2.5}$$

The standard molar entropy changes of a wide variety of electron-transfer reactions have been meaured in this way by use of appropriate combinations of electrodes like Ag|AgCl|Cl$^-$, Hg|HgCl|Cl$^-$, Pt|Cl$_2$|Cl$^-$, Pb|PbCl$_2$|Cl$^-$, Tl|TlCl|Cl$^-$, Ag|AgI|I$^-$, and Pt|I$_2$|PbI$_2$|Pb^{2+}.

§ 15.3   Consequences of Nernst's heat theorem for pure substances

If it is true that the entropy of a solid at very low temperatures is independent of pressure then it follows that at such temperatures all the derivatives of S with respect to p are zero. In particular

$$(\partial S/\partial p)_T = -(\partial V/\partial T)_p \to 0 \quad \text{as } T \to 0, \tag{15.3.1}$$

and

$$(\partial^2 S/\partial p^2)_T = -\{\partial(\partial V/\partial p)_T/\partial T\}_p \to 0 \quad \text{as } T \to 0. \tag{15.3.2}$$

Equations (1) and (2) are tested for Ar(s) in figure 15.1; both $(\partial V_m/\partial T)_p$ and $\{\partial(\partial V_m/\partial p)_T/\partial T\}_p$ tend to zero as the temperature tends to zero, and the flatness of the curves suggests that higher derivatives also tend to zero.

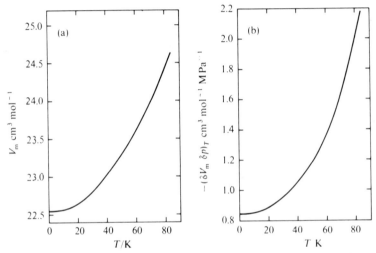

Figure 15.1. Plots of (a), the molar volume V_m and (b), the derivative $(\partial V_m/\partial p)_T$ against temperature for Ar(s).

For helium the liquid can coexist with the solid at temperatures near zero, and at pressures less than $p^{s+l}(T \to 0)$ the liquid remains stable as $T \to 0$. The molar volume of liquid helium at a pressure less than $p^{s+l}(T \to 0)$ also follows equation (1). At the equilibrium pressure $p^{s+l}(T)$, since the change of molar volume, $(V_m^l - V_m^s)$, remains finite it follows from Clapeyron's equation (9.3.3) that if $(S_m^l - S_m^s) \to 0$ as $T \to 0$ then

$$dp^{s+l}/dT \to 0 \quad \text{as } T \to 0. \tag{15.3.3}$$

Equation (3) is in accord with experiment; indeed it has been estimated that at least the first seven derivatives $d^n p^{s+l}/dT^n$ vanish as $T \to 0$.

The agreement with Nernst's heat theorem for liquid helium raises a question. How can a liquid be regarded as a 'perfectly ordered solid' and why then do the supercooled liquids called glasses fail to obey the theorem? If the answer is that at very low temperatures and a pressure less than $p^{s+1}(T \to 0)$ liquid helium is stable, while glasses are metastable with respect to perfectly ordered solids, then another question arises. How then can a glass, or solid CO, fail to obey the theorem while the metastable forms of tin, phosphine, and many other solids do obey the theorem? The answers to such questions are still far from clear.

§ 15.4   Spectroscopic entropies of gaseous substances

The differences $[S_B^\ominus(g, T) - R \ln\{g(\varepsilon_{0,B})\}]$ can be calculated for gaseous substances from spectroscopically determined quantities by the methods described in § 14.10. When the calculated values are compared with calorimetrically determined values of $\{S_B^\ominus(g, T) - S_B(s, T \to 0)\}$ we obtain values of the difference:

$$R \ln\{g(\varepsilon_{0,B})\} - S_B(s, T \to 0). \tag{15.4.1}$$

The values so obtained for the difference (1) are nearly always zero within experimental error. If the lowest energy level of a gaseous molecule is regarded as non-degenerate (though we have no way of knowing that) then when the difference (1) is zero we have $S_B(s, T \to 0) = 0$.

The values so obtained for the difference (1) are a little greater than zero for precisely those substances, CO, N_2O, NO, H_2O, and CH_3D, that are known to lead to failure of Nernst's heat theorem in the form (15.1.2). Moreover, the values obtained for the difference (1) agree within experimental error with the discrepancies from Nernst's heat theorem obtained from purely thermodynamic measurements for reactions involving one or more of the exceptional substances. The difference (1) could be made to disappear for CO and N_2O by calculating the 'spectroscopic entropy' with symmetry numbers of 2 rather than 1, that is to say by assuming that a CO or an N_2O (NNO) behaves as if the two ends of the molecule were identical.

Such considerations lead to the accepted explanation of the failure of Nernst's heat theorem for the few exceptional substances. If the orientational degeneracy of the solid were removed at the temperature T_0, so that the solid were perfectly ordered, then the difference (1) would always be zero and Nernst's heat theorem would always be obeyed in the form of equation (15.1.3). In CO and N_2O the degeneracy is ascribed to random arrangements of neighbouring CO and OC molecules or of NNO and ONN molecules in the solid at the temperature T_0. If the degeneracy were removed that would lead to an extra term $R \ln 2$ in $\{S_B^\ominus(g, T) - S_B(\text{p.o.s.}, T \to 0)\}$. We do not know whether the discrepancy is caused by the lowest temperature T_0 not being low enough, or by the 'freezing in' of the degeneracy, that is to say by the change from the randomly oriented state to the perfectly ordered state having become too slow at the temperature T_0 to be observed within the time of a calorimetric experiment.

Similar explanations have been offered for the other exceptional substances. For NO a discrepancy of $R \ln 2^{1/2}$ is accounted for by assuming random orientations of neighbouring $\overset{NO}{ON}$ dimers. For H_2O (and D_2O) a discrepancy of about $R \ln(3/2)$ is accounted for by assuming random occupation of four O—H—O groups by one approximately linear and three non-linear arrangements. For CH_3D and CHD_3 a discrepancy of $R \ln 4$, and for CH_2D_2 of $R \ln 6$, are accounted for by the persistence of random orientations of the C—H and C—D bonds.

For all these exceptional substances the values of '$S_m^\ominus(B, g, T)$' that appear in primary thermodynamic tables are those calculated from spectroscopic quantities by statistical-thermodynamic formulae with the proper symmetry numbers. Those values can then be used to evaluate $\Sigma_B \nu_B S_B^\ominus$, and thence $K^\ominus$, without any errors arising from the assumption of Nernst's heat theorem.

§ 15.5 Hydrogen

For hydrogen (and for deuterium) a discrepancy with Nernst's heat theorem appeared when the lowest temperature T_0 at which measurements of the heat capacity had been made was 12 K. When T_0 was lowered from 12 K to 2 K the heat capacity was found to pass through a minimum at about 6 K and then to increase again, promising to eliminate the discrepancy. If measurements of the heat capacity were to be made down to a still lower T_0 then there is little doubt that hydrogen would obey Nernst's heat theorem. In primary thermodynamic tables the value tabulated for '$S_m^\ominus(H_2, g, T)$' is that calculated from equation (14.11.9) with the appropriate partition functions q for o-H_2 and for p-H_2.

While the lowest temperature T_0 was 12 K, the calorimetric standard molar entropy of H_2 was apparently about $6.7 \, \mathrm{J \, K^{-1} \, mol^{-1}}$ too small. This corresponds within experimental error to $\frac{3}{4}R \ln 3$ arising from the H_2 at T_0 being ($\frac{1}{4}$p-$H_2 + \frac{3}{4}$o-H_2), the p-H_2 having no orientational degeneracy but the o-H_2 having three ($2j + 1$ with $j = 1$) energetically almost equivalent orientations. When $T_0 \ll 12$ K the apparent degeneracy of the o-H_2 is removed, the discrepancy with Nernst's heat theorem disappears, and there is no need even to know about the existence of nuclear-spin isomers.

§ 15.6 The 'third law of thermodynamics'

In spite of 70 years' effort no 'third law of thermodynamics' independent of statistical-mechanical arguments, universally obeyed, and somehow incorporating Nernst's heat theorem, has yet been formulated.

A close approach was made by the law of the unattainability of zero temperature, but in the words of Münster:† "We are therefore forced to conclude that the Nernst Heat Theorem is at least not necessarily a consequence of the unattainability principle and that the latter cannot assume the role of a more general principle."

Abandoning the restrictions of classical thermodynamics Guggenheim‡ states the third law in the form: "By the standard methods of *statistical thermodynamics* it is

---

† Münster, A. *Statistical Thermodynamics*. Vol. II. Springer-Verlag: Berlin and Academic Press: New York. **1974**. p. 79.

‡ Guggenheim, E. A. *Thermodynamics*. North-Holland: Amsterdam. Fifth edition. **1967**. p. 60.

possible to derive for certain entropy changes general formulae which cannot be derived from the *zeroth, first,* or *second laws of classical thermodynamics.* In particular one can obtain formulae for entropy changes in highly disperse systems (i.e. gases), for those in very cold systems (i.e. when $T \to 0$), and for those associated with the mixing of very similar substances (e.g. isotopes)." If there is a law then it is probably statistical-mechanical rather than classical-thermodynamic, but again in Münster's words:[†] "We are still far from understanding the Nernst heat theorem on a statistical basis and only this could form the foundation of an exact formulation."

Problems for chapter 15

Problem 15.1

You are given the directly measured standard equilibrium constant $K^{\ominus}(500 \text{ K}) = 0.0633$ for the reaction: $C_2H_5CH_2OH(g) = C_2H_5CHO(g) + H_2(g)$, and the following calorimetrically measured standard molar entropy increments and standard molar enthalpies of formation.

B	$C_2H_5CH_2OH(g)$	$C_2H_5CHO(g)$	$H_2(g)$
$\{S_B^{\ominus}(g, 500 \text{ K}) - S_B(s, T \to 0)\}/\text{J K}^{-1} \text{ mol}^{-1}$	376.6	353.4	145.6[a]
$\Delta_f H_B^{\ominus}(500 \text{ K})/\text{kJ mol}^{-1}$	−267.5	−194.5	0

Use these quantities to calculate a value of $\Sigma_B \nu_B S_B(s, T \to 0)$ for the reaction, and compare your value with that predicted by Nernst's heat theorem.

[a] For $H_2(g)$ the value was calculated from spectroscopic quantities by the appropriate statistical formula.

Problem 15.2

In place of the values of $\Delta_f H_B^{\ominus}$ in problem 15.1 you are given that the directly measured values of $K^{\ominus}(T)$ led to $d(RT \ln K^{\ominus})/dT = (122.8 \pm 0.5) \text{ J K}^{-1} \text{ mol}^{-1}$ at 500 K. Use the given quantities to calculate a value of $\Sigma_B \nu_B S_B(s, T \to 0)$ for the reaction, and compare your value with that predicted by Nernst's heat theorem.

Problem 15.3

The standard molar heat capacities $C_{p,m}^{\ominus}(M)$ of monoclinic sulphur and $C_{p,m}^{\ominus}(R)$ of rhombic sulphur were each measured from low temperatures up to the transition temperature 368.5 K. Values of the smoothed difference are given below. The calorimetrically measured molar enthalpy of transition is $(397 \pm 40) \text{ J mol}^{-1}$. Compare these results with that predicted by Nernst's heat theorem.

T/K	60	80	100	120	140	160	180	200	220
$10^3 \{C_{p,m}^{\ominus}(M) - C_{p,m}^{\ominus}(R)\}/\text{J K}^{-1} \text{ mol}^{-1}$	29	84	167	264	364	469	569	665	753

T/K	240	260	280	300	340	360	368.5
$10^3 \{C_{p,m}^{\ominus}(M) - C_{p,m}^{\ominus}(R)\}/\text{J K}^{-1} \text{ mol}^{-1}$	833	908	979	1046	1105	1115	1209

[†] Münster, A. *Classical Thermodynamics.* Wiley-Interscience: London. **1970**. p. 158.

Problem 15.4

Given for the galvanic cell $Hg(l)|HgCl(s)|HCl(aq)|Cl_2(g, p^{\ominus})|Pt$, that the temperature coefficient $dE^{\ominus}/dT$ of the e.m.f. $E^{\ominus}$ is -0.945 mV K^{-1} at 298.15 K, calculate the standard molar entropy change for the cell reaction. Given the calorimetrically determined values of $\{S_B^{\ominus}(s \text{ or } 1, 298.15 \text{ K}) - S_B(s, T \to 0)\}$: $Cl_2(g)$, 223.0 J K^{-1} mol^{-1}; HgCl(s), 96.2 J K^{-1} mol^{-1}; and Hg(l), 76.0 J K^{-1} mol^{-1}, calculate $\Sigma_B \nu_B S_B(s, T \to 0)$ and compare your value with that predicted by Nernst's heat theorem.

Problem 15.5

Calculate the 'spectroscopic' molar entropy of $H_2O(g)$ at 298.15 K given that $M(H_2O) = 0.01802$ kg mol^{-1}, $\theta_R = 22.3$ K, and $\theta_V/K = 2290$, 5250, and 5400. Compare your result with the calorimetrically determined quantity: $\{S_m^{\ominus}(H_2O, g, 298.15 \text{ K}) - S_m(H_2O, s, T \to 0)\} = 185.3$ J K^{-1} mol^{-1}.

Chapter 16

Liquid mixtures

§ 16.1 Introduction and scope

Although the present chapter is entitled 'Liquid mixtures', these may be taken to include pure liquids; they should also be taken to include homogeneous solid mixtures ('solid solutions') where these are relevant. We postpone a discussion of solutions, especially dilute solutions, until chapter 18, reminding the reader in the meantime that a solution is conventionally distinguished from a mixture (see § 2.10) by our choosing to regard one (or more) of the components of a solution as the solvent and the remainder as solutes, and by then treating the solvent differently from the solutes. In a mixture the components are all treated on a par.

We shall confine ourselves to mixtures of two substances; extension to more than two components is straightforward. We shall also confine ourselves in this chapter to liquids (or solids) under pressures within about 100 kPa of atmospheric pressure, that is to say at temperatures not greater than about $0.7T^c$ where T^c is the gas-liquid critical temperature of either component. The behaviour of fluid mixtures over larger ranges of pressure and temperature, and of fluid mixtures of substances having widely different critical temperatures, will be dealt with in chapter 17.

§ 16.2 Mixing functions

We recall from § 2.9 that the molar functions of mixing for a binary mixture $\{(1-x)A + xB\}$, where x denotes the mole fraction of B, are defined by

$$\Delta_{\mathrm{mix}}X_m \stackrel{\mathrm{def}}{=} X_m(T, p, x) - (1-x)X_m(T, p, 0) - xX_m(T, p, 1)$$

$$= (1-x)(X_A - X_A^*) + x(X_B - X_B^*),\qquad (16.2.1)$$

where X denotes any extensive quantity such as G, H, S, or V; X_A and X_B denote partial molar quantities (see § 2.7); and * denotes 'pure'.

§ 16.3 Ideal mixtures

An *ideal mixture* is defined as one for which

$$\lambda_A^{id}(T, p, x) \stackrel{\mathrm{def}}{=} (1-x)\lambda_A^*(T, p), \qquad \lambda_B^{id}(T, p, x) \stackrel{\mathrm{def}}{=} x\lambda_B^*(T, p).\qquad (16.3.1)$$

For an ideal mixture the molar functions of mixing are given by

$$\Delta_{mix}G_m^{id} \overset{\text{def}}{=} RT\{(1-x)\ln(1-x)+x\ln x\}, \tag{16.3.2}$$

$$\Delta_{mix}S_m^{id} \overset{\text{def}}{=} -R\{(1-x)\ln(1-x)+x\ln x\}, \tag{16.3.3}$$

$$\Delta_{mix}H_m^{id} \overset{\text{def}}{=} 0, \tag{16.3.4}$$

$$\Delta_{mix}V_m^{id} \overset{\text{def}}{=} 0. \tag{16.3.5}$$

For an ideal mixture the molar Gibbs function of mixing is negative and the molar entropy of mixing is positive† for any value of x.

No real mixture is exactly ideal, but mixtures of very similar substances such as $\{(1-x)n\text{-}C_6H_{14}+xn\text{-}C_7H_{16}\}$ or $\{(1-x)C_6H_6+xC_6H_5CH_3\}$ do behave more or less ideally.

An ideal mixture is not quite the same as a mixture that obeys Raoult's law; this point will be discussed in § 16.10.

§ 16.4   Excess functions

It has become customary to use instead of the molar function of mixing $\Delta_{mix}X_m$ the excess molar function X_m^E defined by

$$X_m^E \overset{\text{def}}{=} \Delta_{mix}X_m - \Delta_{mix}X_m^{id}, \tag{16.4.1}$$

so that 'excess' means 'excess over ideal'. The excess molar functions are given by

$$G_m^E = \Delta_{mix}G_m - RT\{(1-x)\ln(1-x)+x\ln x\}, \tag{16.4.2}$$

$$S_m^E = \Delta_{mix}S_m + R\{(1-x)\ln(1-x)+x\ln x\}, \tag{16.4.3}$$

$$H_m^E = \Delta_{mix}H_m, \tag{16.4.4}$$

$$V_m^E = \Delta_{mix}V_m. \tag{16.4.5}$$

§ 16.5   Activity coefficients

The *activity coefficients* f_A and f_B of the components *of a mixture*‡ are defined by the relations:

---

† Notwithstanding statements to the contrary in many elementary textbooks, for a real mixture the molar entropy of mixing need not be positive. Thermodynamics inposes no such restriction, and nature sometimes exercises its right to make $\Delta_{mix}S_m(T, p, x)$ negative. It is negative (about -8.8 J K^{-1} mol^{-1}) for example for $\{0.5(C_2H_5)_2NH+0.5H_2O\}$ at 322.25 K and atmospheric pressure.

‡ Compare the use in chapter 18 of the same name, activity coefficient, but a different symbol, γ_B, for the solutes in a solution. Compare also the use in § 12.7 of the same symbol, f_B, but a different name, fugacity, for a property of a gaseous mixture. The notation used here is that internationally recommended.

$$RT \ln f_A \stackrel{\text{def}}{=} G_m^E - x(\partial G_m^E/\partial x)_{T,p} = \mu_A^E = RT \ln\{\lambda_A/(1-x)\lambda_A^*\}, \quad (16.5.1)$$

$$RT \ln f_B \stackrel{\text{def}}{=} G_m^E + (1-x)(\partial G_m^E/\partial x)_{T,p} = \mu_B^E = RT \ln\{\lambda_B/x\lambda_B^*\}. \quad (16.5.2)$$

§ 16.6 Gibbs–Duhem equation

We recall the Gibbs–Duhem equation (5.7.2), which we rewrite for a binary mixture in the form:

$$(H_m/RT^2)\, dT - (V_m/RT)\, dp + (1-x)\, d \ln \lambda_A + x\, d \ln \lambda_B = 0. \quad (16.6.1)$$

At constant temperature and pressure equation (1) becomes

$$(1-x)\, d \ln \lambda_A + x\, d \ln \lambda_B = 0, \quad (T, p \text{ constant}). \quad (16.6.2)$$

We note that ideal mixtures are thermodynamically possible. We further note that equation (2) can be rewritten in the form:

$$(1-x)\, d \ln f_A + x\, d \ln f_B = 0, \quad (T, p \text{ constant}). \quad (16.6.3)$$

By rearrangement and integration of equation (3) we obtain

$$\int_0^1 \ln(f_B/f_A)\, dx = 0, \quad (T, p \text{ constant}). \quad (16.6.4)$$

Equation (4) is a necessary, though decidedly not sufficient, criterion for thermodynamic consistency of separately measured values of f_A and f_B; it is sometimes useful as a first test by allowing us to say that some results are not thermodynamically consistent. The limitations of equation (4) as a test will be discussed in § 16.8.

§ 16.7 Measurement of the excess functions

We have already dealt in § 4.11 with calorimetric methods for the measurement of the excess molar enthalpy H_m^E, and in § 2.9 with dilatometric methods for the measurement of the excess molar volume V_m^E. Here we shall concentrate on methods of measurement of the excess molar Gibbs function G_m^E (or of μ_A^E and μ_B^E, or of f_A and f_B). The excess molar entropy S_m^E can be obtained by use of the relation: $S_m^E = (H_m^E - G_m^E)/T$, and so need not be separately discussed. Alternatively, if G_m^E is measured at each of several temperatures then S_m^E follows as $-(\partial G_m^E/\partial T)_{p,x}$ and H_m^E as $\{G_m^E - T(\partial G_m^E/\partial T)_{p,x}\}$. It is usually preferable, however, to rely on a calorimetric measurement of H_m^E and to measure G_m^E at only one temperature; numerical differentiation is an inaccurate process, especially when the available temperature range is small.

The key to the determination of the molar Gibbs function $G_m^l(T, p, x^l)$ of a binary liquid mixture is the determination of the mole fractions x^l and x^g and the vapour pressure p^{l+g} of coexisting liquid and gaseous phases. Only two of those three quantities are needed since $\mathcal{F} = \mathcal{C} + 2 - \mathcal{P} = 2 + 2 - 2 = 2$; the third can be obtained by use of the Gibbs–Duhem equation (16.6.1) for each phase. Nevertheless all three quantities are sometimes measured and can then be used to test the results for thermodynamic consistency. Indeed, the experimental method in which all three quantities are measured, from a recirculating still, is often so unreliable that it is as well that we can test the results for thermodynamic consistency. Modern methods usually depend on the measurement of x^l and p^{l+g} or of x^g and p^{l+g} at true equilibrium.

So determined, $G_m^l(T, p, x^l)$ contains two standard chemical potentials $\mu_A^{\ominus}(g, T)$ and $\mu_B^{\ominus}(g, T)$. These can be eliminated, and G_m^E can be obtained, by measurements also of the vapour pressures p_A^{l+g} and p_B^{l+g} of the two pure substances. We shall now develop the necessary algebra.

We begin by writing the chemical potential of the substance A in the liquid phase in the form:

$$\mu_A^l(T, p, x^l) = \mu_A^l(T, p^{l+g}, x^l) + \int_{p^{l+g}}^{p} V_A^l(T, p, x^l)\, dp. \qquad (16.7.1)$$

The condition for equilibrium of liquid and gaseous phases is

$$\mu_A^l(T, p^{l+g}, x^l) = \mu_A^g(T, p^{l+g}, x^g), \qquad (16.7.2)$$

which, used with equation (12.6.15):

$$\mu_A^g(T, p^{l+g}, x^g) = \mu_A^{\ominus}(g, T) + RT\, \ln\{(1 - x^g)p^{l+g}/p^{\ominus}\}$$
$$+ \int_0^{p^{l+g}} \{V_A^g(T, p, x^g) - RT/p\}\, dp, \qquad (16.7.3)$$

allows us to rewrite equation (1) in the form:

$$\mu_A^l(T, p, x^l) = \mu_A^{\ominus}(g, T) + RT\, \ln\{(1 - x^g)p^{l+g}/p^{\ominus}\}$$
$$+ \int_0^{p^{l+g}} \{V_A^g(T, p, x^g) - RT/p\}\, dp + \int_{p^{l+g}}^{p} V_A^l(T, p, x^l)\, dp. \qquad (16.7.4)$$

If the vapour pressure p^{l+g} is low enough to ensure that we may neglect higher virial coefficients than the second in the equation of state of the gas then we may use equations (12.6.6), (12.6.7), and (12.6.10) to write

$$\int_0^{p^{l+g}} \{V_A^g(T, p, x^g) - RT/p\}\, dp = B_{AA}p^{l+g} + 2(x^g)^2 \delta_{AB}p^{l+g}. \qquad (16.7.5)$$

If the vapour pressure p^{l+g} is close enough to the pressure p then we may assume that the partial molar volume V_A^l in the liquid phase is independent of pressure and write

$$\int_{p^{l+g}}^{p} V_A^l(T, p, x^l)\, dp = V_A^l(T, x^l)(p - p^{l+g}). \qquad (16.7.6)$$

Substitution of (5) and (6) into (4) yields

$$\mu_A^l(T, p, x^l) = \mu_A^{\ominus}(g, T) + RT\, \ln\{(1 - x^g)p^{l+g}/p^{\ominus}\} + B_{AA}p^{l+g}$$
$$+ 2(x^g)^2 \delta_{AB}p^{l+g} + V_A^l(T, x^l)(p - p^{l+g}). \qquad (16.7.7)$$

By subtracting from (7) the corresponding equation for pure substance A and then using equation (16.5.1) we obtain for the excess chemical potential of A

$$\mu_A^E(T, p, x^l) = RT \ln\{f_A(T, p, x^l)\}$$
$$= RT \ln\{(1-x^g)p^{l+g}/(1-x^l)p_A^{*l}\} + (B_{AA} - V_A^{*l})(p^{l+g} - p_A^{l+g})$$
$$+ 2(x^g)^2 \delta_{AB} p^{l+g} + (V_A^l - V_A^{*l})(p - p^{l+g}). \qquad (16.7.8)$$

For the substance B we similarly obtain

$$\mu_B^E(T, p, x^l) = RT \ln\{f_B(T, p, x^l)\}$$
$$= RT \ln\{x^g p^{l+g}/x^l p_B^{*l}\} + (B_{BB} - V_B^{*l})(p^{l+g} - p_B^{l+g})$$
$$+ 2(1-x^g)^2 \delta_{AB} p^{l+g} + (V_B^l - V_B^{*l})(p - p^{l+g}). \qquad (16.7.9)$$

By use of the relation: $G_m^E = (1-x)\mu_A^E + x\mu_B^E$, we finally obtain

$$G_m^E(T, p, x^l) = (1-x^l)RT \ln\{(1-x^g)p^{l+g}/(1-x^l)p_A^{*l}\} + x^l RT \ln\{x^g p^{l+g}/x^l p_B^{*l}\}$$
$$+ (1-x^l)(B_{AA} - V_A^{*l})(p^{l+g} - p_A^{l+g}) + x^l(B_{BB} - V_B^{*l})(p^{l+g} - p_B^{l+g})$$
$$+ \{(1-x^l)(x^g)^2 + x^l(1-x^g)^2\}2\delta_{AB} p^{l+g}$$
$$+ V_m^E(T, x^l)(p - p^{l+g}). \qquad (16.7.10)$$

The relative magnitudes of the terms in equation (10) vary of course from mixture to mixture; typically those in the first row might be 500 J mol^{-1}, those in the second row 50 J mol^{-1}, that in the third row 5 J mol^{-1}, and that in the last row if $|p - p^{l+g}| <$ 100 kPa then about 0.1 J mol^{-1}. The last term is often negligible; the molar excess Gibbs function $G_m^E(T, p, x^l)$ is approximately independent of the pressure p when p is sufficiently close to the vapour pressures p^{l+g} for all values of x^l. In the absence of measurements of δ_{AB} or of a reliable method of estimating it, the term in the third row is also often ignored.

§ 16.8   The recirculating still for measurements of excess molar Gibbs functions

In a recirculating still a liquid mixture is boiled and the vapour is condensed and returned to the still via a trap where some of it is retained as liquid for analysis. When the still has reached a steady state at the chosen temperature (not an equilibrium state, but one hopes not significantly different) the pressure p^{l+g} is measured and samples of the liquid phase and of the (condensed) gaseous phase are taken for analysis. The analysis may be done by measurement of the density or the refractive index or other suitably varying property and comparison with values obtained for synthetic mixtures of known composition. For rough measurements gas-liquid chromatography is suitable. Occasionally (for example for mixtures of HCl and H_2O) chemical analysis is suitable. The resulting values of x^l, x^g, and p^{l+g} at the temperature T can be used, with values of the second virial coefficients and of the molar volumes of the liquids, to calculate $G_m^E(T, p, x^1)$ directly from equation (16.7.10).

Many kinds of recirculating still have been used. The best of them were carefully designed to ensure steady boiling (without 'bumping' or 'flash boiling'), to avoid premature partial condensation (and consequent enrichment of the gaseous phase with respect to the more volatile component), to avoid entrainment of droplets of the liquid phase in the gaseous phase (with consequent impoverishment of the gaseous phase with respect to the more volatile component), to ensure that the measured temperature is as close as possible to that of equilibrium of liquid and gas at the measured pressure, and to ensure that the condensed sample of the gaseous phase is uniform. One of the most successful recirculating stills is shown in figure 16.1; further details of the design and operation can be found in the literature.†

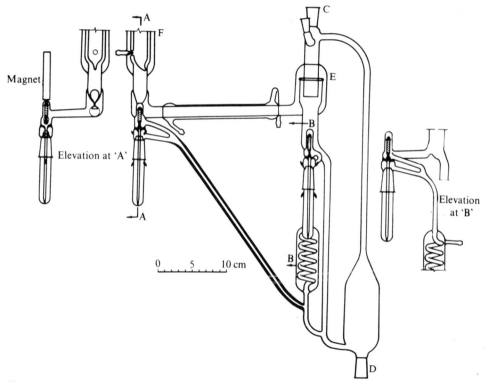

Figure 16.1. A recirculating still for measurements of the molar excess Gibbs function. At C a platinum resistance thermometer was inserted, and at D a heating coil of bare platinum wire. The boiler acted as a Cottrell pump (like a coffee percolator) by spraying an intimate mixture of liquid and gas over the thermometer in the 'disengagement chamber' E. The boiler was surrounded by a nichrome heating coil and was then heavily lagged up to E. The condenser at F was connected through a large volume of air as a pressure ballast and through pressure-adjusting apparatus to a precise manometer.
(Reproduced with permission from Brown, I. *Aust. J. Sci. Res. A* **1952**, 5, 530.)

† Brown, I.; Ewald, A. H. *Aust. J. Sci. Res. A* **1950**, 3, 306; Brown, I. *Ibid.* **1952**, 5, 530.

The 'equal-area' test according to equation (16.6.4), traditionally made on measurements of p^{1+g}, x^1, and x^g from recirculating stills, is severely limited in at least two ways. First, the integral $\int_0^1 \ln(f_B/f_A)\,dx$ may be close to zero even when the results are grossly inconsistent thermodynamically; for example if $\ln f_A = Ax$ and $\ln f_B = A(1-x)$ the integral is zero in spite of those equations failing to satisfy the Gibbs–Duhem equation (16.6.3). Second, in the function $\ln(f_B/f_A)$ the pressure p^{1+g} has almost cancelled out; the 'equal-area' test can sometimes be used to detect errors in the measurements of the compositions x^1 and x^g, but will fail to detect quite serious errors in measurements of the equilibrium pressures p^{1+g}.† Nevertheless an example of the 'equal-area' test is shown in figure 16.2 for a set of results from a recirculating still.

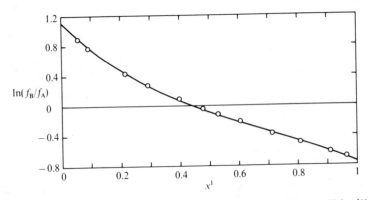

Figure 16.2. A test of equation (16.6.4) for results obtained with a recirculating still for $\{(1-x^1)CCl_4 + x^1CH_3COCH_3\}$ at 318.15 K. ○, Experimental points. The curve is that calculated from the equation: $G_m^E/RT = x^1(1-x^1)\{0.8274 + 0.1764(1-2x^1) + 0.1033(1-2x^1)^2\}$, and so is entirely consistent with the Gibbs–Duhem equation (16.6.3).

Nowadays recirculating stills are used mostly for producing isobaric phase diagrams for chemical-engineering purposes. However, just as 'ebulliometric' methods have found favóur over 'static' methods for measurements of the vapour pressures of pure liquids (see § 9.5), so redesigned recirculating stills might be re-established, with the method of analysis discussed in the following section, for precise measurements of p^{1+g} and x^1 of liquid mixtures.

§ 16.9 Equilibrium methods for measurements of excess molar Gibbs functions

The use of 'static' equilibrium measurements of x^g and p^{1+g} (or of x^1 and p^{1+g}) to obtain excess molar Gibbs functions G_m^E has many advantages and only two possible disadvantages over the use of a recirculating still. First, the calculations are rather heavier; the Gibbs–Duhem equation (16.6.1) must be used to evaluate x^1 (or x^g) which can then be used with x^g and p^{1+g} (or x^1 and p^{1+g}) to evaluate G_m^E from equation

† Marsh, K. N. *SPRCT(II)*, chapter 1, p. 4.

(16.7.10). That disadvantage disappeared with the advent of computers. Second, much greater attention must be paid to the rigorous degassing of the mixture. That disadvantage is by far outweighed by the loss of the uncertainties inherent in any recirculating still, by the much shorter times taken by static measurements, by the much smaller amounts of substance needed, and by the avoidance of analysis by the use of synthetic mixtures of accurately known composition.

Measurements of x^g and p^{l+g} are more suitable for mixtures of liquids having similar volatilities; measurements of x^l and p^{l+g} are more suitable for mixtures of liquids having widely different volatilities. We shall now discuss each method in turn, beginning with that of x^g and p^{l+g}.

A gaseous mixture of accurately known mole fraction x^g is prepared as follows in a reservoir, consisting of a flask enclosed in a suitably heated oven. Two ampoules containing known masses of the substances are sealed into ampoule holders† attached to the flask. The flask is thoroughly evacuated and then the ampoules are broken. The gaseous mixture is thoroughly stirred,‡ for example by prolonged forced convection.

The gaseous mixture is then admitted, all chances of condensation being prevented by adequate heating of connecting tubes, into an evacuated manometer with one closed limb and with mercury withdrawn into the mercury-supply tube. A sketch of the manometer is shown in figure 16.3(a). The whole manometer is enclosed in a glass-fronted thermostat. A sample of the gaseous mixture is then trapped in the closed limb by raising the mercury, the supply line from the reservoir is closed, and the open limb of the manometer is evacuated. A little care and experience are needed to ensure that no condensation takes place and that the sample of gaseous mixture remains trapped in the closed limb when the open limb is evacuated.

The pressure of the mixture is measured in steps through the sharp discontinuity, illustrated in figure 16.4, that occurs when the 'dew pressure' is reached and liquid begins to be formed. At the dew point the gaseous mixture has the mole fraction x^g and the pressure is p^{l+g}. The measurement may be checked by re-expansion through the dew point.

At constant temperature the Gibbs–Duhem equation (16.6.1) may be written for $\mu_A^g = \mu_A^l = \mu_A$, $\mu_B^g = \mu_B^l = \mu_B$, and $p = p^{l+g}$, in the form:

$$(1-x^l)(d\mu_A - V_A^l \, dp^{l+g}) + x^l(d\mu_B - V_B^l \, dp^{l+g}) = 0 , \qquad (16.9.1)$$

which may be solved for x^l to give

$$x^l = (d\mu_A - V_A^l \, dp^{l+g})/(d\mu_A - V_A^l \, dp^{l+g} - d\mu_B + V_B^l \, dp^{l+g}) . \qquad (16.9.2)$$

---

† An account of these, and of many other experimental points, is given by Williamson, A. G. *An Introduction to Non-Electrolyte Solutions.* Oliver and Boyd: Edinburgh. **1967**.

‡ The difficulty of thoroughly mixing two gases, especially when they have rather different densities, should not be underestimated.

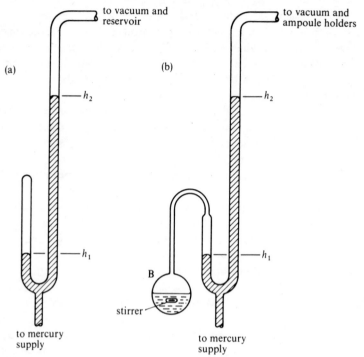

Figure 16.3. Sketches of apparatus suitable for measurements (a), of x^g and p^{l+g} and (b), of x^l and p^{l+g}.

By differentiation with respect to x^g of (16.7.3) with (16.7.5), and of the corresponding equation for $d\mu_B^g(T, p^{l+g}, x^g)$, equation (2) can be expressed in the form:

$$x^l = x^g \frac{RT - (1-x^g)\{RT/p^{l+g} + \beta_A + 2(x^g)^2\delta_{AB}\}\, dp^{l+g}/dx^g - 4x^g(1-x^g)\delta_{AB}p^{l+g}}{RT - x^g(1-x^g)\{\beta_A - \beta_B - 2(1-2x^g)\delta_{AB}\}\, dp^{l+g}/dx^g - 4x^g(1-x^g)\delta_{AB}p^{l+g}},$$
(16.9.3)

where $\beta_C = (\beta_{CC} - V_C^{*l})$ and where we have ignored any distinction between V_C^l and V_C^{*l}, an approximation equivalent to ignoring the last term in equation (16.7.10). A set of measured values of x^g and p^{l+g} is used to produce a corresponding set of values of dp^{l+g}/dx^g. Substitution into (3) then gives the corresponding set of values of x^l, and substitution of x^g, x^l, and p^{l+g} into (16.7.10) gives $G_m^E(T, p, x^l)$.

Alternatively, a particular algebraic form with m disposable parameters may be assumed for $G_m^E(x^l)$. The corresponding forms can be found for $\mu_A^E(x^l)$ and $\mu_B^E(x^l)$ by use of (16.5.1) and (16.5.2). Identification of those forms with (16.7.8) and (16.7.9) gives $2n$ equations for n values of x^l and m parameters. These are sufficient to solve the problem by iteration from trial values of the unknowns if $n \geqslant m$ and if the chosen algebraic form was suitable.[†] The form:

$$G_m^E(T, p, x^l) = x^l(1-x^l)\{A_1 + A_2(1-2x^l) + A_3(1-2x^l)^2 + \cdots\}, \quad (16.9.4)$$

† It is usually desirable to make about 10 measurements spread over the range of mole fractions.

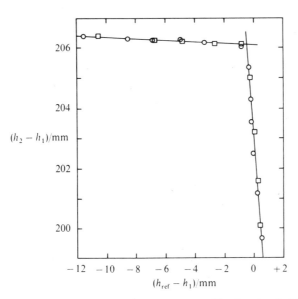

Figure 16.4. Measurement of the dew pressure in an apparatus like that sketched in figure 16.3(a) for $\{(1-x^g)C_6H_6 + x^g C_6H_{12}\}$ at 313.15 K with $x^g \approx 0.4$. The difference in height $(h_2 - h_1)$, proportional to the pressure, is plotted against $(h_{ref} - h_1)$, a linear function of the volume, where h_{ref} is the height of a reference mark on the manometer. $\bigcirc$, Measurements taken on compression; $\square$, measurements taken on re-expansion.

is widely used, and is found to fit the results for most mixtures with at most $m = 4$. For some mixtures the form (4) fails; then some other form must be tried, or better the first method must be used.

We now turn to measurements of x^l and p^{l+g}. An apparatus like that shown in figure 16.3(b), with the mercury withdrawn into the supply line, is thoroughly evacuated. The bulb B, containing a magnetically operated stirrer, is immersed in a beaker of liquid nitrogen. A mixture of mole fraction x is made by transferring known masses of the two substances to the bulb B. The mixture is confined to B by raising the level of mercury; then the liquid nitrogen is removed, and the whole manometer is immersed in a glass-fronted thermostat. Measurements of $(h_2 - h_1)$ then lead to p^{l+g}. The mole fraction x^l in the liquid phase is not known exactly; some of the mixture is in the gaseous phase. The value of x^l can be found by correction from the value of x if the volume and the equation of state of the gaseous mixture are known.

A set of values of x^l and p^{l+g} is analysed by one of the two methods explained for x^g and p^{l+g}. The solution of the Gibbs–Duhem equation (16.9.1) for x^g is now a little more complicated but can be found by numerical integration. Alternatively, some analytical equation like (4) may be assumed for G_m^E, and the n values of x^g and the m parameters can be found iteratively if $n \geqslant m$ and if the chosen equation was suitable.

The $p^{l+g}(x^l)$ method sometimes has a disadvantage compared with the $p^{l+g}(x^g)$ method when the liquids have similar volatilities. Although the interface of liquid-to-gas is designed to be below the level of mercury at h_1, a fluctuation of temperature can cause liquid to appear on the mercury surface. Equilibrium, with all the liquid phase in the bulb B, is then reached only after slowly attenuated oscillations arising from the enthalpies of evaporation. The liquid on the mercury surface begins to move back, under the influence of the earth's gravitational field, through the gaseous phase to recondense in the bulb B; that causes the temperature of the liquid remaining at the mercury surface to fall and that of the liquid in the bulb B to rise, and that causes more liquid to evaporate from the bulb B and condense on the mercury surface.

The disadvantage disappears, and the $p^{l+g}(x^l)$ method becomes much better than the $p^{l+g}(x^g)$ method, when there is a considerable disparity between the volatilities of the two liquids. In the extreme case when the liquid B is almost involatile, $x^g = 0$, and equation (8), with the last term ignored as before, becomes

$$\mu_A^E(T, p, x^l) = RT \ln\{p^{l+g}/(1-x^l)p_A^{*l}\}+(B_{AA}-V_A^{*l})(p^{l+g}-p_A^{*l}). \qquad (16.9.5)$$

The other excess chemical potential $\mu_B^E(T, p, x^l)$ can then be found by integration of the Gibbs–Duhem equation at constant temperature and pressure, and $G_m^E(T, p, x^l)$ follows.

A third method is to determine the 'dew pressure' $p^d (x^g = x)$ and the 'bubble pressure' $p^b (x^l = x)$ for the same mixture of definite but only roughly known x. Equations (16.7.8) and (16.7.9) then provide $4n$ equations in the $3n$ unknown mole fractions and m parameters of a proposed equation for G_m^E, which can again be solved iteratively if $n \geq m$.

Reviews of experimental methods of measurement of excess functions can be found in the literature.† A method, not dealt with here, of measurement of the activity coefficient at infinite dilution of volatile A in a mixture with involatile B by gas-liquid chromatography has also been reviewed.‡

§ 16.10   Raoult's law

According to Raoult's law:

$$(1-x^g)p^{l+g} = (1-x^l)p_A^{*l}, \quad x^g p^{l+g} = x^l p_B^{*l}. \qquad (16.10.1)$$

By comparison with (16.7.8) and (16.7.9) we see that an ideal mixture ($\mu_A^E = 0$ and $\mu_B^E = 0$) obeys Raoult's law only if the gaseous mixture is perfect and if the molar volumes of the liquids are negligibly small.

Elementary textbooks sometimes assert that deviations from ideality must be either everywhere positive for both components or everywhere negative for both

---

† Williamson, A. G. ETd(II), chapter 16 (part 1).
Marsh, K. N. SPRCT(II), chapter 1.
‡ Letcher, T. M. SPRCT(II), chapter 2.

components, and similarly for deviations from Raoult's law. One of those two kinds of behaviour is indeed often observed, but the only thermodyamic restriction is that deviations from ideality must satisfy the Gibbs–Duhem equation (16.6.3), and many examples are known for which one or both of μ_A^E and μ_B^E changes sign between $x^1 = 0$ and $x^1 = 1$. Certainly the deviations from ideality cannot be positive for one component and negative for the other component for all values of x^1, but there is no such restriction on the deviations from Raoult's law. These points are illustrated in figure 16.5 for mixtures of acetone and nitromethane. The deviations from Raoult's law are everywhere negative for acetone and everywhere positive for nitromethane. The deviations from ideality are everywhere negative for acetone, but change sign from negative at low mole fractions of nitromethane to positive at high mole fractions of nitromethane. The curves in figure 16.5(b) are those calculated from an equation for $G_m^E(x^1)$, and so automatically obey the Gibbs-Duhem equation (16.6.3).

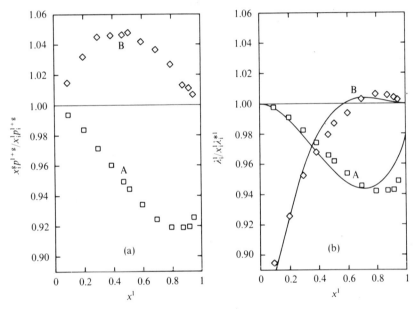

Figure 16.5. Plots against x^1 for $\{(1 - x^1)CH_3COCH_3 + x^1CH_3NO_2\}$ at 318.15 K of (a), deviations from Raoult's law (Raoult's law is obeyed if $x_i^g p^{1+g}/x_i^l p_i^{1+g} = 1; i = A, B$) and (b), deviations from ideality (for an ideal mixture $\lambda_i^l/x_i^l \lambda_i^{*l} = f_i = 1; i = A, B$). The points were obtained from measurements made with a recirculating still: $\square$, A = CH_3COCH_3; $\diamondsuit$, B = CH_3NO_2. The curves in figure 16.5(b) are those calculated from the equation: $G_m^E/RT = -x^1(1 - x^1)\{0.1025 + 0.0799(1 - 2x^1)\}$.

We note that

$$\partial\mu_A^E/\partial x^1 \to 0 \quad \text{as } x^1 \to 0, \quad \text{and} \quad \partial\mu_B^E/\partial x^1 \to 0 \quad \text{as } x^1 \to 1. \tag{16.10.2}$$

In the language of solutions: in any sufficiently dilute solution the *solvent* behaves ideally. It is on that basis that we shall treat dilute solutions differently from mixtures in chapter 18.

§ 16.11   Brief review of results for excess functions

Excess molar functions would be found in secondary thermodynamic tables. However, even for binary mixtures, a prodigious amount of work has resulted in values for only a small fraction of all the possible pairs of liquids, often measured at only one temperature. A bibliography of measured excess functions for binary mixtures has been published.[†] Usually only G_m^E, H_m^E, and V_m^E, or some of those, and occasionally also some of $C_{p,m}^E$, $(\partial V_m^E/\partial T)_p$, and $(\partial V_m^E/\partial p)_T$, have been measured. The excess functions, and their derivatives with respect to temperature and pressure, vary greatly in sign and magnitude.[‡]

   Except for special classes of mixtures carefully chosen to test some statistical-mechanical theory based on a model, measurements of excess functions, though interesting and exacting to make, lead to few reliable generalizations. Many of the vast number of papers reporting new excess functions include a 'rationalization' of the results in terms of naive speculations about the ways in which complex molecules might interact and the effects that such interactions might have on the excess functions. The task of accumulating all the excess functions of all mixtures seems hopeless, and the random acquisition of results seems pointless. That is why the development of statistical-mechanical theories is so important in this field.

§ 16.12   Conformal mixtures

Common to all theories of *conformal mixtures* of substances A and B is the assumption that the pair-interaction energies u_{ij} are given as functions of the intermolecular separation r by

$$u_{ij}(r) = \varepsilon_{ij}\Phi(r/r_{ij}^*)\,, \quad (i, j = A, B)\,, \tag{16.12.1}$$

where ε_{ij} and r_{ij}^* are an energy and a distance characteristic of the pair of molecules i and j, and Φ is the same function for all i, j. Consequently, applications of theories of conformal mixtures strictly are limited to mixtures of spherical molecules that conform with a principle of corresponding states. However, the theories may also be applied usefully to mixtures of nearly spherical molecules that conform roughly with a principle of corresponding states.

   Of the many theories based on equation (1) we shall consider here only a *one-fluid theory*. In a one-fluid theory it is assumed that a liquid mixture may be represented, as far as its configurational properties are concerned, by a hypothetical pure fluid ψ which also conforms with equation (1), so that

$$u_{\psi\psi}(r) = \varepsilon_{\psi\psi}\Phi(r/r_{\psi\psi}^*)\,. \tag{16.12.2}$$

Three further assumptions are needed to specify a particular one-fluid theory: the function Φ must be established; a pair of *one-fluid relations* must be assumed for $\varepsilon_{\psi\psi}$

† Hicks, C. P. SPRCT(II), chapter 9, pp. 275–538.
‡ Rowlinson, J. S. *Liquids and Liquid Mixtures*. Second edition. Butterworth: London. **1969**.
Swinton, F. L. SPRCT(II), chapter 5.
Williamson, A. G. SPRCT(II), chapter 6.

and $r_{\psi\psi}^*$ in terms of ε_{AA}, ε_{AB}, ε_{BB}, r_{AA}^*, r_{AB}^*, and r_{BB}^*; and a pair of *combining rules*, like (12.6.33) and (12.6.34), must be assumed for ε_{AB} and r_{AB}^* in terms of ε_{AA}, ε_{BB}, r_{AA}^*, and r_{BB}^*.

Of the several possible methods of establishing the function Φ we shall choose an explicit equation of state, in particular that of Carnahan and Starling (12.5.18). (Other methods used to establish Φ include the use of the actual physical properties of pure A and pure B, or a statistical-mechanical model, such as the 'cell model' of a liquid.)[†]

For the one-fluid relations we shall follow van der Waals by choosing the forms (12.6.31) and (12.6.32), or in our present notation,

$$r_{\psi\psi}^* = \{(1-x)^2 (r_{AA}^*)^3 + 2x(1-x)(r_{AB}^*)^3 + x^2 (r_{BB}^*)^3\}^{1/3}, \tag{16.12.3}$$

$$\varepsilon_{\psi\psi} = \{(1-x)^2 \varepsilon_{AA}(r_{AA}^*)^3 + 2x(1-x)\varepsilon_{AB}(r_{AB}^*)^3 + x^2 \varepsilon_{BB}(r_{BB}^*)^3\}/(r_{\psi\psi}^*)^3. \tag{16.12.4}$$

For the combining rules we shall use (12.6.33) and (12.6.34), or in our present notation,

$$r_{AB}^* = (r_{AA}^* + r_{BB}^*)/2, \tag{16.12.5}$$

$$\varepsilon_{AB} = \zeta(\varepsilon_{AA}\varepsilon_{BB})^{1/2}. \tag{16.12.6}$$

Although ζ can be determined experimentally, for example from the second virial coefficients of gaseous mixtures, we shall treat it here as a diposable parameter because it has not yet been measured with sufficient accuracy for any mixture.[‡]

The equation (12.5.18) of Carnahan and Starling is

$$p = (RT/V_m)(1 + y + y^2 - y^3)/(1-y)^3 - a/V_m^2, \tag{16.12.7}$$

which, in view of the relation $y = b/4V_m \stackrel{\text{def}}{=} \beta/V_m$, we may rewrite in the form:

$$p = RT(V_m^3 + \beta V_m^2 + \beta^2 V_m - \beta^3)/V_m(V_m - \beta)^3 - a/V_m^2, \tag{16.12.8}$$

where $V_m = V_m(T, p, x)$, $\beta = \beta(x)$, and $a = a(x)$.

When (8) is substituted into equation (12.6.17) we obtain for the molar Helmholtz function:

$$A_m(T, V_m, x) = (1-x)\mu_A^{\ominus}(g, T) + x\mu_B^{\ominus}(g, T) - RT$$

$$+ RT\{(1-x)\ln(1-x) + x \ln x\} - RT \ln(p^{\ominus}V_m/RT)$$

$$- \int_{\infty}^{V_m} \{RT(V_m^3 + \beta V_m^2 + \beta^2 V_m - \beta^3)/V_m(V_m - \beta)^3 - a/V_m^2$$

$$- RT/V_m\} \, dV_m. \tag{16.12.9}$$

† Rowlinson, J. S. *Liquids and Liquid Mixtures*. Second edition. Butterworth: London. **1969**.
   McDonald, I. R. *Equilibrium Theory of Liquid Mixtures*. Specialist Periodical Report: *Statistical Mechanics*. Volume 1. The Chemical Society: London. **1973**, pp. 134–193.
   Münster, A. *Statistical Thermodynamics*. Volume II. Springer-Verlag: Berlin and Academic Press: New York. **1974**, chapter XVII.
   ‡ Another parameter, also in principle accessible from measurements of the second virial coefficients of gaseous mixtures, has sometimes been postulated for the combining rule (5) or (12.6.33).

The integration can be carried out, after resolution of the first term into partial fractions:

$$(V_m^3 + \beta V_m^2 + \beta^2 V_m - \beta^3)/V_m(V_m - \beta)^3$$
$$= 1/V_m + 4\beta/(V_m - \beta)^2 + 2\beta^2/(V_m - \beta)^3, \qquad (16.12.10)$$

and leads to

$$A_m(T, V_m, x)/RT = (1-x)\mu_A^\ominus(g, T)/RT + x\mu_B^\ominus(g, T)/RT - 1$$
$$+ (1-x)\ln(1-x) + x \ln x - \ln(p^\ominus V_m/RT) + 4\beta/(V_m - \beta)$$
$$+ \beta^2/(V_m - \beta)^2 - a/RTV_m. \qquad (16.12.11)$$

In chapter 17 on fluid mixtures equation (11) will be used with $V_m(T, p, x)$ determined by (8); here we shall simplify matters by dealing only with liquids at negligible pressures. We may then substitute $G_m(T, p \to 0, x) = G_m(T, x)$ for $A_m(T, V_m, x)$, and write for the molar excess Gibbs function:

$$G_m^E(T, x)/RT = (1-x)\ln\{V_m(T, 0)/V_m(T, x)\} + x \ln\{V_m(T, 1)/V_m(T, x)\}$$
$$+ 4[\beta(x)/\{V_m(T, x) - \beta(x)\} - (1-x)\beta(0)/\{V_m(T, 0) - \beta(0)\}$$
$$- x\beta(1)/\{V_m(T, 1) - \beta(1)\}]$$
$$+ [\beta(x)/\{V_m(T, x) - \beta(x)\}]^2 - (1-x)[\beta(0)/\{V_m(T, 0) - \beta(0)\}]^2$$
$$- x[\beta(1)/\{V_m(T, 1) - \beta(1)\}]^2$$
$$- a(x)/RTV_m(T, x) + (1-x)a(0)/RTV_m(T, 0)$$
$$+ xa(1)/RTV_m(T, 1), \qquad (16.12.12)$$

with $V_m(T, x)$ determined according to (8) with $p = 0$:

$$V_m(V_m^3 + \beta V_m^2 + \beta^2 V_m - \beta^3) - (a/RT)(V_m - \beta)^3 = 0. \qquad (16.12.13)$$

The molar energy $U_m(T, V_m, x)$ follows after differentiation of (11) with respect to T at constant V_m and x. At $p \to 0$ we may substitute $H_m(T, p \to 0, x) = H_m(T, x)$ for $U_m(T, V_m, x)$, and so obtain for the molar excess enthalpy:

$$H_m^E(T, x) = -a(x)/V_m(T, x) + (1-x)a(0)/V_m(T, 0) + xa(1)/V_m(T, 1). \qquad (16.12.14)$$

For the molar excess volume we have

$$V_m^E(T, x) = V_m(T, x) - (1-x)V_m(T, 0) - xV_m(T, 1). \qquad (16.12.15)$$

Equations (12), (14), and (15) can be used, with $V_m(T, x)$ determined by (13), to evaluate $G_m^E(T, x)$, $H_m^E(T, x)$, and $V_m^E(T, x)$ for given $\beta(x)$ and $a(x)$; so far these equations apply to any one-fluid theory of conformal mixtures based on the equation of state of Carnahan and Starling.

When we use equations (3) and (4) (or, since we are using a two-parameter equation of state, equations 12.6.31 and 12.6.32) as the one-fluid relations, and equations (5) and (6) (or equations 12.6.33 and 12.6.34) as the combining rules, the excess functions can be evaluated for given $\beta(0)$, $a(0)$, $\beta(1)$, $a(1)$, and the single disposable parameter ζ. The four quantities $\beta(0)$, $a(0)$, $\beta(1)$, and $a(1)$ can be expressed as follows in terms of the critical constants of the pure substances.

By solving the equations:

$$(\partial p/\partial V_m)_T = 0, \quad (\partial^2 p/\partial V_m^2)_T = 0, \quad (T = T^c, p = p^c, V_m = V_m^c), \quad (16.12.16)$$

for p given by Carnahan and Starling's equation (8) we obtain

$$\beta = 0.13045 V_m^c, \quad a = 1.3829 RT^c V_m^c, \quad p^c V_m^c/RT^c = 0.35894. \quad (16.12.17)$$

We can now calculate G_m^E, H_m^E, and V_m^E, given ζ and the critical properties T_A^c, V_A^c, T_B^c, and V_B^c (or T_A^c, p_A^c, T_B^c, and p_B^c) of the pure substances A and B.

Comparisons with experiment for $x = 0.5$ are given in table 16.1 for mixtures of simple substances ('condensed gases') that conform with the principle of corresponding states, and in table 16.2 for mixtures of the tetramethyls of carbon, silicon, and tin, which roughly conform among themselves with a principle of corresponding states. In these tables ζ has been chosen so as to fit the experimental value of G_m^E, and the theory is tested by seeing how well H_m^E and V_m^E are then fitted. At least qualitatively the fit is satisfactory. The values of ζ are close to the rough

Table 16.1. Comparison with experiment of the one-fluid theory of conformal mixtures for Carnahan and Starling's equation of state for six mixtures of 'condensed gases'.

Mixture	T/K	V_A^c/V_B^c	T_A^c/T_B^c	ζ	$G_m^E/J\,mol^{-1}$	$H_m^E/J\,mol^{-1}$	$V_m^E/cm^3\,mol^{-1}$
$0.5Ar + 0.5O_2$	84	0.96	0.97	0.99	37	60	+0.14
			Calculated:	0.9874	(37)	59	+0.10
$0.5N_2 + 0.5CH_4$	91	0.90	0.66	—	170	139	−0.35
			Calculated:	0.9791	(170)	138	−1.33
$0.5Ar + 0.5Kr$	116	0.81	0.72	0.99	84	43	−0.52
			Calculated:	0.9901	(84)	1^a	−1.23
$0.5Ar + 0.5CH_4$	91	0.76	0.79	0.99	74	103	+0.18
			Calculated:	0.9736	(74)	86	−0.04
$0.5O_2 + 0.5N_2$	84	0.87	1.23	—	39	42	−0.31
			Calculated:	1.0059	(39)	36	−0.61
$0.5Ar + 0.5N_2$	84	0.83	1.19	1.00	34	51	−0.18
			Calculated:	1.0033	(34)	35	−0.49

^{a} The calculated values change sign near $x = 0.5$.

Table 16.2. Comparison with experiment of the one-fluid theory of conformal mixtures for Carnahan and Starling's equation of state for three mixtures of tetramethyls at $T = 283$ K.

Mixture	p_A^c/p_B^c	T_A^c/T_B^c	ζ	$G_m^E/J\,mol^{-1}$	$H_m^E/J\,mol^{-1}$	$V_m^E/cm^3\,mol^{-1}$
$0.5C(CH_3)_4 + 0.5Si(CH_3)_4$	1.13	0.97	—	7	22	+0.05
		Calculated: 0.9974		(7)	18	+0.02
$0.5Si(CH_3)_4 + 0.5Sn(CH_3)_4$	0.95	0.86	—	71	68	−0.39
		Calculated: 0.9948		(71)	64	−0.46
$0.5C(CH_3)_4 + 0.5Sn(CH_3)_4$	1.07	0.83	—	96	99	−0.31
		Calculated: 0.9857		(96)	93	−0.36

values obtained experimentally where these exist. The calculated and experimental excess molar enthalpies agree quite well. The calculated excess molar volumes are always too small, but lie more or less in the correct sequence. The experimental values for some of these mixtures are difficult to measure, and are rather uncertain.

§ 16.13   Lattice theory of mixtures

In a lattice theory of liquid mixtures it is assumed that each molecule is made up of a number of *elements*, so chosen that each element can plausibly be regarded as occupying one site of a lattice. Any lattice model is unrealistic for liquids, which are distinguished from solids by the absence of any long-range order, but we may hope that it will nevertheless allow us to account approximately for the numbers of *contacts* of elements of one molecule with elements of another molecule. Here we shall limit ourselves to binary mixtures of molecules A and B having between them only two kinds i and j of elements. We shall apply the theory to mixtures of n-alkanes, regarding them as composed of 'end' and 'middle' elements.

Let molecule A have r_A elements and $zq_A = (zr_A - 2r_A + 2)$ contacts with elements of other molecules, and similarly for molecule B, where z is the coordination number of the lattice. Let the number of contacts of the i elements of a molecule A with elements of other molecules (A and B) be $zq_A\beta_A$, and of the j elements be $zq_A(1-\beta_A)$; and similarly for a molecule B. For example let $i = CH_3-$ and $j = -CH_2-CH_2-$ in a mixture of n-alkanes; then for say n-hexane $+ n$-hexadecane putting $z = 8$ we have $r_A = 4$ and $r_B = 9$, $q_A = 26/8$ and $q_B = 56/8$, and $\beta_A = 14/26$, $(1-\beta_A) = 12/26$, $\beta_B = 14/56$, and $(1-\beta_B) = 42/56$.

According to the methods of statistical mechanics† an approximation to the excess molar Helmholtz function A_m^E of such a mixture is given by

$$A_m^E/RT = (1-x)\ln r_A + x \ln r_B - \ln\{(1-x)r_A + xr_B\}$$
$$+ \tfrac{1}{2}z\{(1-x)q_A + xq_B\}\ln[\{(1-x)r_A + xr_B\}/\{(1-x)q_A + xq_B\}]$$
$$+ \tfrac{1}{2}z(1-x)q_A \ln(q_A/r_A) + \tfrac{1}{2}zxq_B \ln(q_B/r_B)$$
$$+ [x(1-x)q_Aq_B(\beta_A - \beta_B)^2/\{(1-x)q_A + xq_B\}]w/kT, \tag{16.13.1}$$

where $2w/z$ is the energy of interchange of an ii contact and a jj contact to form two ij contacts. (We note in passing that when $r_A = r_B = 1$, and $\beta_A = 1$, $\beta_B = 0$, so that A and B are different 'monomers', equation (1) reduces to $A_m^E/RT = x(1-x)w/kT$.)

For the excess molar energy U_m^E we obtain

$$U_m^E = [x(1-x)q_Aq_B(\beta_A - \beta_B)^2/\{(1-x)q_A + xq_B\}]Lu, \tag{16.13.2}$$

where $u = (w - T\,dw/dT)$.

If we are to hope that w or u might be independent of x, and independent of r_A and r_B for a series of binary mixtures of molecules (such as n-alkanes) containing only the elements i and j, then we must at least be satisfied that the lattice remains fixed for

† Guggenheim, E. A. *Mixtures*. Clarendon Press: Oxford. **1952**.

pure A, for pure B, and for any mixture of A and B. We therefore choose the elements so that V_A^*/r_A and V_B^*/r_B are the same, and we do not allow any volume change on mixing. For the series of n-alkanes, for example, V_m^*/r with $r = (n+2)/2$, where n denotes the number of carbon atoms, is almost constant; that is why we choose CH_3- and $-CH_2-CH_2-$ groups as elements. We must not compare equations (1) and (2) with the excess molar Gibbs function $G_m^E(p)$ and excess molar enthalpy $H_m^E(p)$ measured at constant pressure, but must correct those measured quantities to those corresponding to the 'Procrustean bed' implied by the lattice model. Good approximations to the required formulae are

$$A_m^E \approx G_m^E(p),\qquad(16.13.3)$$

$$U_m^E \approx H_m^E(p) - T\{(1-x)\alpha_A^*/\kappa_{T,A}^* + x\alpha_B^*/\kappa_{T,B}^*\}V_m^E(p),\qquad(16.13.4)$$

where α and κ_T denote isobaric expansivity and isothermal compressibility.

In figure 16.6 experimental values of A_m^E and U_m^E, determined according to equations (3) and (4) for a series of mixtures of n-alkanes, are compared with calculated values: for A_m^E from equation (1) with a constant value of w, and for U_m^E from equation (2) with a constant value of u. The agreement of experiment with the theory is highly satisfactory.

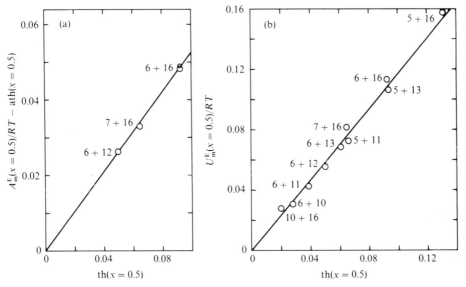

Figure 16.6. $\bigcirc$, Experimental values at 293.15 K for mixtures of n-alkanes: $\{0.5n\text{-}C_aH_{2a+2} + 0.5C_bH_{2b+2}\}$ (a), of A_m^E determined by equation (16.13.3) less the first five terms ath(x) of equation (16.13.1) and (b), U_m^E determined by equation (16.13.4), plotted against the factor th(x) given in square brackets in the last term of equation (16.13.1) or in equation (16.13.2). The values of a and b are shown on the diagram. The straight lines are those calculated for $z = 8$, $r_A = (a+2)/2$, and $r_B = (b+2)/2$: (a), from equation (16.13.1) with $Lw = 1290$ J mol^{-1} and (b), from equation (16.13.2) with $Lu = 2860$ J mol^{-1}.

One of the principal applications of equation (1) is to liquid mixtures of a 'monomer' A and a polymer B. (It seems natural to think of the polymer, often a solid at the temperature of study, as being dissolved as a solute in a solvent. We shall nevertheless continue to use the language of mixtures.) When we put $\alpha_A = \alpha_B = 1$ in equation (1) we obtain for λ_A/λ_A^* the formula:

$$\lambda_A/\lambda_A^* \approx (1-x^g)p^{l+g}/p_A^{l+g} \approx p^{l+g}/p_A^{l+g}$$

$$= (1-\phi)\{1-(2/zq_A)\phi(1-1/\rho)\}^{-zq_A/2}, \tag{16.13.5}$$

where we have used the symbol ϕ for $r_B x/\{(1-x)r_A + xr_B\}$ and the symbol ρ for r_B/r_A. We may call ϕ the (idealized) volume fraction of polymer in the mixture. Equation (5) can be applied as it stands, but can be simplified without much loss by assuming that $zq_A/2 \gg 1$ so as to obtain Flory's formula:

$$p^{l+g}/p_A^{l+g} = (1-\phi)\exp\{\phi(1-1/\rho)\}. \tag{16.13.6}$$

When $\rho \gg 1$, Flory's formula takes the even simpler form:

$$p^{l+g}/p_A^{l+g} = (1-\phi)e^\phi. \tag{16.13.7}$$

Had we assumed not that $\alpha_A = \alpha_B = 1$, but rather that $\alpha_A = 0$ and $\alpha_B = 1$ so that 'monomer' molecules interact as single elements with elements of the polymer, we should have obtained instead of (7) the formula:

$$p^{l+g}/p_A^{l+g} = (1-\phi)\exp(\phi + \phi^2 w/kT). \tag{16.13.8}$$

In figure 16.7 we have plotted experimental values of p^{l+g}/p_A^{l+g} against the volume fraction ϕ of polystyrene for mixtures of toluene and polystyrene at 298.15 K. Curve (a) is that calculated for $\rho = 3000$ according to Raoult's law: $(1-x^g)p^{l+g}/p_A^{l+g} \approx$

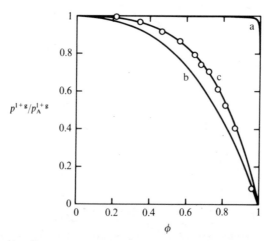

Figure 16.7. Plot of p^{l+g}/p_A^{l+g} against the volume fraction ϕ of polystyrene ($M = 290\,000$ g mol^{-1}) for mixtures of toluene and polystyrene at 298.15 K. The curves are those calculated (a), according to Raoult's law for $\rho = 3000$; (b), according to equation (16.13.7); and (c), according to equation (16.13.8) with $w/kT = 0.37$. $\bigcirc$, Experimental results.

$p^{l+g}/p_A^{l+g} = (1-x^l)$; such a mixture of monomer and polymer is far from ideal. Curve (b) is that calculated according to equation (7), clearly a much better approximation to the experimental results than Raoult's law. Curve (c) is that calculated according to equation (8) with $w/kT = 0.37$; the agreement between experiment and theory is now probably within experimental error.

§ 16.14 Simple ('regular') mixtures

A *simple mixture* is defined as one for which the excess molar Gibbs function is given by

$$G_m^E = x(1-x)Lw ,\tag{16.14.1}$$

where w may depend on temperature and pressure but not on composition. It follows that the activity coefficients in a simple mixture are given by

$$\ln f_A = x^2 w/kT, \qquad \ln f_B = (1-x)^2 w/kT.\tag{16.14.2}$$

According to lattice theory the excess molar Helmholtz function is given by $A_m^E = x(1-x)Lw$ when the molecules A and B may each be regarded as occupying one site on a lattice and the z *contact points* of an A molecule are all of one kind, and of a B molecule are all of another kind. We may therefore expect a mixture to be simple when the two molecules are of similar size and are spherically symmetrical; this is borne out in practice. We may then interpret w as an energy of interchange. When $w > 0$ (the common case) there is a tendency towards 'molecular segregation', which at low enough temperatures might result, if solid phases are not formed first, in separation into two coexisting liquid phases (see § 16.15). When $w < 0$ there is a tendency towards 'molecular integration'; mixtures like $\{(1-x)CHCl_3 + x(C_2H_5)_2O\}$ that show a tendency to 1—1 compound formation (see § 16.19) behave approximately like simple mixtures with $w < 0$.

According to Hildebrand's 'solubility-parameter' theory w can be calculated from the properties of the pure substances by the relation:

$$Lw = V_A^* V_B^* (\delta_A - \delta_B)^2 /\{(1-x)V_A^* + xV_B^*\},\tag{16.14.3}$$

where the 'solubility parameter' $\delta_i = \{(\Delta_1^g H_i^* - RT)/V_i^*\}^{1/2}$. Solubility-parameter theory fails to give even roughly correct values of w for liquid mixtures, though it has been used successfully to correlate the solubilities of iodine, and of some other solutes, in a wide variety of non-polar solvents.[†]

§ 16.15 Phase separation in liquid mixtures

Some liquid mixtures separate into two liquid phases below an *upper critical solution temperature* (UCST) or above a *lower critical solution temperature* (LCST). These

---

[†] Hildebrand, J. H.; Scott, R. L. *Solubility of Nonelectrolytes.* Third edition. Reinhold: New York. **1950**.
Hildebrand, J. H.; Prausnitz, J. M.; Scott, R. L. *Regular and Related Solutions.* Van Nostrand Reinhold: New York. **1970**.

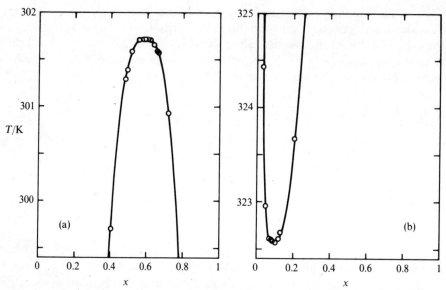

Figure 16.8. Partial-miscibility diagrams of temperature plotted against mole fraction (a), for $\{(1-x)C_6H_{12}+xCH_2I_2\}$ showing a UCST and (b), for $\{(1-x)H_2O+xCH_3(C_2H_5)_2N\}$ showing an LCST. The curves drawn through the experimental results will be discussed in § 16.16.

two kinds of behaviour are illustrated in figure 16.8. Mixtures with an LCST often also have a UCST at a higher temperature, and so have a closed miscibility loop, provided that the mixture does not boil before the UCST is reached. More rarely an LCST is found at a temperature above that of a UCST, leaving an interval of temperature over which the liquids are completely miscible. These two kinds of behaviour are illustrated in figure 16.9.

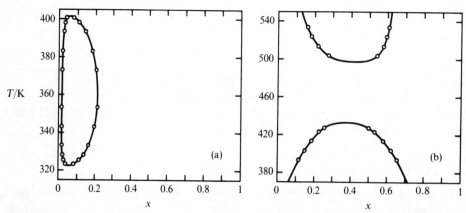

Figure 16.9. Partial-miscibility diagrams of temperature plotted against mole fraction for mixtures having both a UCST and an LCST: (a), for $\{(1-x)H_2O+xCH_2(OH)CH_2OC_4H_9\}$ for which the UCST is at a higher temperature than the LCST and (b), for $\{(1-x)C_6H_6+xS_8\}$ for which the UCST is at a lower temperature than the LCST.

The coexistence of two phases α and β of a binary mixture is implied by the simultaneous existence of two Gibbs–Duhem equations:

$$S_m^\alpha \, dT - V_m^\alpha \, dp + (1 - x^\alpha) \, d\mu_A + x^\alpha \, d\mu_B = 0 \,, \tag{16.15.1}$$

$$S_m^\beta \, dT - V_m^\beta \, dp + (1 - x^\beta) \, d\mu_A + x^\beta \, d\mu_B = 0 \,. \tag{16.15.2}$$

These lead to the equation, with two independent variables instead of three:

$$(S_m^\beta - S_m^\alpha) \, dT - (V_m^\beta - V_m^\alpha) \, dp + (x^\beta - x^\alpha) \, d(\mu_B - \mu_A) = 0 \,. \tag{16.15.3}$$

By use of the relations:

$$(\partial G_m/\partial x)_{T,p} = \mu_B - \mu_A \,, \quad (\partial^2 G_m/\partial x^2)_{T,p} = \{\partial(\mu_B - \mu_A)/\partial x\}_{T,p} \,, \tag{16.15.4}$$

and

$$\{\partial(\mu_B - \mu_A)/\partial T\}_{p,x} = -(S_B - S_A) \,, \tag{16.15.5}$$

with

$$\mu_A = H_A - TS_A \,, \quad \mu_B = H_B - TS_B \,, \quad \mu_A^\alpha = \mu_A^\beta \,, \quad \mu_B^\alpha = \mu_B^\beta \,, \tag{16.15.6}$$

we may rewrite (3) in terms of the variables T, p, x^α, in the form:

$$\{(1 - x^\beta)(H_A^\beta - H_A^\alpha) + x^\beta(H_B^\beta - H_B^\alpha)\} \, d \ln T - (V_m^\beta - V_m^\alpha) \, dp$$
$$+ (x^\beta - x^\alpha)(\partial^2 G_m/\partial x^2)_{T,p}^\alpha \, dx^\alpha = 0 \,. \tag{16.15.7}$$

Equation (7), and its analogue for the variables T, p, x^β, sometimes called the Gibbs–Konovalow equations, inter-relate the properties of two coexisting phases of a binary mixture.

For example, the dependence of the phase-separation temperature T on the mole fraction x^α in the phase α at constant pressure p is given by

$$(\partial T/\partial x^\alpha)_p = -T(x^\beta - x^\alpha)(\partial^2 G_m/\partial x^2)_{T,p}^\alpha / \{(1 - x^\beta)(H_A^\beta - H_A^\alpha)$$
$$+ x^\beta(H_B^\beta - H_B^\alpha)\} \,. \tag{16.15.8}$$

According to (12.6.23) we have $(\partial^2 G_m/\partial x^2)_{T,p} \geq 0$. If we adopt the convention that $x^\beta > x^\alpha$ then the sign of $(\partial T/\partial x^\alpha)_p$ is opposite to that of the denominator of (8).

For a simple mixture we have

$$G_m^E = x(1 - x)Lw \,, \tag{16.15.9}$$

so that

$$G_m = (1 - x)\mu_A^* + x\mu_B^* + RT\{(1 - x)\ln(1 - x) + x \ln x\} + x(1 - x)Lw \,, \tag{16.15.10}$$

$$\mu_A = \mu_A^* + RT \ln(1 - x) + x^2 Lw \,, \tag{16.15.11}$$

$$\mu_B = \mu_B^* + RT \ln x + (1 - x)^2 Lw \,, \tag{16.15.12}$$

and

$$H_A = H_A^* + x^2 L\{w - T(\partial w/\partial T)_p\}, \tag{16.15.13}$$

$$H_B = H_B^* + (1-x)^2 L\{w - T(\partial w/\partial T)_p\}. \tag{16.15.14}$$

Since according to (13) and (14)

$$(1-x^\beta)(H_A^\beta - H_A^\alpha) + x^\beta(H_B^\beta - H_B^\alpha) = -(x^\beta - x^\alpha)^2 L\{w - T(\partial w/\partial T)_p\}, \tag{16.15.15}$$

the denominator of (8) has the same sign as $\{w - T(\partial w/\partial T)_p\}$. If $\{w - T(\partial w/\partial T)_p\} > 0$ then $(\partial T/\partial x^\alpha)_p > 0$; the unmixing temperature increases with increasing x^α towards a UCST. If $\{w - T(\partial w/\partial T)_p\} < 0$ then $(\partial T/\partial x^\alpha)_p < 0$; the unmixing temperature decreases with increasing x^α towards an LCST.

Alternatively, we may use equations (11) and (12) to express the conditions for coexisting phases for a simple mixture in the forms:

$$(1-x^\alpha)/(1-x^\beta) = \exp\{(x^\beta - x^\alpha)(x^\alpha + x^\beta)w/kT\}, \tag{16.15.16}$$

$$x^\alpha/x^\beta = \exp\{(x^\beta - x^\alpha)(x^\alpha + x^\beta - 2)w/kT\}. \tag{16.15.17}$$

Making use of the relation $x^\beta = 1 - x^\alpha$, obvious for a simple mixture from symmetry about $x = \frac{1}{2}$, we can rewrite either of (16) or (17) in the form:

$$(1-x)/x = \exp\{(1-2x)w/kT\}, \tag{16.15.18}$$

which can easily be solved numerically. The curve calculated from (18) does not, however, fit the experimental coexistence curve, especially in the neighbourhood of a critical point, even for a mixture that is otherwise approximately simple. We shall discuss this point further in the following section.

§ 16.16   Critical points in liquid mixtures

We begin by emphasising that there is no formal distinction between a critical solution temperature and a (liquid + gas) critical temperature for a mixture. Either is determined by equations (12.6.24) and (12.6.25):

$$(\partial^2 G_m/\partial x^2)_{T,p} = 0, \quad \text{and} \quad (\partial^3 G_m/\partial x^3)_{T,p} = 0, \tag{16.16.1}$$

with

$$(\partial^4 G_m/\partial x^4)_{T,p} > 0. \tag{16.16.2}$$

Sometimes, as we shall see in the next chapter, (liquid + liquid) critical points and (liquid + gas) critical points become indistinguishable. We shall nevertheless concentrate in this section on UCST's and LCST's in liquid mixtures at constant low pressures. For such mixtures we may rewrite equation (1) in terms of the excess molar Gibbs function:

$$\left.\begin{aligned} (\partial^2 G_m^E/\partial x^2)_T^c &= -RT^c/x^c(1-x^c), \\ (\partial^3 G_m^E/\partial x^3)_T^c &= RT^c(1-2x^c)/(x^c)^2(1-x^c)^2. \end{aligned}\right\} \tag{16.16.3}$$

For a simple mixture it follows at once that

$$-2w = -RT^c/x^c(1-x^c), \quad \text{and} \quad 0 = RT^c(1-2x^c)/(x^c)^2(1-x^c)^2. \quad (16.16.4)$$

From the second of equations (4) we can deduce, if it was not already obvious, that $x^c = 0.5$. Using that result we deduce from the first of equations (4) that $RT^c = Lw/2$. A critical temperature exists only for $w > 0$; it is a UCST if $(w - T\, dw/dT) > 0$ and an LCST if $(w - T\, dw/dT) < 0$. A critical temperature exists only if $(G_m^E/RT)^c = 0.5$, implying at ordinary temperatures a $G_m^E(x = 0.5)$ of about $+1250\ \text{J mol}^{-1}$. This rule, exact for simple mixtures, serves to indicate an approach to critical unmixing for a wide range of real mixtures that do not deviate too far from simplicity.

For mixtures in general the critical temperature is given by putting $x^\alpha \to x^\beta$ in equation (16.15.8). Putting $x^\beta = x^c + |\delta x|$ and $x^\alpha = x^c - |\delta x|$ (and so assuming symmetry about the critical point) at a temperature just below a UCST or just above an LCST we obtain†

$$(\partial T/\partial x^\alpha)_p = -T(\partial^2 G_m/\partial x^2)_{T,p}/(\partial^2 H_m/\partial x^2)_{T,p}|\delta x|. \quad (16.16.5)$$

It follows that at a UCST $(\partial^2 H_m/\partial x^2)_{T,p} = (\partial^2 H_m^E/\partial x^2)_{T,p} < 0$, and at an LCST $(\partial^2 H_m^E/\partial x^2)_{T,p} > 0$. For simple mixtures, and for at least nearly all real mixtures, $(\partial^2 H_m^E/\partial x^2)_{T,p}$ has the opposite sign to H_m^E; mixing is endothermic near a UCST and is exothermic near an LCST.

We now return to the remark at the end of the previous section. No 'classical' theory will give agreement with the experimental coexistence curve near the critical temperature. If we write

$$(x^\beta - x^\alpha) \propto |T^c - T|^\beta, \quad (T \to T^c), \quad (16.16.6)$$

then any theory giving an equation that can be expanded as a Taylor series about the critical point will lead to $\beta = \frac{1}{2}$. We recall the discussion of the critical point of a pure substance in § 9.8, where we conjectured that a critical point for a mixture behaves with respect to 'corresponding' variables like a critical point for a pure substance. In particular when we use the 'dictionary' given on page 135 we expect all the critical exponents and the relations between them to remain the same. It will now be clear why we chose in figure 16.8 to plot points only within a kelvin or two of the critical temperature, and that the exponents $\beta = 0.347$ for $\{(1-x)C_6H_{12} + xCH_2I_2\}$ at a UCST and $\beta = 0.34$ for $\{(1-x)H_2O + xCH_3(C_2H_5)_2N\}$ at an LCST agree within experimental errors with the values obtained for β from the densities of coexisting

---

† $(1-x^\beta)(H_A^\beta - H_A^\alpha) + x^\beta(H_B^\beta - H_B^\alpha) = (1 - x^c - |\delta x|)2|\delta x|(\partial H_A/\partial x)_{T,p} + (x^c + |\delta x|)2|\delta x|(\partial H_B/\partial x)_{T,p}$

$$= 2(\delta x)^2\{(\partial H_B/\partial x)_{T,p} - (\partial H_A/\partial x)_{T,p}\} \quad \text{(a)}$$

$$= 2(\delta x)^2(\partial^2 H_m/\partial x^2)_{T,p}. \quad \text{(b)}$$

(a), From equation (2.7.12): $(1-x)(\partial H_A/\partial x)_{T,p} + x(\partial H_B/\partial x)_{T,p} = 0$.
(b), From equation (2.7.5): $H_m = (1-x)H_A + xH_B$.
   The proof is set out more fully, though with some confusion over signs, in Prigogine, I.; Defay, R. *Chemical Thermodynamics.* Everett, D. H.: translator. Longmans Green: London. **1954**, chapter 18.

phases of a pure substance in § 9.8. Relatively few accurate measurements have yet been made of critical exponents for liquid mixtures, but what has been done both experimentally and theoretically has been extensively reviewed in the literature.†

§ 16.17 Standard thermodynamic functions for a liquid (or solid) substance

The standard chemical potential $\mu_B^\ominus(l, T)$ of a liquid substance B is defined by the relation:

$$\mu_B^\ominus(l, T) \overset{\text{def}}{=} \mu_B^*(l, T, p^\ominus), \tag{16.17.1}$$

so that it is identified with the chemical potential of the pure substance at the same temperature and at the standard pressure $p^\ominus$. All the other standard thermodynamic functions of a liquid substance follow. Thus

$$\lambda_B^\ominus(l, T) = \lambda_B^*(l, T, p^\ominus), \tag{16.17.2}$$

$$S_B^\ominus(l, T) = S_B^*(l, T, p^\ominus), \tag{16.17.3}$$

$$H_B^\ominus(l, T) = H_B^*(l, T, p^\ominus), \tag{16.17.4}$$

$$C_{p,B}^\ominus(l, T) = C_{p,B}^*(l, T, p^\ominus). \tag{16.17.5}$$

Equation (1) can be rewritten in the form:

$$\mu_B^\ominus(l, T) = \mu_B^*(l, T, p) + \int_p^{p^\ominus} V_B^*(l, T, p)\, dp, \tag{16.17.6}$$

where $V_B^*(l, T, p)$ is the molar volume of the pure liquid substance at the temperature T and pressure p. When the standard pressure $p^\ominus$ is chosen as 101.325 kPa (as it usually is by thermochemists) and provided that the pressure p is reasonably close to atmospheric pressure, the integral in equation (6) makes only a small contribution and is often omitted.

Equation (1) can also be rewritten in the form, relevant to a liquid mixture:

$$\mu_B^\ominus(l, T) = \mu_B(l, T, p, x_C) + \{\mu_B^*(l, T, p) - \mu_B(l, T, p, x_C)\}$$

$$+ \int_p^{p^\ominus} V_B^*(l, T, p)\, dp, \tag{16.17.7}$$

where $\mu_B(l, T, p, x_C)$ is the chemical potential of the substance B in a liquid mixture of composition specified by the set of mole fractions x_C at the temperature T and pressure p. Alternatively, by use of the defining equation (16.5.1) or (16.5.2), we may write (7) in the form:

$$\mu_B^\ominus(l, T) = \mu_B(l, T, p, x_C) - RT \ln(x_B f_B) + \int_p^{p^\ominus} V_B^*(l, T, p)\, dp, \tag{16.17.8}$$

where f_B is the activity coefficient of the substance B in the mixture.

† Scott, R. L. *SPRCT(II)*, chapter 8.

For liquid mixtures we have no general result that we can use, as we did for the corresponding expression for gaseous mixtures, to express the chemical-potential difference $\{\cdots\}$ in (7), or the activity coefficient f_B in (8), in terms of some known function of the composition plus an integral of partial molar volumes with respect to the pressure. We have, however, learnt how to measure the chemical-potential difference $\{\cdots\}$ or the activity coefficient f_B.

All the formulae of this section are applied also to solids merely by change of the state symbol from l to s; $\mu_B(s, T, p, x_C)$ is then the chemical potential of the substance B in a solid single-phase† mixture of composition specified by the set of mole fractions x_C at the temperature T and pressure p.

§ 16.18   Chemical equilibrium in liquid (or solid) mixtures

For chemical equilibrium of the reaction $0 = \Sigma_B \nu_B B$ in a liquid (or solid) mixture, or between a liquid mixture and one or more pure solids (or pure liquids), the standard equilibrium constant is given by

$$K^\ominus(T) = \prod_B \{\lambda_B^\ominus(\text{l or s}, T)\}^{-\nu_B}$$

$$= \prod_B \{\lambda_B^*(\text{l or s}, T, p^\ominus)\}^{-\nu_B}$$

$$\approx \prod_B (x_B^{eq} f_B^{eq})^{\nu_B}, \tag{16.18.1}$$

the approximation in the last expression arising from omission of the integrals from equation (16.17.8). If any of the substances B is a pure substance then $x_B^{eq} = 1$ and $f_B^{eq} = 1$. That is the origin of the advice, sound if we are neglecting the effect of any difference between the pressure p and the standard pressure $p^\ominus$, to leave out factors relating to pure solids (or pure liquids) in the formulation of equilibrium constants.

For an ideal mixture equation (1) becomes

$$K^\ominus(T) \approx \prod_B (x_B^{eq})^{\nu_B}, \tag{16.18.2}$$

the 'classical', though little used, form for the standard equilibrium constant of a reaction in a liquid mixture. There exists no limiting condition under which all liquid mixtures become ideal, comparable with the condition $p \to 0$ under which all gaseous mixtures become perfect. Moreover, few real mixtures are even approximately ideal so that equation (2) is only rarely of useful accuracy. Nevertheless equation (2) is surprisingly accurate for the very reaction in a liquid mixture, the hydrolysis of ethyl ethanoate, which was studied by Berthelot and St. Gilles in 1862 and used by

---

† Such a single-phase solid mixture is often called a 'solid solution'. We wish, however, to reserve the word 'solution' for the special case of a mixture in which we choose to distinguish between the solvent and the solutes, and especially for dilute solutions (see § 2.10 and chapter 18).

Horstmann and by van't Hoff in the earliest formulations of the law of chemical equilibrium. It was lucky, and remains surprising, that mixtures of ethyl ethanoate, water, ethanol, and ethanoic acid are nearly ideal.

§ 16.19   Ideal associated mixtures

When the excess functions of a mixture are dominated by the effects of compound formation it is sometimes reasonable to assume that the mixture is an ideal mixture of A, B, and of the compounds $A_m B_n$.

We shall illustrate the theory for $\{(1-x)p\text{-}C_4H_8O_2 \text{ (dioxan)} + x\text{CHCl}_3\}$ which probably forms two hydrogen-bonded compounds AB and AB_2. Let the 'actual' mole fractions of A, AB, AB_2, and B, be $(1-y_{11}-y_{12}-y)$, y_{11}, y_{12}, and y, respectively. Then x is related to the y's by the formula:

$$x = (y_{11} + 2y_{12} + y)/(1 + y_{11} + 2y_{12}). \tag{16.19.1}$$

We assume the chemical equilibria:

$$A + B = AB, \quad \text{and} \quad A + 2B = AB_2, \tag{16.19.2}$$

so that

$$\lambda_A \lambda_B = \lambda_{AB}, \quad \text{and} \quad \lambda_A \lambda_B^2 = \lambda_{AB_2}, \tag{16.19.3}$$

and we further assume that the four species form an ideal mixture so that

$$\lambda_A = (1 - y_{11} - y_{12} - y)\lambda_A^*, \quad \lambda_{AB} \propto y_{11}, \quad \lambda_{AB_2} \propto y_{12}, \quad \lambda_B = y\lambda_B^*. \tag{16.19.4}$$

When we use (4) in (3) we obtain

$$\left.\begin{aligned} y_{11}/(1 - y_{11} - y_{12} - y)y &= K_1^\ominus, \\ y_{12}/(1 - y_{11} - y_{12} - y)y^2 &= K_2^\ominus, \end{aligned}\right\} \tag{16.19.5}$$

where $K_1^\ominus$ and $K_2^\ominus$ are the standard equilibrium constants of reactions (2).

Using equations (5) in the first and last of equations (4) we obtain

$$\left.\begin{aligned} \lambda_A/\lambda_A^* &= (1 - y_{11} - y_{12} - y) = (1-y)/(1 + K_1^\ominus y + K_2^\ominus y^2), \\ \lambda_B/\lambda_B^* &= y, \end{aligned}\right\} \tag{16.19.6}$$

and using equations (5) in (1) we obtain

$$x = \{(1 - K_1^\ominus)y + K_2^\ominus y^2(2-y)\}/\{1 + K_1^\ominus y(2-y) + K_2^\ominus y^2(3-2y)\}. \tag{16.19.7}$$

We have now expressed all the experimental variables in terms of y and of $K_1^\ominus$ and $K_2^\ominus$.

We can eliminate y from equations (6) and obtain

$$(1 - \lambda_A/\lambda_A^* - \lambda_B/\lambda_B^*)/(\lambda_A/\lambda_A^*)(\lambda_B/\lambda_B^*) = K_1^\ominus + K_2^\ominus (\lambda_B/\lambda_B^*), \tag{16.19.8}$$

an equation that can be used as a sensitive test of the theory by plotting experimental values of the left-hand side against λ_B/λ_B^*. If the theory is valid the points should lie

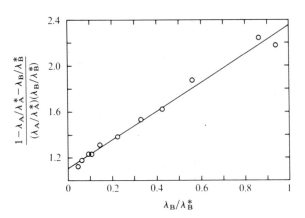

Figure 16.10. Test of equation (16.19.8) for $\{(1-x)p\text{-}C_4H_8O_2 + xCHCl_3\}$ at 323.15 K. The results were measured by means of a recirculating still. The straight line gives $K_1^\ominus = 1.11$ and $K_2^\ominus = 1.24$.

on a straight line, and if they do then the standard equilibrium constants can be determined from the intercept and slope. Such a plot is shown in figure 16.10 for $\{(1-x)p\text{-}C_4H_8O_2 + xCHCl_3\}$; equation (8) is obeyed within experimental error. The experimental values of λ_A/λ_A^* and λ_B/λ_B^* are plotted against x in figure 16.11(a) with the curves calculated from equations (6) with y determined for given x by equation (7).

The excess molar Gibbs function G_m^E is given, whether association actually occurs or not, by the formula:

$$G_m^E/RT = (1-x)\ln\{\lambda_A/(1-x)\lambda_A^*\} + x\ln\{\lambda_B/x\lambda_B^*\}. \qquad (16.19.9)$$

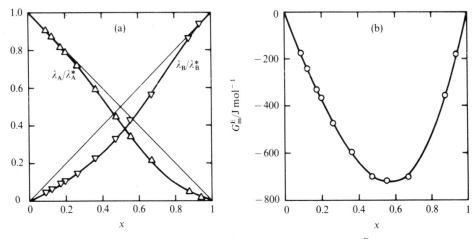

Figure 16.11. Experimental values (a), of λ_A/λ_A^* and λ_B/λ_B^* and (b), of G_m^E, plotted against x for $\{(1-x)p\text{-}C_4H_8O_2 + xCHCl_3\}$ at 323.15 K. The curves are those calculated (a), from equations (16.19.6) and (b), from equation (16.19.10), with equation (16.19.7) for $K_1^\ominus = 1.11$ and $K_2^\ominus = 1.24$.

Using equations (6) we obtain for G_m^E the formula:

$$G_m^E/RT = (1-x)\ln\{(1-y)/(1-x)(1+K_1^\ominus y + K_2^\ominus y^2)\} + x\ln\{y/x\}, \quad (16.19.10)$$

with y again determined for given x by equation (7). The experimental values of G_m^E are plotted against x in figure 16.11(b) with the curve calculated from equation (10) with (7).

§ 16.20   Equilibria of liquid mixtures and pure solid phases

If the solid phase that coexists with a liquid mixture of A and B is pure A then the freezing temperature of the mixture is lower than that of pure A; if the solid phase is pure B then the freezing temperature is lower than that of pure B. When there is no other effect, a freezing-temperature diagram like that shown in figure 16.12(a) is obtained. The intersection E of the two curves determines the *eutectic* temperature and *eutectic* composition; the eutectic temperature is the lowest temperature at which any liquid mixture of A and B can exist. At the eutectic temperature, for any mole fraction, the system contains three coexisting phases: liquid mixture, pure solid A, and pure solid B. At a given pressure such a system is invariant; the temperature remains constant until one or more of the phases disappears.

Such a diagram can be constructed from 'cooling curves' like that sketched for $x = 0.5$ in figure 16.12(b). The liquid cools continuously until $T \approx 261$ K; after a discontinuity the mixture continues to cool, but less rapidly, while solid B separates,

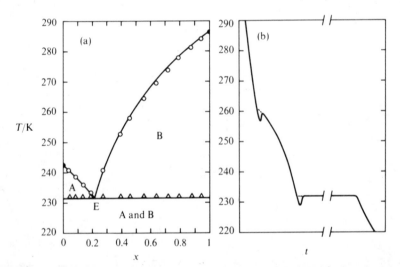

Figure 16.12. (a), Freezing temperatures of $\{(1-x)C_6H_5Br + xp\text{-}C_6H_4(CH_3)_2\}$ plotted against mole fraction x. The areas are labelled according to the substances present as solid phases. $\bigcirc$, Freezing temperatures; $\triangle$, eutectic temperature. The curves and the eutectic line are those calculated for an ideal liquid mixture from equations (16.20.1) and (16.20.2) with $\Delta_s^l H(C_6H_5Br) = 10.64$ kJ mol^{-1} and $\Delta_s^l H\{p\text{-}C_6H_4(CH_3)_2\} = 19.10$ kJ mol^{-1}. (b), Sketch of the cooling curve of a mixture with overall mole fraction $x = 0.5$.

until $T \approx 232$ K; after another discontinuity the temperature of the system remains constant (there is an 'arrest' in the cooling curve) while three phases are present, until another discontinuity marks the disappearance of the last trace of liquid; the solid phases then begin to cool towards the temperature of the surroundings. The discontinuities in cooling curves are often masked by supercooling but can be found by extrapolation as is shown in figure 16.12(b).

If the liquid mixtures are ideal (see § 16.3) we may use equation (10.6.9), ignoring terms of $O(1 - T/T^*)^2$, to write

$$-\ln(1-x) = (\Delta_s^l H_A^* / R T_A^*)(1 - T_A/T_A^*), \tag{16.20.1}$$

$$-\ln x = (\Delta_s^l H_B^* / R T_B^*)(1 - T_B/T_B^*). \tag{16.20.2}$$

Equations (1) and (2) can be used to calculate the freezing temperatures $T_A(x)$ and $T_B(x)$, and from their intersection the eutectic temperature and composition. The curves in figure 16.12(a) are those calculated in that way. In spite of the assumptions implicit in equations (1) and (2) the agreement between theory and experiment is quite good.

If A and B form a solid compound then the corresponding phase diagram is like one of those shown in figure 16.13. In figure 16.13(a) the compound melts at a *congruent* melting temperature T_C. In figure 16.13(b) the compound decomposes

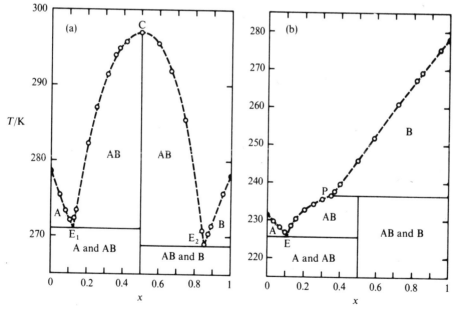

Figure 16.13. Phase diagrams for mixtures that form a solid compound. (a), Mixtures $\{(1-x)C_6H_6 + xC_6F_6\}$ form a solid compound $C_6H_6 \cdot C_6F_6$ that melts congruently at $T \approx 297$ K; (b), mixtures $\{(1-x)C_5H_5N + xC_6F_6\}$ form a compound $C_5H_5N \cdot C_6F_6$ that melts incongruently at $T \approx 237$ K. The areas are labelled according to the substances present as solid phases.

into liquid of mole fraction x_P and pure solid B at an *incongruent* melting temperature or *peritectic* temperature T_P.

If liquid mixtures of A and B are only partially miscible with an upper critical
solution temperature then, if no solid compound is formed, the phase diagram is like
that shown in figure 16.14, where we have this time chosen a pair of metals for our
example.

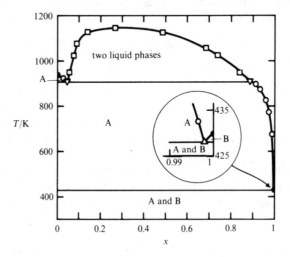

Figure 16.14. Freezing temperatures of $\{(1-x)\text{Al} + x\text{In}\}$ plotted against mole fraction x. The areas are
labelled according to the substances present as solid phases or as two coexisting liquid phases. ○, Freezing
temperatures; □, temperatures of coexisting liquid phases; △, eutectic temperature; ▽, coexistence
temperature of two liquid phases and pure solid B.

§ 16.21 Equilibria of liquid mixtures and mixed-crystal phases

A mixed-crystal phase ('solid solution') is a one-phase solid mixture. Two examples
are shown in figure 16.15, both for a complete series of mixed-crystal phases from
$x = 0$ to $x = 1$.

The minimum Az in figure 16.15(b) is called an *azeotrope*, a term that we shall meet
again in chapter 17 in connexion with two fluid phases. When two binary mixtures
are in equilibrium we have according to equation (16.6.1):

$$(H_m^\alpha/RT^2)\,\mathrm{d}T - (V_m^\alpha/RT)\,\mathrm{d}p + (1-x^\alpha)\,\mathrm{d}\ln\lambda_A + x^\alpha\,\mathrm{d}\ln\lambda_B = 0, \qquad (16.21.1)$$

and

$$(H_m^\beta/RT^2)\,\mathrm{d}T - (V_m^\beta/RT)\,\mathrm{d}p + (1-x^\beta)\,\mathrm{d}\ln\lambda_A + x^\beta\,\mathrm{d}\ln\lambda_B = 0. \qquad (16.21.2)$$

If $x^\alpha = x^\beta$ then it follows that $(\partial T/\partial x)_p = 0$. It also follows that $(\partial p/\partial x)_T = 0$. Systems
with an azeotrope at a maximum in the freezing-temperature diagram, or at a
horizontal point of inflexion, are possible but apparently much less common. A solid
azeotropic mixture melts to give a liquid mixture having the same composition.

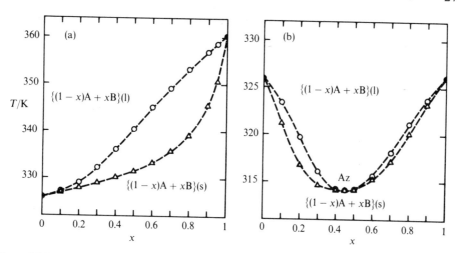

Figure 16.15. Phase diagrams for liquid mixtures that form single-phase solid mixtures. (a), $\{(1-x)p\text{-}$ $C_6H_4Cl_2 + xp\text{-}C_6H_4Br_2\}$; (b), $\{(1-x)p\text{-}C_6H_4Cl_2 + xp\text{-}C_6H_4ClI\}$. $\bigcirc$, Freezing temperatures of the liquid mixtures; $\triangle$, melting temperatures of the solid mixtures.

Similarly a gaseous azeotropic mixture condenses to give a liquid mixture having the same composition. In this respect an azeotrope behaves like a pure substance. Its composition depends, however, on the pressure and so it is not a compound.

§ 16.22   Other equilibria between liquid mixtures and solid phases

The phase diagrams shown in figures 16.11 to 16.15 are among the simplest of those involving solid phases. Many other, more complex, kinds of phase diagram result from the formation of more than one solid compound, from partial miscibility of the solid mixture giving rise to two mixed-crystal phases, from polymorphic transitions in one or more solid phases, and from all kinds of combinations of these. Further complexities arise when we consider the effect of changing the pressure. All these complexities are compounded for systems having more than two components. We shall not pursue equilibria involving solid phases any further here but refer the reader to the specialist literature.†

Problems for chapter 16

Problem 16.1

From the following results obtained from a recirculating still for mixtures of $A = CCl_4$ and $B = CHCl_3$ at 313.15 K calculate values of the activity coefficients f_A

---

† Haase, R.; Schönert, H. *Solid-Liquid Equilibrium*. Pergamon: Oxford. **1969**.
Ricci, J. E. *The Phase Rule and Heterogeneous Equilibrium*. Van Nostrand: New York. **1951**.
Zernike, J. *Chemical Phase Theory*. Kluwer: Deventer. **1955**.

and f_B, of the excess chemical potentials μ_A^E and μ_B^E, and of the excess molar Gibbs function G_m^E.

$$p_A^{l+g} = 28.443 \text{ kPa}; \quad B_{AA} = -1465 \text{ cm}^3 \text{ mol}^{-1}; \quad V_A^{*l} = 99 \text{ cm}^3 \text{ mol}^{-1};$$

$$p_B^{l+g} = 48.064 \text{ kPa}; \quad B_{BB} = -1040 \text{ cm}^3 \text{ mol}^{-1}; \quad V_B^{*l} = 82 \text{ cm}^3 \text{ mol}^{-1};$$

$$\delta_{AB} = 0; \quad x_B^l = 0.5242; \quad x_B^g = 0.6456; \quad p^{l+g} = 40.242 \text{ kPa}.$$

Problem 16.2

Measurements of the dew pressure p^{l+g} of gaseous mixtures of $A = Si(CH_3)_4$ and $B = Sn(CH_3)_4$ at 279.95 K as a function of the mole fraction x_B^g gave for $x_B^g = 0.10660$: $p^{l+g} = 28.703 \text{ kPa}$ and $dp^{l+g}/dx_B^g = -113.08 \text{ kPa}$. Given that $(B_{AA} - V_A^{*l}) = -1452 \text{ cm}^3 \text{ mol}^{-1}$, $(B_{BB} - V_B^{*l}) = -2251 \text{ cm}^3 \text{ mol}^{-1}$, $\delta_{AB} = +69 \text{ cm}^3 \text{ mol}^{-1}$, $p_A^{l+g} = 48.285 \text{ kPa}$, and $p_B^{l+g} = 6.087 \text{ kPa}$, calculate the mole fraction x_B^l of the coexisting liquid phase and thence the excess molar Gibbs function G_m^E.

Problem 16.3

A measurement of the vapour pressure p^{l+g} of a liquid mixture $\{(1-x^l)C_6H_{14} + x^l C_{16}H_{34}\}$ at 293.15 K gave $p^{l+g} = 8.224 \text{ kPa}$ for $x^l = 0.47362$. At the same temperature the vapour pressure p_A^{l+g} of pure hexane was 16.145 kPa. That of pure hexadecane was less than 10^{-4} kPa. Given that $(B_{AA} - V_A^{*l}) = -2197 \text{ cm}^3 \text{ mol}^{-1}$ calculate the activity coefficient f_A and the excess chemical potential μ_A^E of C_6H_{14} in the mixture. How could a series of such measurements be used to obtain the excess molar Gibbs function G_m^E?

Problem 16.4

Calculate the values of G_m^E, H_m^E, and V_m^E for $(0.5N_2 + 0.5CH_4)$ at $T = 91$ K according to the one-fluid theory of conformal mixtures based on Carnahan and Starling's equation of state. (Solve equation 16.12.13 for V_m by trial and error.) Take $\zeta = 0.98$. The critical temperatures are 126.2 K for N_2 and 190.6 K for CH_4, and the critical molar volumes are 89.5 cm^3 mol^{-1} for N_2 and 99.0 cm^3 mol^{-1} for CH_4. Compare your values with those in table 16.1.

Problem 16.5

Use the lattice theory to predict the values of G_m^E and of H_m^E for $(0.5C_6H_{14} + 0.5C_{13}H_{28})$ at 293.15 K given that for mixtures of normal alkanes $Lw = 1290 \text{ J mol}^{-1}$ and $Lu = 2860 \text{ J mol}^{-1}$, that $T\alpha^*/\kappa_T^*$ has the values 251.5 MPa for pure hexane and 304.6 MPa for pure tridecane, and that $V_m^E = -0.370 \text{ cm}^3 \text{ mol}^{-1}$ for the mixture in question.

Problem 16.6

For liquid mixtures of A = rubber and B = benzene the following values of p^{l+g}/p_A^{l+g} were obtained as a function of the volume fraction ϕ of B at 298.15 K. Find the value of w/kT that best fits these results to equation (16.13.8). Use that value to calculate values of p^{l+g}/p_A^{l+g} corresponding to the experimental ones.

ϕ	0.312	0.441	0.524	0.710	0.823
p^{l+g}/p_A^{l+g}	0.979	0.948	0.916	0.729	0.534

Problem 16.7

Relative activities λ_A/λ_A^* and λ_B/λ_B^* were obtained as follows for $\{(1-x)(C_2H_5)_3N + xCHCl_3\}$ at 283 K.

x	0.2500	0.5000	0.7500
λ_A/λ_A^*	0.6920	0.3147	0.0857
λ_B/λ_B^*	0.0858	0.3201	0.6979

Assuming an ideal associated mixture of A, AB, and B, find the best value of the standard equilibrium constant $K^\ominus$. Use that value of $K^\ominus$ to compare calculated values of λ_A/λ_A^*, λ_B/λ_B^*, and G_m^E with the experimental values.

Problem 16.8

Liquid mixtures $\{(1-x)C_6H_6 + x(C_6H_5)_2\}$ are approximately ideal. The two substances neither mix nor form any compound in the solid. Calculate and plot against the mole fraction x the freezing temperatures of the mixtures. Estimate the temperature and mole fraction of the eutectic. C_6H_6 melts at 278.6 K with molar enthalpy of melting 9.92 kJ mol^{-1}; $(C_6H_5)_2$ melts at 343.6 K with molar enthalpy of melting 16.82 kJ mol^{-1}.

Problem 16.9

Sketch cooling curves for compositions as follows for the mixtures whose phase diagrams are given in figures 16.13 to 16.15. Figure 16.13(a): $x = 0.4$ and $x = 0.5$; figure 16.13(b): $x = 0.25$, x_P, $x = 0.45$, $x = 0.5$, and $x = 0.7$; figure 16.14: $x = 0.02$ and $x = 0.5$; figure 16.15(a): $x = 0.5$; figure 16.15(b): x_{Az}.

Chapter 17

Fluid mixtures

§ 17.1 Introduction

It will have become obvious in the preceding chapter why the excess molar functions are useful only for mixtures of liquids of similar volatility at pressures within 100 kPa or so of atmospheric pressure. For a mixture like $\{(1-x)C_6H_6+xN_2\}$ at 300 K the nitrogen is well above its gas-liquid critical temperature and so cannot exist as a pure liquid; for a mixture like $\{(1-x)C_6H_6+xC_{14}H_{10}$ (anthracene)$\}$ at 300 K the anthracene is well below its melting temperature and so cannot exist as a pure liquid. For a mixture of liquids of comparable volatility at a pressure of say 1 MPa we do not usually know enough about the equation of state of the coexisting gaseous mixture to be able to use equation (16.7.3) as a step towards the determination of G_m^E; that severely limits the range of temperatures over which G_m^E is accessible.

In the present chapter we shall give an introductory account of phase diagrams for fluid mixtures.[†] In particular we shall consider $p(T)$ *projections*, and $p(x)$ *sections* (isotherms), of the three-dimensional (p, T, x) diagram. The corresponding $T(x)$ sections (isobars) are left as an exercise for the reader. We shall not deal at all with the molar volumes of the phases. Nor shall we deal at all with the separation of solid phases. For fuller accounts of the field the reader should turn to the literature.[‡]

§ 17.2 Experimental methods

In the most usual methods of constructing (p, T, x) diagrams, either the pressure p of a sample of known mole fraction x and temperature T is progressively increased through the dew pressure p^d and the bubble pressure p^b, or samples of coexisting phases at pressure p and temperature T are removed and analysed. The second of these methods can be used with a recirculating still or with a static apparatus. Sometimes the critical loci $T^c(x)$ and $p^c(x)$ are found directly by adjusting the temperature and pressure of a sample of known mole fraction x until two coexisting phases just become identical. All these methods are also used for pure fluids, except that the dew pressure p^d should coincide with the bubble pressure p^b and that sampling the coexisting phases is unnecessary. Full accounts of the experimental

---

[†] Even the elementary taxonomy of $p(T, x)$ diagrams is still unfamiliar to most physical chemists. For example no text-book of physical chemistry known to the author deals at all with the $p(x)$ isotherm for a mixture at a temperature above the critical temperature of one of the components.

[‡] Zernike, J. *Chemical Phase Theory*. Kluwer: Deventer. **1955**.
Rowlinson, J. S. *Liquids and Liquid Mixtures*. Butterworth: London. Second edition. **1969**, chapter 6.
Schneider, G. M. *SPRCT(II)*, chapter 4.

methods, and especially of the precautions needed to make measurements at high
pressures, can be found in the literature.†

§ 17.3 The simplest kind of phase diagram

The simplest kind of behaviour of a binary fluid mixture is that in which the gas-liquid
critical points of the two pure substances are joined by a continuous curve and in
which there is no azeotropy and no coexistence of three fluid phases. An example of
this kind of behaviour is shown in figure 17.1, where we have plotted on the left the
$p(T)$ projection c of the critical line joining the critical points A^c and B^c of the two
pure substances A and B, and the vapour-pressure curves $p^{1+g}(T)$, marked A and B,

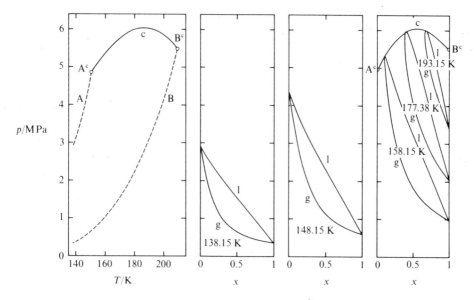

Figure 17.1. The $p(T)$ projection c of the critical line and the $p^{1+g}(T)$ curves A and B for the pure
substances, and some $p(x)$ sections each at a constant temperature T, for fluid mixtures $\{(1-x)Ar + xKr\}$.

of the two pure substances; and on the right some $p(x)$ sections each at a constant
temperature T. We shall keep to this pattern of diagram when we turn to mixtures
having more complicated behaviour. We have not included in the $p(T)$ diagram any
sections at constant values of x (except for $x = 0$ and for $x = 1$), but we note that these
would be curves of dew pressure p^d and bubble pressure p^b meeting on the critical
line at the critical pressure p^c, and that the critical line c is their envelope.‡ The
critical point $p^c(x)$ and $T^c(x)$ does not usually occur at an extremum of the $p(T)$
projection for fixed mole fraction x. In the right-hand part of the diagram the critical

† Young, C. L. *SPRCT(II)*, chapter 3.
Schneider, G. M. *ETd(II)*, chapter 16, part 2.
‡ Such a $p(T)$ section at constant x is called an *isopleth*.

line c is the locus of maxima of the $p(x)$ isothermal sections, horizontal lines joining the two branches of each isotherm being 'tie lines' joining the mole fractions of coexisting fluid phases. For the simple behaviour shown in figure 17.1 we have been able to label the two branches of each $p(x)$ curve g for gas and l for liquid. For more complicated behaviour we shall be unable to distinguish 'gas' from 'liquid'.

§ 17.4   Retrograde condensation

The phenomenon called *retrograde condensation* (or retrograde evaporation) occurs whenever the dew curve, or the bubble curve, is intersected twice by a pathway. The term is usually confined to the two-fold intersection of the dew-pressure curve of an isothermal section, or of the bubble-temperature curve of an isobaric section, by a path of constant composition. As an example, in figure 17.2 we have plotted on a

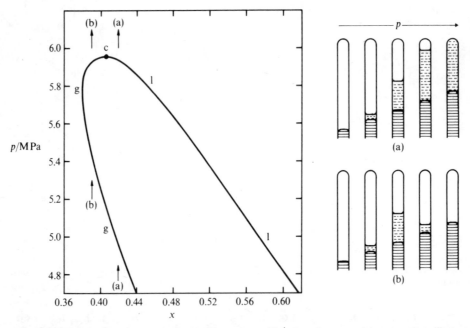

Figure 17.2. Part of the isotherm at 177.38 K from figure 17.1 plotted on a much larger scale to illustrate retrograde condensation. The curve marked g is the dew-pressure curve and that marked l is the bubble-pressure curve. When the gaseous mixture (a) is compressed it condenses in the ordinary way as shown diagrammatically on the upper right-hand side. When the gaseous mixture (b) is compressed it begins to condense to liquid but as the compression is continued the liquid phase at first grows but then shrinks and finally disappears, as shown diagrammatically on the lower right-hand side. In (a) and (b) the denser phase is shown as the shaded one, but that does not imply any difference of kind between the two final fluids.

much larger scale part of the isotherm at 177.38 K from figure 17.1; the retrograde condensation is explained in the legend. It will be obvious that there will be pathways of retrograde condensation, even if the term is used in its restricted sense, for any

fluid mixture. It is easy to understand the bewilderment that must have greeted the
first observation of this phenomenon. A gaseous mixture was expected to condense
to a liquid when the pressure was increased at constant temperature or when the
temperature was decreased at constant pressure. Moreover, at first sight the series of
observations labelled (b) on the right-hand side of figure 17.2 appears to the
experimentalist to defy the rule that at given temperature an increase of pressure
must result in a decrease of volume.

§ 17.5   The barotropic effect

The *barotropic effect* occurs when the molar masses of the pure substances and the
molar volumes of two coexisting phases happen to lead to nearly equal densities:

$$\{(1-x^{\alpha})M_A+x^{\alpha}M_B\}/V^{\alpha}_m \approx \{(1-x^{\beta})M_A+x^{\beta}M_B\}/V^{\beta}_m . \qquad (17.5.1)$$

A small change of pressure or of temperature causes the two phases to change places
in the earth's gravitational field; the hitherto more dense phase fountains up and
takes the place of the hitherto less dense phase. Which is then the 'liquid' and which
the 'gas'? An example is argon + ammonia for which the barotropic effect has been
observed at temperatures between 300 and 400 K and pressures between 0.4 and
0.5 MPa.

§ 17.6   Azeotropy

We recall how an azeotrope differs from a critical point: for an azeotrope $x^{\alpha} = x^{\beta}$ but
$V^{\alpha}_m \neq V^{\beta}_m$; for a critical point $x^{\alpha} = x^{\beta}$ *and* $V^{\alpha}_m = V^{\beta}_m$. Azeotropy may, but need not,
persist up to the critical line. An example of 'positive' azeotropy persisting up to the
critical line is shown in figure 17.3. A positive azeotrope has a maximum vapour
pressure at given temperature (or a minimum boiling temperature at given pressure).
The converse behaviour is called negative azeotropy; it is relatively uncommon.

   The region of figure 17.3 around the point of intersection Z of the azeotropic line
Az and the critical line c is drawn on a larger scale in figure 17.4. The isotherm at T_3,
passing through Z, meets the critical line with a horizontal cusp.

   When azeotropy does not persist up to the critical line the maxima (or minima) of
curves like those in figure 17.3 occur at mole fractions that vary so as to reach $x = 1$
(or $x = 0$) at a temperature below the critical temperature of pure substance B (or A).

§ 17.7   The coexistence of three fluid phases

Of the many binary mixtures for which three fluid phases can coexist there are still
very few for which the phase equilibria have been studied over the whole of the
relevant range of p, T, and x. Only rarely has it been possible to complete such a
study in a single apparatus; different parts of the phase diagram have been studied in
different laboratories, but they have usually been studied for different pairs of
substances. The presentation of such diagrams is beset by related problems of scale,

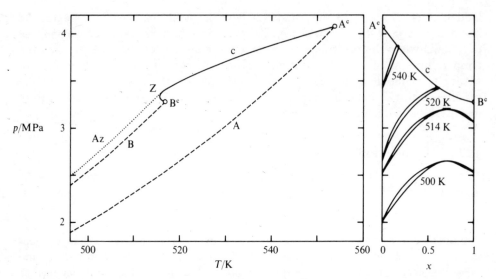

Figure 17.3. The $p(T)$ projection c of the critical line and the $p^{l+g}(T)$ curves A, B, and Az for the pure substances and for azeotropic mixtures, and some $p(x)$ sections each at a constant temperature T, for fluid mixtures $\{(1-x)C_6H_{12} + xC_6F_6\}$ showing positive azeotropy persisting up to the critical line.

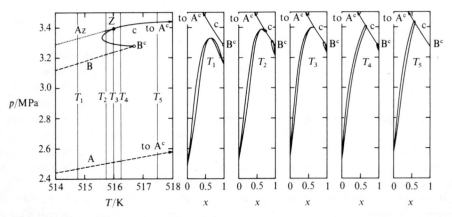

Figure 17.4. Part of figure 17.3 drawn on a larger scale around the point of intersection Z of the azeotropic and critical lines. In the isotherms the mole fractions of coexisting phases have been spread a little so as to show the separate phases distinctly.

problems of the kind already illustrated by our need to redraw a small part of figure 17.3 on a larger scale in figure 17.4, and within figure 17.4 by our need sometimes to exaggerate the differences between the compositions of coexisting phases so as to be able to see them separately.

A phase diagram for one kind of mixture having a region of coexistence of three fluid phases is shown, for $\{(1-x)CF_4 + xCHF_3\}$, in figure 17.5. At temperatures above about 131 K the diagram is of the same form as figure 17.1. At temperatures

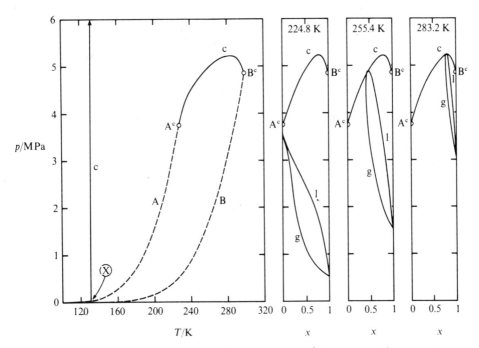

Figure 17.5. The $p(T)$ projections c of the critical lines and the $p^{l+g}(T)$ curves A and B for the pure substances, and some $p(x)$ sections each at a constant temperature T, for fluid mixtures $\{(1-x)CF_4 + xCHF_3\}$. The region marked X is drawn on a larger scale in figure 17.6.

below 130.5 K, however, three fluid phases can coexist. The three-phase curve, intermediate between curves A and B, terminates at an upper critical end point (UCEP), at a temperature of 130.5 K, from which a second critical line begins and rises very steeply with a positive slope: $dp^c/dT^c = 35$ MPa K^{-1}. The critical line is known to continue upwards at least to $p = 100$ MPa. The region around the UCEP, marked X on figure 17.5, is shown on a much larger scale in figure 17.6. The slope of the critical line c in figure 17.6 has been deliberately reduced (it would appear to be vertical on this scale if it were drawn correctly) so as to show the lower critical pressure on the same scale of pressure on the right-hand $p(x)$ section. Actually, the temperature called '131 K' in figure 17.6 would be only 0.0008 K greater than the temperature of the UCEP (about 130.5 K). The pressure of the UCEP has been guessed on the basis of a single $p^{l+g}(x)$ isotherm at 145.2 K. The shape of the three-fluid line 3F has also been guessed. In the $p(x)$ sections, the shapes have all been guessed on the basis of the single $p^{l+g}(x)$ isotherm at 145.2 K and the shapes of the $p(x)$ liquid-liquid coexistence curves at much higher pressures and of the $T(x)$ liquid-liquid coexistence curve at pressures near atmospheric.

If the slope of the critical curve c in figure 17.6 had been negative rather than positive (it is negative for example for $\{(1-x)CH_4 + x2,6,10,15,19,23\text{-}(CH_3)_6C_{24}H_{44}$ (squalane)$\}$ for which $dp^c/dT^c = -1.25$ MPa K^{-1}) then the phase diagram would be

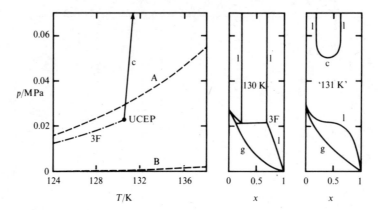

Figure 17.6. The region of figure 17.5 around the UCEP for $\{(1-x)CF_4 + xCHF_3\}$. The critical line c should appear vertical on this scale, but its slope has been greatly reduced for clarity.

like that sketched in figure 17.7. An upper critical pressure would appear at a temperature slightly less than that of the UCEP, and no liquid-liquid immiscibility would occur at temperatures greater than that of the UCEP.

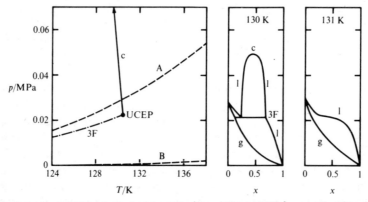

Figure 17.7. Behaviour of a hypothetical mixture like $\{(1-x)CF_4 + xCHF_3\}$ except that the critical curve c has a negative slope.

The $T(x)$ sections, left as exercises for the reader, corresponding to figures 17.6 and 17.7 would both show upper critical temperatures like those in figures 16.8(a) and 16.14.

For other mixtures the UCEP and the three-fluid line can lie above both the curves A and B for the pure substances. The reader is invited to sketch the diagram corresponding to figure 17.6 for such a mixture.

A $p(T)$ projection for another kind of mixture having a region of coexistence of three fluid phases is shown, for $\{(1-x)CH_4 + xC_6H_{14}\}$, in figure 17.8. Here the range of coexistence of three fluid phases is bounded not only above (that is on increasing

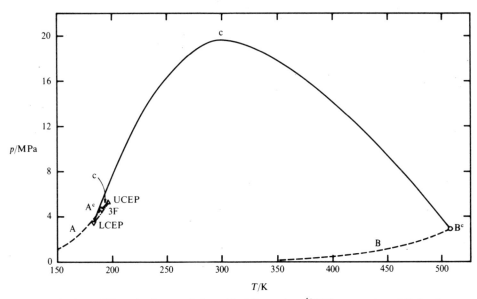

Figure 17.8. The $p(T)$ projections c of the critical lines, the $p^{l+g}(T)$ curves A and B for the pure substances, and the three-fluid curve 3F, for fluid mixtures $\{(1-x)CH_4 + xC_6H_{14}\}$. The three-fluid curve is bounded above by a UCEP and below by an LCEP.

the temperature and pressure) at a UCEP, but also below at a lower critical end point (LCEP). The critical line that starts from the critical point B^c of the pure substance B goes, not to A^c, but to the LCEP, while the UCEP is joined to A^c by another critical line. The impossibility of distinguishing between a liquid and a gas is especially evident; the 'gas-liquid' critical line that starts from B^c is continuous with the 'liquid-liquid' critical line that starts from LCEP. The part of figure 17.8 that includes LCEP, UCEP, and A^c is shown on a larger scale, with $p(x)$ isothermal sections, in figure 17.9. At 186.2 K there is a pressure (about 3.7 MPa) at which three fluid phases coexist and there is an upper critical pressure (on the line LCEP to B^c); at 193.2 K there is a pressure (about 4.75 MPa) at which three fluid phases coexist and there are two upper critical pressures (one on the line A^c to UCEP and one on the line LCEP to B^c); and at 198.0 K there is no pressure at which more than two fluid phases coexist and there is one upper critical pressure (on the line LCEP to B^c).

Other mixtures show variations of the pattern exemplified in figure 17.9. Such variations can be generated by regarding A^c as fixed with UCEP a varying distance away and making any angle with a horizontal line through A^c, the three-fluid line 3F being of any length but obeying the same kind of restriction (for $\mathscr{C} = 2, \mathscr{P} = 3: \mathscr{F} = 1$) as the lines A and B (for $\mathscr{C} = 1, \mathscr{P} = 2: \mathscr{F} = 1$), namely obeying a Clapeyron equation. The reader is invited to sketch diagrams corresponding to figure 17.9 for some of those variations.

Another variation can arise from the superposition of two diagrams like figure 17.6 and figure 17.9. As the temperature is increased, a three-fluid line terminates at

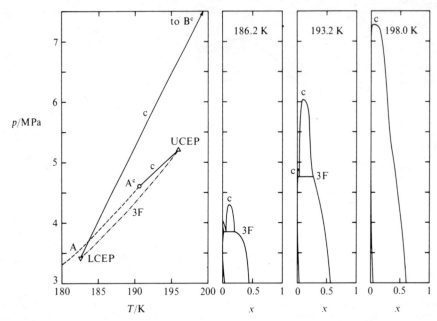

Figure 17.9. Part of the $p(T)$ projection from figure 17.8 for $\{(1-x)CH_4+xC_6H_{14}\}$ drawn on a larger scale, with a $p(x)$ section at each of three temperatures: one between those of LCEP and A^c, one between those of A^c and UCEP, and one greater than that of UCEP.

a UCEP, from which a critical line goes upwards indefinitely, as in figure 17.6, and then after a range of temperatures at which no more than two fluid phases coexist, a second three-fluid line begins at an LCEP and terminates at a UCEP, as in figure 17.9. Such a variation appears to be uncommon but a few examples are known such as $\{(1-x)CH_4+xCH_2{:}CH{\cdot}C_4H_9$ (hex-1-ene)$\}$. The reader is invited to sketch the phase diagram for a mixture of this kind.

A phase diagram for another kind of mixture having a region of coexistence of three fluid phases is shown, for $\{(1-x)C_6H_6+xH_2O\}$, in figure 17.10. The critical line c that starts from B^c, after a minimum in the pressure, now goes at least to the highest accessible pressures (unless it is terminated by intersection with a line for the coexistence of a solid phase and two fluid phases).

Many variations exist of the kind of phase diagram shown in figure 17.10. The critical line c between A^c and UCEP can have any slope (of either sign) and the lines A and B (that is the relative volatilities of the two pure substances) can be reversed. The critical line c that starts from B^c can overlap the range of temperatures of the critical line c from A^c to UCEP, can be single-valued with respect to temperature (so that the lower critical pressure in the $p(x)$ diagram at 570 K is not present), and can be single-valued with respect to pressure. Finally, the critical line c that starts from B^c can, after a minimum with respect to temperature, proceed at accessible pressures to temperatures greater than the critical temperature of pure substance B, as shown

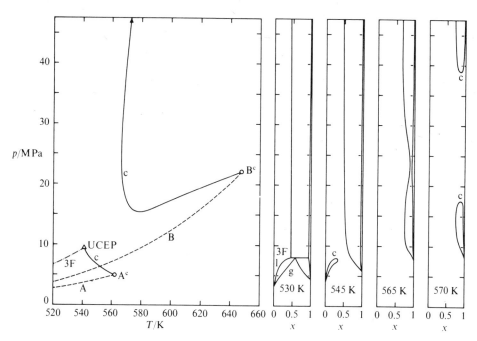

Figure 17.10. The $p(T)$ projections c of the critical lines, the $p^{l+g}(T)$ curves A and B for the pure substances, and the three-fluid curve 3F, and some $p(x)$ sections each at a constant temperature T, for fluid mixtures $\{(1-x)C_6H_6 + xH_2O\}$.

for fluid mixtures $\{(1-x)He + xAr\}$ in figure 17.11, or can even proceed directly to such higher temperatures by starting from B^c with a positive slope, as shown for fluid mixtures $\{(1-x)He + xCH_4)\}$ in figure 17.12. For the examples shown in figures 17.11 and 17.12 it is believed that the other critical line that would start from A^c would terminate in a UCEP as in the example shown in figure 17.10. These two are examples of so-called 'gas-gas immiscibility', a term that was used to describe the coexistence of two fluid phases at temperatures greater than the critical temperature of either pure substance. The term is misleading, however, if only in view of the high densities of such coexisting phases.

§ 17.8   Application of the theory of conformal mixtures

For pairs of molecules that can be expected to conform reasonably well with a principle of corresponding states the theory outlined in § 16.12 has been successfully used, with plausible values of the parameter ζ, to discriminate among some of the kinds of phase diagram to be expected of fluid mixtures having different ratios T_A^c/T_B^c and p_A^c/p_B^c. As an example, the version of the theory based on the equation of state of Carnahan and Starling correctly distinguishes between behaviour like that shown in figure 16.11 for $\{(1-x)He + xAr\}$ and like that shown in figure 16.12 for

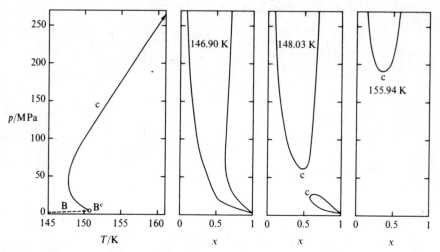

Figure 17.11. The $p(T)$ projection c of the critical line and the $p^{l+g}(T)$ curve B for the pure substance B, and some $p(x)$ sections each at a constant temperature T, for fluid mixtures $\{(1-x)\text{He}+x\text{Ar}\}$.

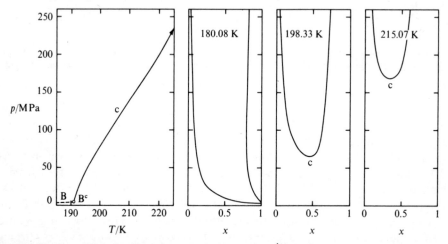

Figure 17.12. The $p(T)$ projection c of the critical line and the $p^{l+g}(T)$ curve B for the pure substance B, and some $p(x)$ sections each at a constant temperature T, for fluid mixtures $\{(1-x)\text{He}+x\text{CH}_4\}$.

$\{(1-x)\text{He}+x\text{CH}_4\}$, and the computed phase diagrams are qualitatively correct.[†] We still have a good deal to learn, however, about more efficient computation of such results; fuller accounts can be found in the literature.[‡]

[†] McGlashan, M. L.; Stead, K.; Warr, C. J. *Chem. Soc. Faraday Trans. II* **1977**, 73, 1889.
[‡] Scott, R. L.; Van Konynenburg, P. H. *Disc. Faraday Soc.* **1970**, 49, 87.
   Hicks, C. P.; Young, C. L. *Chem. Rev.* **1975**, 75, 119; *J. Chem. Soc. Faraday Trans. II* **1977**, 73, 597;
Hurle, R. L.; Jones, F.; Young, C. L. *ibid.* 613; Hurle, R. L.; Toczylkin, L.; Young, C. L. *ibid.* 618.

Chapter 18

Solutions, especially dilute solutions

§ 18.1 Introduction

A *solution* is a special description of a mixture for which it is convenient to distinguish sharply between the *solvent* A on the one hand and the *solutes* B, C, $\cdots$, on the other hand. The solvent is often present in great excess; the solution is then called a *dilute solution*. We shall be especially concerned with dilute solutions both in this chapter and in chapter 20 on electrolyte solutions.

The special properties ('colligative' properties) of dilute solutions follow from a theorem: *in any sufficiently dilute solution the solvent behaves ideally*. The solutes do not usually behave ideally in any 'sufficiently dilute' solution but, whether they do or not, the solvent always approaches ideal behaviour as infinite dilution is approached. We shall now prove that theorem.

We write the Gibbs–Duhem equation (5.7.2) at constant temperature and pressure in the form:

$$\left\{1 - \sum_B x_B\right\} d \ln \lambda_A + \sum_B x_B \, d \ln \lambda_B = 0, \quad (T, p \text{ constant}). \qquad (18.1.1)$$

For each solute B there is obviously some range, however small, between 0 and $\sum_B x_B$ over which we may write

$$\lambda_B \propto x_B, \quad \left(\sum_B x_B \ll 1\right). \qquad (18.1.2)$$

We particularly note that we have not assumed that the proportionality factor is λ_B^*, as it would be for an ideal mixture. Equation (1) then takes the form after some rearrangement:

$$d \ln \lambda_A = d \ln\left(1 - \sum_B x_B\right), \quad \left(\sum_B x_B \ll 1\right). \qquad (18.1.3)$$

Integration of (3) from $\sum_B x_B = 0$ to some small finite value then leads to

$$\ln(\lambda_A/\lambda_A^*) = \ln\left(1 - \sum_B x_B\right), \quad \left(\sum_B x_B \ll 1\right), \qquad (18.1.4)$$

or

$$\lambda_A = \left(1 - \sum_B x_B\right)\lambda_A^*, \quad \left(\sum_B x_B \ll 1\right), \qquad (18.1.5)$$

which is what we set out to prove. An example of (5), and of (2) with the proportionality factor not equal to λ_B^*, can be seen near the left-hand edge ($x \ll 1$) in figure 16.11.

Equation (5) is true in any 'sufficiently dilute' solution, but how dilute is 'sufficiently dilute'? The answer to that question depends strikingly on whether we are dealing with a solution of a non-electrolyte or with a solution of an electrolyte. This, after the conduction of electricity by electrolyte solutions, is the most important distinction between the two kinds of solution. For a solution of a non-electrolyte, equation (5) is obeyed within about 5 per cent up to about $\Sigma_B x_B = 0.1$, an easily accessible range of compositions for experimental work. For a solution of a 1–1 electrolyte dissolved in water as solvent, equation (5) is obeyed within about 5 per cent to about $\Sigma_B x_B = 0.0001$, a barely accessible range of compositions for experimental work. For electrolytes with higher charges on the ions and for solvents of lower electric permittivity the situation is much worse.

In this chapter we shall introduce the general formulae for any solution, whether of non-electrolytes or of electrolytes, and shall develop some formulae that are especially relevant to dilute solutions of non-electrolytes. Solutions of electrolytes will be discussed in chapters 19 and 20.

Both here and in chapters 19 and 20 we shall follow customary practice by using the molality m_B of a solute in place of its mole fraction x_B. Molalities were introduced in § 2.11. We remind the reader that they are related to mole fractions by the formulae:

$$x_B = M_A m_B \bigg/ \left(1 + M_A \sum_B m_B\right),\qquad(18.1.6)$$

or

$$m_B = x_B \bigg/ \left(1 - \sum_B x_B\right) M_A,\qquad(18.1.7)$$

where M_A denotes the molar mass of the solvent A.

By use of (6) we can rewrite (5) in the form:

$$\lambda_A = \left(1 + M_A \sum_B m_B\right)^{-1} \lambda_A^*, \quad \left(M_A \sum_B m_B \ll 1\right).\qquad(18.1.8)$$

§ 18.2   Osmotic coefficient of the solvent

The osmotic coefficient ϕ of the solvent A in a solution is defined by the relation:

$$\ln(\lambda_A/\lambda_A^*) \overset{\text{def}}{=} -\phi M_A \sum_B m_B.\qquad(18.2.1)$$

That definition arises as follows. For a sufficiently dilute solution, according to equation (18.1.8), $\ln(\lambda_A/\lambda_A^*) = -\ln(1 + M_A \Sigma_B m_B)$. For $M_A \Sigma_B m_B \ll 1$ we may expand the logarithm on the right-hand side and keep only the first term. We then use ϕ to express deviations from the resulting formula. Whatever its historical origin, equation (1) is now regarded as the exact definition of ϕ.

The osmotic coefficient ϕ can be determined by any of the methods used for measurement of the difference of chemical potentials of a substance between two phases at the same temperature and pressure (see chapter 10). In particular osmotic coefficients are determined in practice by measurements of vapour pressure (including the important relative method called the isopiestic method), lowering of the freezing temperature of the solvent, and of the osmotic pressure whenever a membrane permeable only to the solvent can be found. Measurements of osmotic pressure are used especially for solutions of macromolecules. The other methods are used especially for electrolyte solutions and will be reviewed in chapter 20.

The dependence of ϕ on temperature is given by the relation:

$$(\partial \phi/\partial T)_p = -(H_A^* - H_A)/RT^2 M_A \sum_B m_B , \qquad (18.2.2)$$

and the dependence of ϕ on pressure by the relation:

$$(\partial \phi/\partial p)_T = (V_A^* - V_A)/RTM_A \sum_B m_B , \qquad (18.2.3)$$

each of which follows immediately on differentiation of equation (1). The differences $(H_A^* - H_A)$ and $(V_A^* - V_A)$ can be found by the methods explained in chapter 2.

§ 18.3   Activity coefficient of a solute

The activity coefficient γ_B of a solute B in a solution is defined by the relation:

$$\lambda_B \overset{\text{def}}{=} m_B \gamma_B (\lambda_B/m_B)^\infty , \qquad (18.3.1)$$

where the superscript $^\infty$ denotes 'infinite dilution' or $\Sigma_B\, m_B \to 0$.

The activity coefficient γ_B can in principle be determined by any of the methods used for measurement of the difference of chemical potentials of a substance between two phases at constant temperature and pressure. In practice the only two methods commonly used are confined to solutions of electrolytes; they will be reviewed in chapter 20.

The dependence of $\ln \gamma_B$ on temperature is given by the relation:

$$(\partial \ln \gamma_B/\partial T)_p = (H_B^\infty - H_B)/RT^2 , \qquad (18.3.2)$$

and its dependence on pressure by the relation:

$$(\partial \ln \gamma_B/\partial p)_T = -(V_B^\infty - V_B)/RT, \qquad (18.3.3)$$

each of which follows immediately on taking logarithms and differentiating equation (1). The differences $(H_B^\infty - H_B)$ and $(V_B^\infty - V_B)$ can be found by the methods explained in chapter 2; they are the changes in the partial molar enthalpy and in the partial molar volume that follow dilution to infinite dilution.

§ 18.4   Relation between the osmotic coefficient of the solvent and the activity coefficients of the solutes

The Gibbs–Duhem equation, obtained in terms of molalities by substitution of (18.1.6) in (18.1.1), has the form:

$$d \ln \lambda_A + M_A \sum_B m_B \, d \ln \lambda_B = 0, \quad (T, p \text{ constant}) . \tag{18.4.1}$$

Substitution of (18.2.1) and (18.3.1) into equation (1) leads after a little manipulation to the relation:

$$d\left\{ (1-\phi) \sum_B m_B \right\} + \sum_B m_B \, d \ln \gamma_B = 0, \quad (T, p \text{ constant}) . \tag{18.4.2}$$

If all but one of ϕ and the γ_B's have been measured over the range of molalities from 0 to $\Sigma_B m_B$ then the remaining one can be calculated.

In particular, for a solution of a single solute B, equation (2) becomes

$$d\{(1-\phi)m_B\} + m_B \, d \ln \gamma_B = 0, \quad (T, p \text{ constant}) . \tag{18.4.3}$$

If ϕ has been measured over the appropriate range of molalities then γ_B can be determined from a plot of $(1-\phi)$ against $\ln(m_B/m^\ominus)$ by the relation:

$$-\ln \gamma_B = (1-\phi) + \int_{m_B=0}^{m_B=m_B} (1-\phi) \, d \ln(m_B/m^\ominus) , \tag{18.4.4}$$

where $m^\ominus$ is a standard molality. Similarly, if γ_B has been measured then ϕ can be determined from the relation:

$$1 - \phi = -m_B^{-1} \int_{m_B=0}^{m_B=m_B} m_B \, d \ln \gamma_B . \tag{18.4.5}$$

§ 18.5   Ideal-dilute solutions

A solution is called *ideal-dilute* if $\phi = 1$ and if $\gamma_B = 1$ for each of the solutes B. For an ideal-dilute solution we thus have

$$\lambda_A = \lambda_A^* \exp\left(-M_A \sum_B m_B \right) , \tag{18.5.1}$$

$$\lambda_B = (\lambda_B/m_B)^\infty m_B , \quad \text{(all B)} . \tag{18.5.2}$$

Since in a sufficiently dilute solution $-\ln(1 + M_A \Sigma_B m_B) \approx -M_A \Sigma_B m_B$, any sufficiently dilute solution is ideal-dilute. Equation (2) is known as Henry's law.

§ 18.6   Determination of the molar mass of a solute

Any of the methods for the measurement of the osmotic coefficient can be used, *for ideal-dilute solutions*, to determine the molar mass M_B of the solute. For non-electrolyte solutions this is the main use of those methods. Since we assume ideal-diluteness as an approximation, and since we usually want only an approximate value of the molar mass, for example to determine N for a solute that we already know from elemental analysis to be say $(C_5H_6)_N$, we may use further approximations and simpler apparatus.

Making use of (18.5.1) we rewrite equation (10.3.11), for the osmotic pressure Π of an ideal-dilute solution of a mass w_B of a solute B dissolved in a volume V of the solvent A, in the form:[†]

$$M_B = w_B RT/\Pi V. \qquad (18.6.1)$$

For macromolecular solutes it is relatively easy to find a membrane permeable only to the solvent, and a reasonable mass of solute dissolved in a reasonable volume of solvent gives an osmotic pressure that is convenient to measure. As an example, if $w_B/V = 0.5$ g/100 cm^3 then $\Pi \approx 5$ kPa for $M_B = 2500$ g mol^{-1}.

We can similarly, though after considerable simplification, rewrite equation (10.5.8), for the relative lowering of the vapour pressure p_A of a mass w_A of solvent A by a mass w_B of an *involatile* solute B in an ideal-dilute solution, in the form:

$$M_B = M_A(w_B/w_A)p_A^*/(p_A^* - p_A), \qquad (18.6.2)$$

where M_A denotes the molar mass of the solvent A and where we have assumed among the other approximations that the gaseous solvent is perfect. As an example, if $M_A = 50$ g mol^{-1} and $w_B/w_A = 0.04$ then $(p_A^* - p_A)/p_A^* \approx 0.01$ for $M_B = 200$ g mol^{-1}. The method, though straightforward, appears not to be widely used.

The most commonly used method for a solute of ordinary molar mass is measurement of the lowering of the freezing temperature of a solvent. We rewrite equation (10.6.9), for the relative lowering $(T_A^* - T_A)/T_A^*$ of the freezing temperature of a mass w_A of solvent A by a mass w_B of a solute B in an ideal-dilute solution, in the form:

$$M_B = M_A(w_B/w_A)(RT_A^*/\Delta_s^l H_A^*)T_A^*/(T_A^* - T_A), \qquad (18.6.3)$$

where we have assumed among other approximations that $(T_A^* - T_A) \ll T_A^*$. As an example, if $M_A = 50$ g mol^{-1}, $w_B/w_A = 0.04$, and $RT_A^*/\Delta_s^l H_A^* = 0.25$, then $(T_A^* - T_A)/T_A^* \approx 0.0025$ for $M_B = 200$ g mol^{-1}. The smaller the molar enthalpy of melting the greater is the lowering of the freezing temperature; camphor, a globular molecule with $M_A = 152$ g mol^{-1} and $RT_A^*/\Delta_s^l H_A^* = 0.55$, is used in Rast's method to give a relative lowering of the freezing temperature nearly seven times greater than that in our example.

[†] We here use the symbol w for mass (*not* 'weight') so as to avoid confusion with our use of m for molality.

Measurements of the elevation of the boiling temperature are also sometimes used to determine the molar mass of an *involatile* solute. We now rewrite equation (10.6.9) in the approximate form:

$$M_B = M_A(w_B/w_A)(RT_A^*/\Delta_l^g H_A^*)T_A^*/(T_A - T_A^*),\qquad (18.6.4)$$

where T_A^* and T_A now denote the boiling temperatures of the pure solvent and of the solution. As an example, if $M_A = 50$ g mol^{-1}, $w_B/w_A = 0.04$, and $RT_A^*/\Delta_l^g H_A^* \approx 0.1$ (Trouton's rule), then $(T_A - T_A^*)/T_A^* \approx 0.001$ for $M_B = 200$ g mol^{-1}.

§ 18.7   Determination of purity from measurements of freezing temperature

Although supplemented nowadays by gas-liquid chromatography, measurement of the dependence of the freezing temperature on the fraction χ that has crystallized is still the best method for determining the purity of a liquid A. The method is valid whenever the solid phase is pure A.

When the external temperature is kept constant and a constant rate of stirring is maintained, χ is proportional to time t; we put $\chi = kt$ where k is a constant. We may then use equation (18.6.3) to write for the mole fraction x_B^i of impurities in the original liquid,

$$x_B^i = (\Delta_s^l H_A^*/RT_A^{*2})(1 - kt)\{T_A^* - T_A(t)\},\qquad (18.7.1)$$

where $T_A(t)$ is the freezing temperature at time t when the fraction $(1 - kt)$ of the liquid remains. In the factor $\{T_A^* - T_A(t)\}$ the freezing temperature T_A^* of pure A is treated as an unknown. Three or more measurements of $T_A(t)$ can be used for the numerical solution of equation (1) for k, T_A^*, and x_B^i. The measurements should begin well after the effect of any supercooling at the beginning of crystallization and should be spread over as wide a range of $\chi = kt$ as is consistent with adequate stirring of the mixture of solid and liquid.

The principle of the calculations can be seen as follows. If measurements are made of $T_1 = T(t_1)$, $T_2 = T(t_1 + \tau)$, and $T_3 = T(t_1 + 2\tau)$, then equation (1) can be solved, and

$$x_B^i = \frac{(\Delta_s^l H_A^*/RT_A^{*2})2\tau(T_1 - T_2)(T_2 - T_3)(T_1 - T_3)}{\{(t_1 + 2\tau)(T_2 - T_3) - t_1(T_1 - T_2)\}\{(T_2 - T_3) - (T_1 - T_2)\}}.\qquad (18.7.2)$$

An example of the use of equation (2) is given in problem 18.5.

§ 18.8   Standard thermodynamic functions for the solvent

The standard chemical potential $\mu_A^\ominus(l, T)$ of the solvent A in a solution is defined by the relation:

$$\mu_A^\ominus(l, T) \overset{\text{def}}{=} \mu_A^*(l, T, p^\ominus),\qquad (18.8.1)$$

a definition identical with that, (16.17.7), of the standard chemical potential of a substance in a liquid mixture or in a pure liquid. Equations (16.17.2) to (16.17.7) therefore apply to the solvent in a solution, the only change being of notation in (16.17.7) which we now write in the form:

$$\mu_A^{\ominus}(l, T) = \mu_A(l, T, p, m_C) + \{\mu_A^*(l, T, p) - \mu_A(l, T, p, m_C)\} + \int_p^{p^{\ominus}} V_A^*(l, T, p)\, dp$$

$$= \mu_A(l, T, p, m_C) + RT\phi M_A \sum_B m_B + \int_p^{p^{\ominus}} V_A^*(l, T, p)\, dp, \qquad (18.8.2)$$

where m_C denotes the set of molalities m_B, m_C, $\cdots$.

The integral in equation (2) makes only a small contribution when $p \approx p^{\ominus}$, and is often omitted; it must of course be retained if (2) is used to derive (18.2.2) or (18.2.3).

§ 18.9   Standard thermodynamic functions for a solute

The standard chemical potential $\mu_B^{\ominus}(\text{solute}, T)$ of a solute B in a solution is defined by the relation:

$$\mu_B^{\ominus}(\text{solute}, T) \overset{\text{def}}{=} \{\mu_B(\text{solute}, T, p^{\ominus}, m_C) - RT \ln(m_B/m^{\ominus})\}^{\infty}$$

$$= \{\mu_B(\text{solute}, T, p, m_C) - RT \ln(m_B/m^{\ominus})\}^{\infty}$$

$$+ \int_p^{p^{\ominus}} V_B^{\infty}(\text{solute}, T, p)\, dp, \qquad (18.9.1)$$

where m_C again denotes the set of molalities m_B, m_C, $\cdots$; $m^{\ominus}$ is a standard molality (usually chosen as 1 mol kg^{-1} in chemical thermodynamics); and the superscript $^{\infty}$ denotes 'infinite dilution' or $\Sigma_B m_B \to 0$.

We rewrite equation (1) in the form:

$$\mu_B^{\ominus}(\text{solute}, T) = \mu_B(\text{solute}, T, p, m_C) - RT \ln(m_B/m^{\ominus}) + \int_p^{p^{\ominus}} V_B^{\infty}(\text{solute}, T, p)\, dp$$

$$+ [\{\mu_B(\text{solute}, T, p, m_C) - RT \ln(m_B/m^{\ominus})\}^{\infty}$$

$$- \{\mu_B(\text{solute}, T, p, m_C) - RT \ln(m_B/m^{\ominus})\}]$$

$$= \mu_B(\text{solute}, T, p, m_C) - RT \ln(m_B \gamma_B/m^{\ominus})$$

$$+ \int_p^{p^{\ominus}} V_B^{\infty}(\text{solute}, T, p)\, dp. \qquad (18.9.2)$$

We can now see why equation (1) was written as it was: the difference $[\cdots]$ in equation (2) can be measured,† vanishes for an ideal-dilute solution, and is known to vanish for any real solution that is sufficiently dilute.

† The term $-RT \ln(m_B/m^{\ominus})$ is included in $\{\cdots\}^{\infty}$ because μ_B diverges at infinite dilution but $\{\cdots\}$ does not.

The integral in equations (1) or (2) makes only a small contribution when $p \approx p^{\ominus}$ and is often omitted; it must of course be retained if (2) is used to derive (18.3.2) or (18.3.3).

§ 18.10 Standard equilibrium constant for a reaction in solution

Remembering that we are treating the solvent A differently from the solutes B, C, $\cdots$, we write the expression for a chemical reaction in solution in the form:

$$0 = \nu_A A + \sum_B \nu_B B .$$ (18.10.1)

The standard equilibrium constant $K^{\ominus}(T) = (\lambda_A^{\ominus})^{-\nu_A} \Pi_B (\lambda_B^{\ominus})^{-\nu_B}$ then takes the form:

$$K^{\ominus}(T) = \exp\left(-\nu_A \phi^{eq} M_A \sum_B m_B^{eq}\right) \prod_B (m_B^{eq} \gamma_B^{eq}/m^{\ominus})^{\nu_B} \exp\left[\int_{p^{\ominus}}^{p^{eq}} \left\{\left(\nu_A V_A^*\right.\right.\right.$$

$$\left.\left.\left.+ \sum_B \nu_B V_B^{\infty}\right)\Big/RT\right\} dp\right].$$ (18.10.2)

When the equilibrium pressure p^{eq} is close to the standard pressure $p^{\ominus}$ the integral in equation (2) becomes negligible; the last factor is often omitted.

The factor $\exp(-\nu_A \phi^{eq} M_A \sum_B m_B^{eq})$ in equation (2) is often close to unity and is also often omitted. For example for H_2O as solvent so that $M_A = 0.018$ kg mol^{-1}, and for $\nu_A = -1$, $\phi^{eq} = 1$, and $\sum_B m_B^{eq} = 0.5$ mol kg^{-1} say, $\exp(-\nu_A \phi^{eq} M_A \sum_B m_B^{eq}) = e^{0.009} \approx 1.01$. That is the origin of the advice commonly given in elementary textbooks to leave out factors relating to the solvent, even when it is one of the reacting substances, in the formulation of equilibrium constants. That is all very well in elementary or approximate treatments, but such factors may not be omitted in accurate treatments.

For an ideal-dilute solution with $p^{eq} = p^{\ominus}$ equation (2) reduces to the form:

$$K^{\ominus}(\text{ideal-dilute}, T) = \exp\left(-\nu_A M_A \sum_B m_B^{eq}\right) \prod_B (m_B^{eq}/m^{\ominus})^{\nu_B}.$$ (18.10.3)

For any real solution with $p^{eq} = p^{\ominus}$ equation (2) can be written in the form:

$$K^{\ominus}(T) = \lim_{\sum_B m_B^{eq} \to 0} \prod_B (m_B^{eq}/m^{\ominus})^{\nu_B}.†$$ (18.10.4)

Equation (4) provides an experimental route from measurements of the molalities m_B^{eq} at chemical equilibrium to the standard equilibrium constant; the measurements must be made in more and more dilute solutions so that $K^{\ominus}(T)$ can be obtained by extrapolation of the 'classical' equilibrium constant $\Pi_B (m_B^{eq}/m^{\ominus})^{\nu_B}$ to $\sum_B m_B^{eq} \to 0$.

† When $m^{\ominus}$ is omitted from $\Pi_B (m_B^{eq}/m^{\ominus})^{\nu_B}$ the quantity is sometimes denoted by K_m. Whereas $K^{\ominus}$ is a number, K_m has the dimension: (molality)$^{\sum_B \nu_B}$. We shall not use the quantity K_m in this book.

The route implied by equation (4) is usually a much easier one than that implied by equation (2) in which in addition to the m_B^{eq}'s the osmotic coefficient ϕ^{eq} and the activity coefficients γ_B^{eq} must be measured.

For non-electrolyte solutions few chemical equilibria have been studied. One example is the reaction:

$$N_2O_4(\text{solute}) = 2NO_2(\text{solute}),\qquad\qquad (18.10.5)$$

which has been studied spectrophotometrically in dilute solution in a solvent such as tetrachloromethane. We shall return to equation (2), as we shall to other equations of this chapter, when we discuss electrolyte solutions in chapter 20.

Problems for chapter 18

Problem 18.1

Solutions of mass concentration w_B/V of a sample of poly(γ-benzyl glutamate) in 1,1,2,2-tetrachloroethane supported a column of height h of the pure solvent (density: 1.566 g cm^{-3}) at osmotic equilibrium at 310.15 K in a laboratory where the acceleration of free fall was 9.81 m s^{-2}. Given the following values of w_B/V and h calculate the average molar mass of the polypeptide.

$(w_B/V)/$kg m^{-3}	3.09	4.64	6.96	10.44
$h/$cm	2.14	3.23	4.92	7.45

Problem 18.2

Benzoic acid is largely dimerized in solution in benzene. Benzoic acid and phenanthrene can each be treated as an involatile solute in benzene. Solutions of benzoic acid and of phenanthrene in benzene were brought to isopiestic equilibrium (equality of the chemical potentials of the solvent; to the approximations of chapter 18, equality of vapour pressures) at 317.05 K, and were found to have the following molalities m of benzoic acid estimated as the monomer $C_6H_5CO_2H$ and m_P of phenanthrene $C_{14}H_{10}$.

$m/$mol kg^{-1}	0.3787	0.3344	0.2622	0.2596	0.0852
$m_P/$mol kg^{-1}	0.2035	0.1807	0.1417	0.1412	0.0489

Calculate the degree of dissociation of $(C_6H_5CO_2H)_2$ in each solution and the standard equilibrium constant ($m^\ominus = 1$ mol kg^{-1}) for the dissociation:

$$(C_6H_5CO_2H)_2(\text{solute}) = 2C_6H_5CO_2H(\text{solute}).$$

Problem 18.3

For $A = C_6H_6$ the freezing temperature $T_A^* = 278.55$ K and the cryoscopic constant $M_A R T_A^{*2}/\Delta_s^1 H_A^* = 5.085$ K mol^{-1} kg. Given the following differences $(T_A^* - T_A)$ of freezing temperature for solutions containing molalities m of $B = C_6H_5CO_2H$ in C_6H_6, calculate the fraction of B_2 dissociated in each solution and the standard equilibrium constant for the dissociation: $B_2(\text{solute}) = 2B(\text{solute})$.

$\Delta T/\text{K}$	0.0189	0.0414	0.0736	0.1205	0.1911	0.2540
$m/\text{mol kg}^{-1}$	0.00621	0.01418	0.02616	0.04372	0.07072	0.09462

Use the mean value of $K^{\ominus}(278.55 \text{ K})$ with the mean value of $K^{\ominus}(317.05 \text{ K})$ from problem 18.2 to estimate $\Delta H_m^{\ominus}$ for the dissociation.

Problem 18.4

The mole-fraction solubility of air, $\{0.8N_2 + 0.2O_2\}$, in water at 273.15 K and atmospheric pressure is 2.4×10^{-5}. Calculate the lowering of the freezing temperature of water caused by the presence of air at atmospheric pressure. The molar enthalpy of melting of $H_2O(s)$ is 6007 J mol^{-1}. Compare your value with the results of problem 9.2.

Problem 18.5

The following freezing temperatures $T(t)$ of analytical-reagent benzene were measured at times t after the onset of crystallization:

t/s	1388	2468	3548
$\{T(t) - T_0\}/\text{K}$	4.1476	3.7346	2.9344

where T_0 is the constant temperature of a bath containing one arm of a difference thermometer. For C_6H_6, $\Delta_s^l H_m^*/RT^{*2} = 0.01542 \text{ K}^{-1}$. Calculate the mole-fraction purity of the sample of C_6H_6.

Chapter 19

Digression on galvanic cells

§ 19.1 Introduction

This chapter is needed to identify the conditions that allow galvanic cells to be used to obtain thermodynamic quantities. No galvanic cell can be at equilibrium, even when the external circuit is open; gradients of chemical potential within the cell always ensure that diffusion is happening. There is strictly no such thing as 'the thermodynamics of galvanic cells'. Nevertheless in certain special cases the electromotive force† (e.m.f.) E of a galvanic cell can be used to calculate the affinity A of a chemical reaction.

§ 19.2 The galvanic cell

We shall now discuss the general galvanic cell:

$$Cu^L \left| \begin{matrix} Re^L \\ Ox^L \end{matrix} \right| \begin{matrix} \text{bridging solutions} \\ \text{of electrolytes} \end{matrix} \left| \begin{matrix} Re^R \\ Ox^R \end{matrix} \right| Cu^R . \qquad (19.2.1)$$

On the left-hand side a copper lead Cu^L is in contact with the reduced form Re^L and the oxidized form Ox^L of an electron-transfer pair of substances related by

$$Re^L = Ox^L + e^- . \qquad (19.2.2)$$

Examples of such pairs are

$$Pt(s) + \tfrac{1}{2}H_2(g) = Pt(s) + H^+(aq) + e^- ,$$

$$\tfrac{1}{2}Cu(s) = \tfrac{1}{2}Cu^{2+}(aq) + e^- ,$$

$$Ag(s) = Ag^+(aq) + e^- ,$$

$$Ag(s) + Cl^-(aq) = AgCl(s) + e^- ,$$

$$Pt(s) + Cl^-(aq) = Pt(s) + \tfrac{1}{2}Cl_2(g) + e^- ,$$

$$Pt(s) + \tfrac{1}{5}Mn^{2+}(aq) + \tfrac{4}{5}H_2O(l) = Pt(s) + \tfrac{1}{5}MnO_4^-(aq) + \tfrac{8}{5}H^+(aq) + e^- ,$$

$$Pt(s) + Fe^{2+}(aq) = Pt(s) + Fe^{3+}(aq) + e^- .$$

---

† 'Electromotive force' is the name used for the electric potential difference measured at virtually zero electric current. The electric potentiometer used to measure the e.m.f. of a galvanic cell may be either a classical bridge with which the e.m.f. can be balanced against an external source of electric potential difference, or a modern electronic meter having an extremely high input impedance, so as to ensure that virtually zero electric current flows through the galvanic cell.

(When platinum is present its function is to act as an electron conductor. In the first example it also acts as a catalyst.) On the right-hand side a copper lead Cu^R is similarly in contact with some other electron-transfer pair Ox^R and Re^R. We shall assume that there is equilibrium (and that the equilibrium is quickly re-established after a small electric current) between the electrons in copper and the electron-transfer pair in contact with it. We then have

$$-F(V^R - V^L) = \mu(e^- \text{ in } Cu^R) - \mu(e^- \text{ in } Cu^L)$$

$$= \mu(Re^R) - \mu(Ox^R) - \mu(Re^L) + \mu(Ox^L), \quad (19.2.3)$$

where $(V^R - V^L)$ is the electric potential difference between the right-hand and left-hand leads and F is the Faraday constant.

It is easy to prove that the combination of chemical potentials on the right-hand side of equation (3) cannot be expressed in terms of the chemical potentials only of pure substances and of electrically neutral dissolved electrolytes. The chemical potential of a single ion has no physical meaning. Thus equation (3), though formally correct, leads nowhere. Nor have we introduced the condition that the electric current shall be zero; the electric potential difference $(V^R - V^L)$ may not be identified with the electromotive force E of the galvanic cell.

By the introduction of an extra-thermodynamic expression for $I = 0$, where I denotes electric current, and the use of Onsager's reciprocal relations,[†] it can be shown[‡] that the e.m.f. of cell (1) is given exactly by the equation:

$$-FE = \mu(Re^R) - \mu(Ox^R) - \mu(Re^L) + \mu(Ox^L) + \sum_i \int_{\mu_i^L}^{\mu_i^R} (t_i/z_i) \, d\mu_i, \quad (19.2.4)$$

where t_i denotes the transport number and z_i the charge number (positive for a cation and negative for an anion) of the ion i, and where the summation extends over every species i of ion present anywhere between the ends L and R of the cell. The transport number t_i of an ion i is defined as the fraction of the electric current arising from the flow of the ion i; it is not of course a thermodynamic quantity. We may use the identity $\sum_i t_i = 1$, so that $t_j = 1 - \sum_{i \neq j} t_i$, to rewrite equation (4) in the form:

$$-FE = \mu(Re^R) - \mu(Ox^R) - \mu(Re^L) + \mu(Ox^L) + z_j^{-1}\mu_j^R - z_j^{-1}\mu_j^L$$

$$+ \sum_{i \neq j} \int_L^R t_i(z_i^{-1} \, d\mu_i - z_j^{-1} \, d\mu_j). \quad (19.2.5)$$

† Onsager's reciprocal relations apply to 'coupled' processes in which two or more 'forces' drive two or more 'fluxes', and lead to the subject called 'the thermodynamics of irreversible processes'. In a thermocouple for example the temperature difference drives a flow of energy and the electric potential difference drives an electric current. Irreversible processes will not be treated in this book; the interested reader is referred to one or more of the following.

Miller, D. G. Chem. Rev. 1960, 60, 20.

Callen, H. B. Thermodynamics. Wiley: New York. 1960. Chapters 15 to 17.

Denbigh, K. G. The Thermodynamics of the Steady State. Methuen: London. 1951.

De Groot, S. R. Thermodynamics of Irreversible Processes. North-Holland: Amsterdam. 1951.

De Groot, S. R.; Mazur, P. Non-Equilibrium Thermodynamics. North-Holland: Amsterdam. 1960.

‡ See for example Guggenheim, E. A. Thermodynamics. North Holland: Amsterdam. 1967. Fifth edition. Chapter 13.

It can be shown, most easily by considering examples, that the ion j can always be chosen so as to reveal that equation (5) contains the chemical potentials only of pure substances and of electrically neutral dissolved electrolytes. We take as a specific example the cell:

$$\text{Pt}|\text{Ag}\left|\begin{array}{c}\text{solution}\\\text{of AgNO}_3\end{array}\right|\begin{array}{c}\text{bridging solutions}\\\text{of AgNO}_3,\text{ Fe(NO}_3)_2,\\\text{and Fe(NO}_3)_3\end{array}\left|\begin{array}{c}\text{solution}\\\text{of Fe(NO}_3)_2\\\text{and Fe(NO}_3)_3\end{array}\right|\text{Pt},\qquad(19.2.6)$$

and choose $j = NO_3^-$, so that $z_j = -1$ and equation (5) becomes

$$-FE = \mu(Fe^{2+}, R) - \mu(Fe^{3+}, R) - \mu(Ag, L) + \mu(Ag^+, L)$$

$$-\mu(NO_3^-, R) + \mu(NO_3^-, L) + \sum_{i \neq NO_3^-}\int_L^R (t_i/z_i)\{d\mu_i + z_i\, d\mu(NO_3^-)\}$$

$$= \{\mu(Fe^{2+}, R) + 2\mu(NO_3^-, R)\} - \{\mu(Fe^{3+}, R) + 3\mu(NO_3^-, R)\} - \mu(Ag, L)$$

$$+ \{\mu(Ag^+, L) + \mu(NO_3^-, L)\} + \int_L^R t(Ag^+)\, d\{\mu(Ag^+) + \mu(NO_3^-)\}$$

$$+ \int_L^R \{t(Fe^{2+})/2\}d\{\mu(Fe^{2+}) + 2\mu(NO_3^-)\}$$

$$+ \int_L^R \{t(Fe^{3+})/3\}\, d\{\mu(Fe^{3+}) + 3\mu(NO_3^-)\}$$

$$= \mu\{Fe(NO_3)_2, R\} - \mu\{Fe(NO_3)_3, R\} - \mu(Ag, L) + \mu(AgNO_3, L)$$

$$+ \int_L^R t(Ag^+)\, d\mu(AgNO_3) + \int_L^R \{t(Fe^{2+})/2\, d\mu\{Fe(NO_3)_2\}$$

$$+ \int_L^R \{t(Fe^{3+})/3\}\, d\mu\{Fe(NO_3)_3\}. \qquad(19.2.7)$$

The integrals in equation (7) depend on the distribution of electrolytes in the bridging solutions; consequently the value of E depends on how the bridging solutions are arranged. Whatever the initial arrangement of the bridging solutions, interdiffusion will be happening; consequently the value of E depends also on time. In certain special cases both kinds of dependence become vanishingly small; it is only for those special cases that galvanic cells are thermodynamically useful.

§ 19.3　Examples of useful galvanic cells

Our first example differs from cell (19.2.6) only by the presence throughout the cell of a uniform molality of some non-reacting electrolyte, for which we choose KNO_3, the molalities of all other electrolytes being kept as small as possible. The e.m.f. of such a cell is given by an equation that differs from (19.2.7) only by the presence of the additional integral:

$$+\int_L^R t(K^+)\, d\mu(KNO_3)\,. \tag{19.3.1}$$

In the limit of 'infinite swamping' each of the transport numbers $t(Ag^+)$, $t(Fe^{2+})$, and $t(Fe^{3+})$, and hence each of the integrals in the last form of equation (19.2.7), becomes vanishingly small. The integral (1) also becomes vanishingly small because $\mu(KNO_3)$ is virtually constant throughout the cell. We then have in place of (19.2.7) the equation:

$$-FE = \mu\{Fe(NO_3)_2, \boldsymbol{S}\} - \mu\{Fe(NO_3)_3, \boldsymbol{S}\} - \mu(Ag, s) + \mu(AgNO_3, \boldsymbol{S})\,, \tag{19.3.2}$$

where R and L have both been replaced by $\boldsymbol{S}$ (for 'swamping'). Cells of this kind thus satisfy three conditions that cells like (19.2.6) do not satisfy: E is independent of the arrangement of the bridging solutions; E is independent of time; and E is a thermodynamic quantity. In particular

$$-FE = -\boldsymbol{A} = (\partial G/\partial \xi)_{T,p}\,, \tag{19.3.3}$$

where $\boldsymbol{A}$ is the affinity (see § 5.9) of the electron-transfer reaction:

$$Fe(NO_3)_3 + Ag(s) = Fe(NO_3)_2 + AgNO_3\,, \tag{19.3.4}$$

when it is carried out with vanishingly small molalities of the reactants in a solution having a high molality of KNO_3. Such an affinity is not an especially interesting quantity. We shall see, however, that this cell can also be used with a double extrapolation to find the standard equilibrium constant $K^\ominus$ of reaction (4). We shall discuss three such applications of this kind of cell in chapter 20.

Our second example is the generally more useful kind of cell:

$$Pt\bigg|H_2(g)\bigg|\begin{array}{c}\text{solution containing}\\ H^+ \text{ and } Cl^-\end{array}\bigg|AgCl(s)\bigg|Ag\bigg|Pt\,, \tag{19.3.5}$$

in which there is apparently only one uniform electrolyte solution. Strictly cell (5) should be written in the form:

$$Pt|H_2(g)\left|\begin{array}{c}\text{solution containing}\\ H^+ \text{ and } Cl^- \text{ and}\\ \text{saturated with } H_2\end{array}\right|\begin{array}{c}\text{solution}\\ \text{containing}\\ H^+ \text{ and } Cl^-\end{array}\left|\begin{array}{c}\text{solution containing}\\ H^+ \text{ and } Cl^- \text{ and}\\ \text{saturated with AgCl}\end{array}\right|AgCl(s)|Ag|Pt,$$

$$\tag{19.3.6}$$

but now it is obvious that $\int_L^R t(Ag^+)\, d\mu(AgCl)$ is negligible because $t(Ag^+)$ is so small, and that $\int_L^R t(H^+)\, d\mu(HCl)$ is negligible because the solution of HCl is virtually uniform, so that $\mu(HCl)$ is virtually constant, throughout the cell. There is no need to add any swamping electrolyte to dispose of the integrals in equation (19.2.5); the three conditions are already satisfied, and the e.m.f. E is given by

$$-FE = \mu(Ag, s) - \mu(AgCl, s) + \mu(HCl, \text{solute}) - \tfrac{1}{2}\mu(H_2, g)\,. \tag{19.3.7}$$

Cells like (5) depend for the simplicity of their behaviour on the choice of two pure-substance electrodes of which one forms an electron-transfer pair with a cation, and the other with an anion, in a uniform electrolyte solution. The use of cells like (5) as chemical potentiometers has already been outlined in § 10.7. That and other applications of cells like (5) will be discussed in chapter 20.

Our third example is the kind of cell:

$$Pt|Ag|AgCl(s)|solution\ containing\ Cl^-|HgCl(s)|Hg(l)|Pt\,, \qquad (19.3.8)$$

which has already been discussed in §15.2 in connexion with Nernst's heat theorem. It should now be obvious, provided that the solution is not too dilute, that the e.m.f. E of cell (8) is given by

$$-FE = \mu\,(Hg, l) - \mu\,(HgCl, s) - \mu\,(Ag, s) + \mu\,(AgCl, s)\,, \qquad (19.3.9)$$

and is independent of the nature and the molalities of the electrolyte solution containing chloride ion.

Our fourth and last example is the kind of cell:

$$Pt|Ag|AgNO_3(m_1)|AgNO_3(m_2)|Ag|Pt\,, \qquad (19.3.10)$$

having two identical electrodes and two solutions of the same single electrolyte differing only in its molality. According to equation (19.2.5), when we choose $j = Ag^+$, the e.m.f. E is given by

$$FE = \int_L^R t(NO_3^-)\,d\mu\,(AgNO_3)\,. \qquad (19.3.11)$$

The e.m.f. of cell (10) is independent of the sharpness or the shape of the liquid-to-liquid junction and is independent of time; it depends only on the molalities m_1 and m_2 of the two solutions in contact with the electrodes. The right-hand side of (11) is not, however, a thermodynamic quantity. If the dependence of the transport number on the molality is known over the range m_1 to m_2 (for example from 'moving-boundary' experiments) equation (11) can be integrated, and cell (10) may nevertheless be used as another kind of chemical potentiometer. We shall not, however, discuss cells like (10) further in this book.

We conclude this section with a remark about the 'salt bridge', a concentrated solution of a non-reacting electrolyte (sometimes set as a gel in a solution of agar) traditionally inserted in place of the bridging solutions in cells like (19.2.6) under the illusion that it will annul the integrals in equation (9.2.5). It seldom if ever does. Nor is there any justification for the claim that it will if the non-reacting electrolyte is chosen as one for which the transport numbers are equal.

§ 19.4   Standard electromotive force

We may use the relations between the chemical potential and the standard chemical potential (given for gases in § 12.8, for pure liquids and solids in § 16.17, for the solvent in § 18.8, and for solutes in § 18.9) to rewrite any of our equations for $-FE$. In particular, equation (19.3.7) becomes

$$-FE = \mu^{\ominus}(\text{Ag, s}) - \mu^{\ominus}(\text{AgCl, s}) + \mu^{\ominus}(\text{H}^+, \text{solute}) + \mu^{\ominus}(\text{Cl}^-, \text{solute}) - \tfrac{1}{2}\mu^{\ominus}(\text{H}_2, \text{g})$$
$$+ 2RT \ln(m\gamma_{\pm}/m^{\ominus}) - \tfrac{1}{2}RT \ln\{x(\text{H}_2, \text{g})p/p^{\ominus}\}, \qquad (19.4.1)$$

where $m^{\ominus}$ denotes a standard molality (usually 1 mol kg^{-1}) and $p^{\ominus}$ a standard pressure (usually 101.325 kPa), where $\gamma_{\pm}$ denotes the activity coefficient of the electrolyte HCl, and where we have assumed with usually trivial inaccuracy that any difference between p and $p^{\ominus}$ may be neglected for liquid and solid phases and that the gaseous phase is a perfect gas mixture. Equation (1) may be written in the more compact form:

$$E = E^{\ominus} - (2RT/F) \ln(m\gamma_{\pm}/m^{\ominus}) + (\tfrac{1}{2}RT/F) \ln\{x(\text{H}_2, \text{g})p/p^{\ominus}\}, \qquad (19.4.2)$$

where $E^{\ominus}$ is called the *standard electromotive force* (of the reaction, not of the cell) and is defined by

$$E^{\ominus} \stackrel{\text{def}}{=} (RT/F) \ln K^{\ominus}$$
$$= -\{\mu^{\ominus}(\text{Ag, s}) - \mu^{\ominus}(\text{AgCl, s}) + \mu^{\ominus}(\text{HCl, solute}) - \tfrac{1}{2}\mu^{\ominus}(\text{H}_2, \text{g})\}/F. \qquad (19.4.3)$$

For historical reasons values of $E^{\ominus}$ are often tabulated instead of values of $K^{\ominus}$ (or of $\ln K^{\ominus}$) for electron-transfer reactions in solution.

§ 19.5 'Standard electrode potentials'

To write, for example,

$$E^{\ominus}\{\tfrac{1}{2}\text{H}_2(\text{g}) + \text{Ag}^+(\text{aq}) = \text{Ag(s)} + \text{H}^+(\text{aq})\} = +0.8 \text{ V}, \qquad (19.5.1)$$

should cause no difficulties. It is merely a way of writing the value of $(RT/F) \ln K^{\ominus}$ where $K^{\ominus}$ is the standard equilibrium constant of the specified reaction. In this case $K^{\ominus}(298.15 \text{ K}) = \exp(E^{\ominus}F/RT) = 3 \times 10^{13}$.

Misunderstanding is unlikely to follow the abbreviation of statements like (1) to

$$E^{\ominus}\{\text{Ag}^+(\text{aq}) + e^- = \text{Ag(s)}\} = +0.8 \text{ V}, \qquad (19.5.2)$$

in which 'e^-' conventionally replaces $\{\tfrac{1}{2}\text{H}_2(\text{g}) - \text{H}^+(\text{aq})\}$, but is more likely to follow the more drastic abbreviation to

$$E^{\ominus}(\text{Ag}^+/\text{Ag}) = +0.8 \text{ V}, \qquad (19.5.3)$$

and seems certain to follow a statement like

'the standard electrode potential of Ag is $+0.8 \text{ V}$'. $\qquad (19.5.4)$

Apart from leading the innocent to the absurd idea that a 'single electrode potential' could ever be measured, such abbreviations often face the reader with a dilemma about the sign. Does 'the standard electrode potential of Ag is $+0.8$ V' imply equation (1), or does it imply equation (1) with the chemical equation reversed? This particular example presents no problem to anyone who knows that silver 'is more noble than' hydrogen, so that $K^{\ominus} > 1$ and $E^{\ominus} > 0$ for the reaction:

$$\tfrac{1}{2}H_2(g) + Ag^+(aq) = Ag(s) + H^+(aq) \,. \tag{19.5.5}$$

It is more difficult, however, to translate 'the standard electrode potential of the Sn^{4+}/Sn^{2+} couple is 0.15 V', especially when one finds in other books 'the standard electrode potential of the Sn^{2+}/Sn^{4+} couple is 0.15 V'. In an attempt to resolve such misunderstandings the International Union of Pure and Applied Chemistry recommended in 1953 (the 'Stockholm Convention') that the term 'standard electrode potential' be used only for the value of $E^{\ominus} = (RT/F) \ln K^{\ominus}$ of the reaction:

$$\tfrac{1}{2}H_2(g) + Ox^z = H^+ + Re^{(z-1)} \,. \tag{19.5.6}$$

Happily, most modern books now follow that recommendation. It might have been better to have persuaded authors to avoid abbreviations that do more harm than good and to write in full what they mean. The term 'standard electrode potential' will not be used again in this book.

Chapter 20

Solutions of electrolytes

§ 20.1 Electrical neutrality

There is always one restriction on our regarding the ions i in a solution of electrolytes as independent components.† That is the condition of electrical neutrality:

$$\sum_i m_i z_i = 0 , \qquad (20.1.1)$$

where m_i denotes the molality, and z_i the charge number (positive for a cation and negative for an anion), of the ion i.

Further restrictions arise as usual from the equilibria of one or more chemical reactions among the ions, the neutral ('undissociated') electrolytes or other non-electrolyte solutes, and the solvent.

§ 20.2 Gibbs–Duhem equation for a solution containing electrolytes

Except that it is subject to the restriction of equation (20.1.1), the Gibbs–Duhem equation (18.4.1):

$$\mathrm{d} \ln \lambda_A + M_A \sum_i m_i \, \mathrm{d} \ln \lambda_i = 0 , \quad (T, p \text{ constant}) , \qquad (20.2.1)$$

or (18.4.2):

$$\mathrm{d}\left\{ (1 - \phi) \sum_i m_i \right\} + \sum_i m_i \, \mathrm{d} \ln \gamma_i = 0 , \quad (T, p \text{ constant}) , \qquad (20.2.2)$$

applies equally well to a solution containing electrolytes when we treat the ions as independent cômponents. The absolute activities λ_i and the activity coefficients γ_i of individual ions are, however, inaccessible experimentally; they are convenient mathematical fictions of which only combinations like $(z_+^{-1} \ln \lambda_+ + |z_-^{-1}| \ln \lambda_-)$, corresponding to electrically neutral electrolytes, can be measured.

§ 20.3 Gibbs–Duhem equation for a solution of a single electrolyte

For a solution of molality m of a single electrolyte $C_{\nu_+} A_{\nu_-}$ we have

$$m_+ = \nu_+ m , \quad \text{and} \quad m_- = \nu_- m . \qquad (20.3.1)$$

† We remind the reader of our discussion on p. 120 of such restrictions in connexion with the phase rule.

Equation (20.2.1) can then be rewritten in the form:

$$d \ln \lambda_A + M_A(\nu_+ m \ d \ln \lambda_+ + \nu_- m \ d \ln \lambda_-) = 0, \quad (T, p \text{ constant}), \quad (20.3.2)$$

and equation (20.2.2) in the form:

$$(\nu_+ + \nu_-) \ d\{(1 - \phi)m\} + \nu_+ m \ d \ln \gamma_+ + \nu_- m \ d \ln \gamma_- = 0, \quad (T, p \text{ constant}). \quad (20.3.3)$$

We define the absolute activity $\lambda_\pm$ of the electrolyte by

$$\ln \lambda_\pm \overset{\text{def}}{=} \nu_+ \ln \lambda_+ + \nu_- \ln \lambda_-, \quad (20.3.4)$$

so that equation (2) reduces to the same form:

$$d \ln \lambda_A + M_A m \ d \ln \lambda_\pm = 0, \quad (T, p \text{ constant}), \quad (20.3.5)$$

as does (18.4.1) for a single solute.

We define the activity coefficient $\gamma_\pm$ of the electrolyte by

$$(\nu_+ + \nu_-) \ln \gamma_\pm \overset{\text{def}}{=} \nu_+ \ln \gamma_+ + \nu_- \ln \gamma_-, \quad (20.3.6)$$

so that equation (3) reduces to the same form:

$$d\{(1 - \phi)m\} + m \ d \ln \gamma_\pm = 0, \quad (T, p \text{ constant}), \quad (20.3.7)$$

as (18.4.3) for a single solute.[†]

§ 20.4   Debye–Hückel formulae for a solution of electrolytes

For a dilute (say $\Sigma_B \ m_B < 1 \text{ mol kg}^{-1}$) solution of non-electrolytes we can safely put $\phi = 1$, and $\gamma_B = 1$ for each solute B. Even for an extremely dilute (say $\Sigma_i \ m_i < 10^{-4} \text{ mol kg}^{-1}$) solution of electrolytes we may not.[‡] It is therefore fortunate that we know the exact form of ϕ and the $\gamma_\pm$'s in the limit as $\Sigma_i \ m_i \to 0$ and approximate forms at finite molalities.

The Debye-Hückel limiting law has the form:[§]

$$-\ln \gamma_\pm = \alpha |z_+ z_-| I^{1/2}, \quad (20.4.1)$$

where I, called the ionic strength, is defined by

$$I \overset{\text{def}}{=} \tfrac{1}{2} \sum_i m_i z_i^2, \quad (20.4.2)$$

[†] The activity coefficient $\gamma_\pm$ has sometimes been called the 'mean' activity coefficient or the 'stoichiometric' activity coefficient.

[‡] The difference arises from the different ranges of the interaction energies of solute particles: in non-electrolyte solutions the pair-interaction energy falls off on dilution roughly as the square of the concentration; in electrolyte solutions the pair-interaction energy falls off on dilution roughly as the cube root of the concentration.

[§] For a derivation of the Debye-Hückel formulae see for example Guggenheim, E. A.; Stokes, R. H. *Equilibrium Properties of Aqueous Solutions of Single Strong Electrolytes*. Pergamon: Oxford. **1969**.

and α is given according to the theory by

$$\alpha = (2\pi L \rho_A^*)^{1/2}(e^2/4\pi\varepsilon kT)^{3/2}, \tag{20.4.3}$$

where ρ_A^* denotes the density of the pure solvent, e the charge on a proton, and ε the electric permittivity ($\varepsilon = \varepsilon_0 \varepsilon_r$ where ε_r is the relative permittivity or dielectric constant) of the solvent. For water as solvent $\alpha(298.15\ \mathrm{K}) = 1.171\ \mathrm{kg}^{1/2}\ \mathrm{mol}^{-1/2}$ and $\alpha(273.15\ \mathrm{K}) = 1.123\ \mathrm{kg}^{1/2}\ \mathrm{mol}^{-1/2}$.

The Debye–Hückel approximation for finite molalities has the form:

$$-\ln \gamma_\pm = \alpha |z_+ z_-| I^{1/2}/(1 + \beta d I^{1/2}), \tag{20.4.4}$$

where β is given according to the theory by

$$\beta = 2(2\pi L \rho_A^*)^{1/2}(e^2/4\pi\varepsilon kT)^{1/2}, \tag{20.4.5}$$

and d, the mean diameter of the ions, is an adjustable parameter. For water as solvent $\beta(298.15\ \mathrm{K}) = 3.282\ \mathrm{nm}^{-1}\ \mathrm{kg}^{1/2}\ \mathrm{mol}^{-1/2}$ and $\beta(273.15\ \mathrm{K}) = 3.239\ \mathrm{nm}^{-1}\ \mathrm{kg}^{1/2}\ \mathrm{mol}^{-1/2}$.

Since a typical mean ionic diameter is about 0.3 nm, the adjustable parameter d is sometimes eliminated from (4) to give

$$-\ln \gamma_\pm = \alpha |z_+ z_-| I^{1/2}/\{1 + (I/\mathrm{mol\ kg}^{-1})^{1/2}\}. \tag{20.4.6}$$

Equation (6) is in turn often extended empirically to higher molalities by the addition of a term proportional to I, and sometimes terms proportional to $I^{3/2}$, I^2, and so on.

The osmotic coefficient ϕ of the solvent corresponding to equation (4) is given by

$$1 - \phi = \tfrac{1}{3}\alpha |z_+ z_-| I^{1/2} \sigma(\beta d I^{1/2}), \tag{20.4.7}$$

where σ is a function defined by

$$\sigma(X) \overset{\text{def}}{=} 3X^{-3}\{1 + X - (1+X)^{-1} - 2\ln(1+X)\}. \tag{20.4.8}$$

When $\beta d I^{1/2} \ll 1$ equation (7) reduces to the limiting law corresponding to equation (1):

$$1 - \phi = \tfrac{1}{3}\alpha |z_+ z_-| I^{1/2} = -\tfrac{1}{3}\ln \gamma_\pm. \tag{20.4.9}$$

In figure 20.1 values of $\ln \gamma_\pm$ are plotted against $|z_+ z_-| I^{1/2}$ for five electrolytes including two having $z_+ > 1$. The curves in figure 20.1 are those plotted for the Debye–Hückel limiting law (1), and for the Debye–Hückel approximation (4) for finite molalities with $d = 0.3046$ nm, that is to say for equation (6). For the five chosen electrolytes the limiting law is obeyed within about 5 per cent of $\gamma_\pm$ for $(z_+ z_-)^2 I \leqslant 0.01\ \mathrm{mol\ kg}^{-1}$, and equation (6) within about 5 per cent of $\gamma_\pm$ for $(z_+ z_-)^2 I \leqslant 0.1\ \mathrm{mol\ kg}^{-1}$. Those conclusions apply roughly to any strong electrolyte. At values of $(z_+ z_-)^2 I$ greater than about 0.1 mol kg^{-1} the experimental values usually diverge progressively from the theory. Nor can the agreement be much improved by the choice of any realistic values for the 'mean ionic diameter' d. The

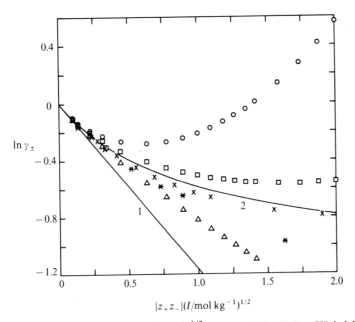

Figure 20.1. Values of ln $\gamma_\pm$ plotted against $|z_+z_-|I^{1/2}$. Curve 1, the Debye–Hückel limiting law, equation (20.4.1); curve 2, the Debye–Hückel approximation, equation (20.4.6), for finite molalities. O, HCl; □, KCl; △, KNO₃; ×, CaCl₂; *, LaCl₃.

Debye–Hückel theory applies with useful accuracy only to "slightly polluted water". It is nevertheless extremely valuable for making theoretically based extrapolations to infinite dilution.

§ 20.5 Solutions of more than one electrolyte

Much work has been devoted to the expression of the activity coefficients of solutions of more than one electrolyte in terms of the activity coefficients of corresponding solutions each of a single electrolyte. Most of this work has been based on Brønsted's principle of *specific interactions of ions* according to which, in addition to the Debye–Hückel expression, the interactions only of oppositely charged ions need to be taken into account. Thus, for a mixture of 1–1 electrolytes ($z_+ = 1$ and $z_- = -1$) we have

$$-\ln\{\gamma_\pm(C', A')\} = \alpha I^{1/2}/(1+\beta dI^{1/2}) - \sum_C \beta_{C,A}m_C - \sum_A \beta_{C,A}m_A, \quad (20.5.1)$$

for each electrolyte C'A', and

$$1 - \phi = \tfrac{1}{3}\alpha I^{1/2}\sigma(\beta dI^{1/2}) - \sum_C\sum_A \beta_{C,A}m_Cm_A/m, \quad (20.5.2)$$

where $m = \Sigma_C \, m_C = \Sigma_A \, m_A$ and C denotes a cation and A an anion in the mixed-electrolyte solution. Full accounts of the thermodynamics of mixed-electrolyte solutions can be found in the literature.[†]

§ 20.6 Measurement of the osmotic coefficient of the solvent

The osmotic coefficient of the solvent can be obtained from measurements of the vapour pressure p^{l+g} of the solution by use of the relation (compare equation 16.9.5):

$$-\phi M_A \sum_i m_i = \ln(p^{l+g}/p_A^{*l+g}) + (B_{AA} - V_A^{*l})(p^{l+g} - p_A^{*l+g})/RT, \qquad (20.6.1)$$

where p_A^{*l+g} denotes the vapour pressure of the pure solvent and where we have assumed that the solutes are involatile. An apparatus like that shown in figure 16.3(b) may be used, or a differential apparatus in which the difference $(p_A^{*l+g} - p^{l+g})$ is measured directly.

Once $\phi(Y, m_Y)$ has been sufficiently established as a function of molality m_Y for any solute Y, the osmotic coefficient $\phi(X, m_X)$ of any other solute X can be measured by the *isopiestic method*. In an isopiestic experiment a known mass of X dissolved in the solvent A is brought to diffusional equilibrium, by allowing transfer of solvent through the vapour phase, with a known mass of Y dissolved in the solvent A. At isopiestic equilibrium $\lambda_A(X, m_X) = \lambda_A(Y, m_Y)$ so that

$$\phi(X, m_X) \sum_i m_i(X) = \phi(Y, m_Y) \sum_i m_i(Y), \qquad (20.6.2)$$

and m_X and m_Y can be found from measurements of the masses of the containers at the end of the experiment together with the predetermined masses of the empty containers and of the electrolytes X and Y. In practice it is necessary to ensure the best possible thermal contact between the two solutions; otherwise evaporation from the solution with the greater λ_A will cool that solution and cause transport of solvent in the wrong direction leading to a very slowly attenuated oscillation about the equilibrium position. Good thermal contact is obtained for example by placing the solutions in silver dishes with their bottom surfaces machined flat on a thick copper block with its upper surface machined flat. Equilibrium is also greatly hastened by evacuation of a vessel (enclosed in a thermostat) containing the dishes on their copper block, and by stirring the solutions for example by slowly rocking the vessel. Even so, at room temperature the time needed to ensure equilibrium is about a day for solutions of molality 1 mol kg^{-1}, several days for solutions of molality 0.1 mol kg^{-1}, and too long for solutions of lower molalities. The dishes should be fitted with hinged lids which can be closed without disturbing the equilibrium before the dishes are removed for weighing at the end of an experiment. Sucrose, sodium chloride, potassium chloride, calcium chloride, or sulphuric acid is commonly used as the reference solute Y.

[†] Harned, H. A.; Robinson, R. A. *Multicomponent Electrolyte Solutions*. Pergamon: Oxford. **1968**.
Guggenheim, E. A. *Applications of Statistical Mechanics*. Clarendon: Oxford. **1966**, chapter 10.

Measurements of the freezing temperature T_A of the solvent from a solution can be used to obtain the osmotic coefficient $\phi(T_A)$, but only at a temperature close to (but not exactly equal to) the freezing temperature T_A^* of the pure solvent. According to equation (10.6.9) we have

$$\phi(T_A)M_A \sum_i m_i = (\Delta_s^l H_A^*/RT_A^*)(1 - T_A/T_A^*)$$

$$+ (\Delta_s^l H_A^*/RT_A^* - \Delta_s^l C_{p,A}^*/2R)(1 - T_A/T_A^*)^2$$

$$+ O(1 - T_A/T_A^*)^3. \tag{20.6.3}$$

A series of values of $\phi(T_A)$ can in principle be corrected to the common temperature T_A^* by use of equation (18.2.2) with calorimetrically determined values of $(H_A^* - H_A)$. The second term on the right-hand side of equation (3) is not quite negligible in the most accurate work. For the solvent water, for example, equation (3) becomes

$$\phi \sum_i m_i/\text{mol kg}^{-1}$$

$$= 0.5376\{(T_A^* - T_A)/\text{K}\}[1 + 4.8 \times 10^{-4}\{(T_A^* - T_A)/\text{K} + \cdots]. \tag{20.6.4}$$

For a 1–1 electrolyte with $m = 1$ mol kg^{-1} and $\phi = 1$, $(T_A^* - T_A) \approx 3.7$ K so that the second term in equation (4) is 0.002. Higher terms in equation (3) are negligible. Accurate measurements of $(T_A^* - T_A)$ are made with a thermopile or two-probe quartz thermometer between two Dewar flasks, one containing a well stirred mush of pure solid solvent and liquid solution and the other of pure solid solvent and pure liquid solvent. When equilibrium has been reached and the temperature difference has been recorded a sample of liquid solution is withdrawn and analysed, for example conductimetrically. Great care must be taken not to include any solid solvent in the sample withdrawn; this would melt and dilute the solution. When there is any doubt about the matter, a considerable effort is needed to provide convincing evidence that the solid phase is pure solvent (and not a solid solution or a solid compound such as a hydrate). The best way is probably to include a second solute, to construct the relevant part of the triangular phase diagram by analysis of saturated solutions and the wet solids ('wet residues') with which they are in equilibrium, and so to identify the solid phases.

Beckmann's apparatus commonly used to determine the approximate molar mass of a solute from depression of the freezing temperature of the solvent is of course no use for measurements of the osmotic coefficient; the solid that separates before the freezing temperature of the solution can be measured changes the predetermined composition of the liquid solution, and a Beckmann thermometer is neither sensitive enough nor reproducible enough for that purpose.

§ 20.7   Measurement of the activity coefficient of a solute

The most important method of determining the activity coefficient $\gamma_\pm$ of an electrolyte is by use of a galvanic cell like (19.3.5). The e.m.f. E of that cell is given by equation (19.4.2):

$$E = E^{\ominus} - (2RT/F) \ln(m\gamma_{\pm}/m^{\ominus}) + (\tfrac{1}{2}RT/F) \ln\{x(H_2, g)p/p^{\ominus}\}. \quad (20.7.1)$$

From a series of measurements of $E(m)$, $\gamma_{\pm}$ can be obtained. For example, if (1) is rewritten in the form:

$$E + (2RT/F) \ln(m/m^{\ominus}) - (\tfrac{1}{2}RT/F) \ln\{x(H_2, g)p/p^{\ominus}\}$$

$$= E^{\ominus} - (2RT/F) \ln \gamma_{\pm}, \quad (20.7.2)$$

then a plot of the left-hand side of (2) against the square root $m^{1/2}$ of the molality will give a nearly straight line with intercept $E^{\ominus}$ at $m^{1/2} = 0$.

Better, if equation (1) is rewritten in the form:

$$E + (2RT/F) \ln(m/m^{\ominus}) - (2RT/F)\alpha m^{1/2}/\{1 + (m/\text{mol kg}^{-1})^{1/2}\}$$

$$- (\tfrac{1}{2}RT/F) \ln\{x(H_2, g)p/p^{\ominus}\}$$

$$= E^{\ominus} - (2RT/F)[\ln \gamma_{\pm} + \alpha m^{1/2}/\{1 + (m/\text{mol kg}^{-1})^{1/2}\}], \quad (20.7.3)$$

so that what is known about $\gamma_{\pm}$ is included on each side, then a plot of the left-hand side of (3) against the molality m will give a more nearly straight line with a smaller slope and consequently a more reliable extrapolation to give $E^{\ominus}$ at $m = 0$.

Given $E^{\ominus}$, values of $\gamma_{\pm}$ can be obtained from equation (1). Alternatively, a computer can be used to obtain $E^{\ominus}$ and $\gamma_{\pm}(m)$ by numerical analysis of $E(m)$ in terms of equation (1). (The value of $E^{\ominus}$ so obtained will be discussed in § 20.8.) The method is limited to electrolytes for which we can find two suitable electrodes, one reacting with the cation and the other with the anion. In practice that excludes for example nitrates and perchlorates.

Another method of determining $\gamma_{\pm}$ is from measurements of the molality s of a solution saturated with respect to the electrolyte X (ionizing to give ν_+ cations and ν_- anions) as a function of the molality m of a more soluble electrolyte Y (ionizing to give ν'_+ cations and ν'_- anions). The electrolyte Y might or might not have an ion in common with the electrolyte X. If the cation is common, then for any solution saturated with X we have the relation:

$$\mu^*(X, s) = \mu(X, \text{solute, saturated})$$

$$= \mu^{\ominus}(X, \text{solute}) + RT \ln[(\nu_+ s$$

$$+ \nu'_+ m)^{\nu_+}(\nu_- s)^{\nu_-}\{\gamma_{\pm}(s, m)/m^{\ominus}\}^{(\nu_+ + \nu_-)}]. \quad (20.7.4)$$

In view of the independence of $\mu^*(X, s)$ and $\mu^{\ominus}(X, \text{solute})$ on the composition of the solution, and of the relation $\gamma_{\pm} \to 1$ as $I \to 0$ (where I denotes the ionic strength), equation (4) leads to

$$\gamma_{\pm}(s, m) = [\lim_{I \to 0} \{(\nu_+ s + \nu'_+ m)^{\nu_+}(\nu_- s)^{\nu_-}\}/(\nu_+ s + \nu'_+ m)^{\nu_+}(\nu_- s)^{\nu_-}]^{1/(\nu_+ + \nu_-)}. \quad (20.7.5)$$

When there is no common ion, equation (5) simplifies to

$$\gamma_{\pm}(s, m) = \lim_{I \to 0} \{s(m)\}/s(m). \quad (20.7.6)$$

In particular, the value $\gamma_\pm(s, m = 0)$ of the activity coefficient of X in a saturated solution in the absence of added electrolyte Y can be obtained. The success of the method, like that dealt with in the previous paragraph, depends on the unambiguity of the extrapolation to $I = 0$. A reliable extrapolation function based on the Debye–Hückel theory again helps, but the method is inevitably limited to sparingly soluble electrolytes. It is also limited to saturated solutions of those electrolytes.

§ 20.8   Determination of the standard equilibrium constant of an electron-transfer reaction

We have already dealt in the previous section with the determination of the standard electromotive force $E^\ominus$ and so have already dealt with the determination of the standard equilibrium constant $K^\ominus$ of an electron–transfer reaction like

$$AgCl(s) + \tfrac{1}{2}H_2(g) = Ag(s) + HCl(solute),\tag{20.8.1}$$

since

$$E^\ominus \stackrel{\text{def}}{=} (RT/F) \ln K^\ominus.\tag{20.8.2}$$

For a reaction such as

$$Fe(NO_3)_3 + Ag(s) = Fe(NO_3)_2 + AgNO_3,\tag{20.8.3}$$

(where we omit the state label 'solute') we cannot construct a galvanic cell for which the e.m.f. is given by an equation like (20.7.1) having no integrals containing transport numbers. To determine $K^\ominus$ for such a reaction we have to resort to double extrapolation. We construct the cell:

$$Pt|Ag(s)\left|\begin{array}{c}AgNO_3(xm)\\KNO_3(m)\end{array}\right|KNO_3(m)\left|\begin{array}{c}Fe(NO_3)_3(xm)\\Fe(NO_3)_2(xm)\\KNO_3(m)\end{array}\right|Pt,\tag{20.8.4}$$

with molalities arranged as shown, where x denotes a fraction. The e.m.f. of the cell (4) is given according to § 19.2 by the expression:

$$-FE = \mu\{Fe(NO_3)_2\} - \mu\{Fe(NO_3)_3\} - \mu(Ag, s) + \mu(AgNO_3)$$

$$+ \sum_{i \neq NO_3^-} \int_L^R (t_i/z_i)\{d\mu_i + z_i\, d\mu(NO_3^-)\},\tag{20.8.5}$$

where the summation includes a term for each of Ag^+, K^+, Fe^{2+}, and Fe^{3+}. Making use of (20.3.1) and (20.3.6), writing $-FE^\ominus = \Sigma_X \nu_X \mu_X^\ominus$, replacing $\mu(Ag, s)$ by $\mu^\ominus(Ag, s)$ (that is to say ignoring the integral in 16.17.6), and rearranging, we obtain

$$FE/RT + \ln(4xm/27m^\ominus)$$

$$= FE^\ominus/RT - 3 \ln[\gamma\{Fe(NO_3)_2\}] + 4 \ln[\gamma\{Fe(NO_3)_3\}]$$

$$- 2 \ln[\gamma(AgNO_3)] - \sum_{i \neq NO_3^-} \int_L^R (t_i/z_i RT)\{d\mu_i + z_i\, d\mu(NO_3^-)\}.\tag{20.8.6}$$

A series of measurements of E with varying x at constant m leads, as we saw in § 19.3, to the relation:

$$\lim_{x \to 0} \{FE/RT + \ln(4xm/27m^{\ominus})\}$$

$$= FE^{\ominus}/RT - 3 \ln[\gamma\{Fe(NO_3)_2\}] + 4 \ln[\gamma\{Fe(NO_3)_3\}] - 2 \ln[\gamma(AgNO_3)].$$
(20.8.7)

A series of such series for different values of m then leads to the relation:

$$\lim_{m \to 0} [\lim_{x \to 0}\{FE/RT + \ln(4xm/27m^{\ominus})\}] = FE^{\ominus}/RT,$$
(20.8.8)

since the activity coefficients of all electrolytes tend to unity at zero ionic strength. The method of determination of $E^{\ominus}$ depends for its success on the unambiguity of the extrapolations. The extrapolations can be made less ambiguous by including a function based on the Debye–Hückel theory in the expression on the left-hand side of (6).

For a few reactions, including reaction (3), the standard equilibrium constant $K^{\ominus}$ can be found by chemical analysis of equilibrium solutions. There are two conditions for the success of this method: $K^{\ominus}$ must be sufficiently close to unity (say between 10^{-2} and 10^2) to ensure that there is an analytically measurable amount of each substance present at equilibrium; it must be possible to analyse the solution without disturbing the extent of reaction at equilibrium. The second condition can be satisfied for reaction (3) by removing all the Ag(s) by filtration before analysis. Spectrophotometry can sometimes be used for analysis effectively with no disturbance of the equilibrium. For example, the reaction:

$$Fe^{3+} + I^- = Fe^{2+} + \tfrac{1}{2}I_2,$$
(20.8.9)

might be studied spectrophotometrically by choosing a wavelength at which only I_2 absorbs, and then assuming Beer's law:

$$\ln(1/T) = \kappa c l,$$
(20.8.10)

where T is the internal transmittance, c is the concentration of the light-absorbing species, l is the length of the optical cell, and κ is a constant found from a separate series of measurements on solutions containing known concentrations of I_2. In this example, however, the algebra would be complicated by the existence also of the complex-ion-forming reaction:

$$I^- + I_2 = I_3^-.$$
(20.8.11)

The results of chemical analysis of a series of solutions progressively more dilute lead to the standard equilibrium constant, for reaction (3) for example:

$$K^{\ominus} = \lim_{I \to 0} \left(4[m\{Fe(NO_3)_2\}]^3 [m(AgNO_3)]^2 / 27[m\{Fe(NO_3)_3\}]^4 m^{\ominus}\right),$$
(20.8.12)

where $I = \tfrac{1}{2}[4m(Fe^{2+}) + 9m(Fe^{3+}) + m(Ag^+) + \{2m(Fe^{2+}) + 3m(Fe^{3+}) + m(Ag^+)\}].$

Table 20.1. Values of the standard equilibrium constant $K^\ominus$ (298.15 K) for a few electron-transfer reactions in aqueous solution.

$$(m^\ominus = 1 \text{ mol kg}^{-1}; p^\ominus = 101.325 \text{ kPa})$$

Reaction	$K^\ominus$	$\ln K^\ominus = FE^\ominus/RT$
$K^+ + \frac{1}{2}H_2(g) = K(s) + H^+$	3.78×10^{-50}	-113.8
$\frac{1}{2}Mg^{2+} + \frac{1}{2}H_2(g) = \frac{1}{2}Mg(s) + H^+$	9.08×10^{-41}	-92.2
$\frac{1}{2}Zn^{2+} + \frac{1}{2}H_2(g) = \frac{1}{2}Zn(s) + H^+$	1.26×10^{-13}	-29.7
$\frac{1}{2}Fe^{2+} + \frac{1}{2}H_2(g) = \frac{1}{2}Fe(s) + H^+$	3.75×10^{-8}	-17.10
$\frac{1}{2}Cd^{2+} + \frac{1}{2}H_2(g) = \frac{1}{2}Cd(s) + H^+$	1.52×10^{-7}	-15.7
$\frac{1}{2}Sn^{2+} + \frac{1}{2}H_2(g) = \frac{1}{2}Sn(s) + H^+$	4.99×10^{-3}	-5.30
$\frac{1}{2}Hg_2^{2+} + \frac{1}{2}H_2(g) = Hg(l) + H^+$	7.43×10^{-2}	-2.600
$\frac{1}{2}Sn^{4+} + \frac{1}{2}H_2(g) = \frac{1}{2}Sn^{2+} + H^+$	3.30×10^2	5.80
$AgCl(s) + \frac{1}{2}H_2(g) = Ag(s) + H^+ + Cl^-$	5.745×10^3	8.656
$\frac{1}{2}Hg_2Cl_2(s) + \frac{1}{2}H_2(g) = Hg(l) + H^+ + Cl^-$	3.383×10^4	10.429
$\frac{1}{2}Cu^{2+} + \frac{1}{2}H_2(g) = \frac{1}{2}Cu(s) + H^+$	4.89×10^5	13.10
$\frac{1}{2}I_2(s) + \frac{1}{2}H_2(g) = I^- + H^+$	1.08×10^9	20.8
$Fe^{3+} + \frac{1}{2}H_2(g) = Fe^{2+} + H^+$	1.07×10^{13}	30.0
$Ag^+ + \frac{1}{2}H_2(g) = Ag(s) + H^+$	3.21×10^{13}	31.1
$Hg^{2+} + \frac{1}{2}H_2(g) = \frac{1}{2}Hg_2^{2+} + H^+$	3.53×10^{15}	35.8
$\frac{1}{2}Br_2(l) + \frac{1}{2}H_2(g) = Br^- + H^+$	1.05×10^{18}	41.5
$\frac{1}{2}Cl_2(g) + \frac{1}{2}H_2(g) = Cl^- + H^+$	9.42×10^{22}	52.9
$\frac{1}{5}MnO_4^- + \frac{8}{5}H^+ + \frac{1}{2}H_2(g) = \frac{1}{5}Mn^{2+} + \frac{4}{5}H_2O$	1.14×10^{26}	60.0
$Ce^{4+} + \frac{1}{2}H_2(g) = Ce^{3+} + H^+$	1.70×10^{27}	62.7

In table 20.1 we give values for a few standard equilibrium constants of electron-transfer reactions in aqueous solution at 298.15 K. In the customary manner each electron-transfer ('redox') pair has been associated with the oxidation of $\frac{1}{2}H_2(g)$ to H^+. The table will be recognized as the 'electrochemical series', but we have refrained from the customary abbreviations (see § 19.5). Further values can be found in the literature.[†]

§ 20.9   Proton-transfer reactions. Acids and bases

An *acid* is a proton donor; a *base* is a proton acceptor. Any acid A is related to its *conjugate* base B by the relation: $A = B + H^+$. A few examples follow.

$$H_3O^+ = H_2O + H^+,$$

$$CH_3CO_2H = CH_3CO_2^- + H^+,$$

$$NH_4^+ = NH_3 + H^+,$$

$$H_2O = OH^- + H^+,$$

[†] Sillén, L. G.; Martell, A. E. *Stability Constants of Metal-Ion Complexes.* Chemical Society: London. **1964**. Supplement, **1971**.

$$H_2SO_4 = HSO_4^- + H^+ ,$$

$$HSO_4^- = SO_4^{2-} + H^+ .$$

Substances like H_2O and HSO_4^- that can function as an acid or as a base are called *amphiprotic*.

In a proton-transfer reaction a proton H^+ is transferred from an acid A' to a base B'', the acid A' being converted to its conjugate base B', and the base B'' being converted to its conjugate acid A''. A few examples follow of proton-transfer reactions.

$$H_3O^+ + OH^- = H_2O + H_2O , \qquad \text{('neutralization')},$$

$$CH_3CO_2H + H_2O = CH_3CO_2^- + H_3O^+ , \quad \text{('dissociation of weak acid')},$$

$$H_2O + NH_3 = OH^- + NH_4^+ , \qquad \text{('dissociation of weak base')},$$

$$NH_4^+ + H_2O = NH_3 + H_3O^+ , \qquad \text{('hydrolysis of weak acid')},$$

$$H_2O + CH_3CO_2^- = OH^- + CH_3CO_2H , \quad \text{('hydrolysis of weak base')},$$

$$NH_4^+ + CH_3CO_2^- = NH_3 + CH_3CO_2H , \quad \text{('reaction of weak acid and weak base')},$$

$$H_2O + H_2O = OH^- + H_3O^+ , \qquad \text{('ionization of water')} .$$

The names given in parentheses are descriptive of the particular reactions (though 'dissociation' is a misleading name for the second and third examples) but mask the similarities: each is an example of a proton-transfer reaction: $A' + B'' = B' + A''$.

§ 20.10 Determination of the standard equilibrium constant of a proton-transfer reaction

It is sufficient to discuss and to tabulate the standard equilibrium constant $K_A^\ominus$, or *standard acidity constant*, for the reaction of each acid A with H_2O: $A + H_2O = B + H_3O^+$.† The $K^\ominus$ for any other proton-transfer reaction can then be obtained from the $K_A^\ominus$'s by multiplication and division.

The principal method of obtaining accurate standard acidity constants is by means of a galvanic cell such as

$$Pt|H_2(g) \left| \begin{array}{c} NaCl(m_1) \\ CH_3CO_2H(m_A) \\ CH_3CO_2Na(m_B) \end{array} \right| AgCl(s)|Ag(s)|Pt . \qquad (20.10.1)$$

A quinhydrone (quinone, $C_6H_4O_2(s)$, Q + quinol, $C_6H_4(OH)_2(s)$, H_2Q: $\frac{1}{2}Q + H_3O^+ + e^- = \frac{1}{2}H_2Q + H_2O$) electrode, or a glass electrode, can be used in place of the

---

† In the most accurate work a distinction must be made between H_3O^+ and H^+; for example if $m(H_3O^+) = 1.000$ mol kg^{-1} then $m(H^+) = 0.982$ mol kg^{-1}. The activity coefficient of the electrolyte $(H_3O^+ + Cl^-)$ is not quite the same as that of $(H^+ + Cl^-)$. The $E^\ominus$ of equation (20.10.2) is not quite the same as that of the reaction: $AgCl(s) + \frac{1}{2}H_2(g) = Ag(s) + H^+ + Cl^-$.

hydrogen electrode. Other left-hand electrodes can be used, of the general type: $Ag(s)|AgCl(s)|NaCl$. For the cell (1) the e.m.f. E is given by

$$-FE/RT = -FE^\ominus/RT + \ln\left[m(H_3O^+)m(Cl^-) \right.$$
$$\left. \times \{\gamma(H_3O^+, Cl^-)/m^\ominus\}^2 \exp\left\{\phi M(H_2O) \sum_i m_i\right\}\right], \quad (20.10.2)$$

where $E^\ominus$ is the standard e.m.f. of the reaction:

$$AgCl(s) + H_2O + \tfrac{1}{2}H_2(g) = Ag(s) + Cl^- + H_3O^+, \quad (20.10.3)$$

which can be determined from the results of separate experiments with for example only HCl in solution in a cell otherwise like (1). In equation (2) we have also assumed either that the fugacity of hydrogen is equal to the standard pressure $p^\ominus$ or if it is not that the appropriate correction has been included in E. Equilibrium of the proton-transfer reaction:

$$CH_3CO_2H + H_2O = CH_3CO_2^- + H_3O^+, \quad (20.10.4)$$

is determined by the equation:

$$K_A^\ominus = m(CH_3CO_2^-)m(H_3O^+)\{\gamma(CH_3CO_2^-, H_3O^+)\}^2$$
$$\times \exp\left\{\phi M(H_2O) \sum_i m_i\right\}\bigg/ m(CH_3CO_2H)\gamma(CH_3CO_2H)m^\ominus, \quad (20.10.5)$$

where $\sum_i m_i$ is the sum of the molalities of all solutes in the solution. The activity coefficient $\gamma(CH_3CO_2H)$ of a non-electrolyte solute can safely be put equal to unity. By use of the relations:

$$m(Cl^-) = m_1, \quad m(H_3O^+) \overset{\text{def}}{=} m_H, \quad m(CH_3CO_2H) = m_A - m_H,$$
$$m(CH_3CO_2^-) = m_B + m_H, \quad (20.10.6)$$

and a Debye–Hückel extrapolation function for each $\gamma_\pm$ (in this example they cancel), equations (2) and (5) can be solved iteratively for $K_A^\ominus$. Explicitly, the relation:

$$\lim_{I\to 0} \left[-FE/RT + FE^\ominus/RT - \ln\{m_1(m_A - m_H)/(m_B + m_H)m^\ominus\}\right] = \ln K_A^\ominus, \quad (20.10.7)$$

can be solved for $K_A^\ominus$ with $m_H = 0$, and then the value of $K_A^\ominus$ can be improved by calculation of m_H from equation (5), and so on until the value of $K_A^\ominus$ converges.

A similar method can be used with a cell such as

$$Pt|H_2(g)\begin{vmatrix} NaCl(m_1) \\ NaOH(m_2) \end{vmatrix}AgCl(s)|Ag(s)|Pt, \quad (20.10.8)$$

to determine the particular standard acidity constant, called the *standard ionization product* of water and denoted by $K_w^\ominus$, for the reaction:

$$H_2O + H_2O = OH^- + H_3O^+. \quad (20.10.9)$$

The e.m.f. of cell (8) is again given by equation (2), and for the proton-transfer equilibrium we have

$$K_w^\ominus = \{m(OH^-)m(H_3O^+)/(m^\ominus)^2\}\{\gamma(H_3O^+, OH^-)\}^2 \exp\left\{2\phi M(H_2O)\sum_i m_i\right\}.$$

$$(20.10.10)$$

Equations (10) and (2) can be solved for $K_w^\ominus$ given that $m(Cl^-) = m_1$, $m(H_3O^+) \overset{\text{def}}{=} m_H$, and $m(OH^-) = m_2 + m_H \approx m_2$.

Other methods used for the determination of acidity constants include spectrophotometry (for example, 2,4-dinitrophenolate ion absorbs light of wavelength 436 nm while its conjugate acid, 2,4-dinitrophenol, does not; or by addition of an indicator to a solution of a non-absorbing conjugate pair and measurement of the 'indicator ratio' m_{HIn}/m_{In}), and measurements of electric conductivity. Full accounts of such methods of measurement of acidity constants, and of proton-transfer reactions in general including multiprotic reactions and reactions in solvents other than water, can be found in the literature.[†]

Standard acidity constants usually pass through a maximum on changing the temperature by a few tens of kelvins around room temperature. That fact should be enough to warn the reader against attempts to 'rationalize' the relative values of standard acidity constants at any one temperature by discussion of relative 'electron-withdrawing' effects and the like. One might for example argue that diethylacetic acid will be stronger than acetic acid, that is to say have a larger standard acidity constant, because of the greater 'electron-withdrawing' effect of the extra alkyl groups. That 'explanation' would 'work' at 290 K, but how would one then explain the relative strengths at say 310 K? The standard acidity constants of those two acids are plotted against temperature in figure 20.2.

Compilations of standard acidity constants can be found in the literature.[‡]

§ 20.11   The standard solubility product

Consider again (see p. 308) a solution saturated with an electrolyte X (ionizing to give ν_+ cations and ν_- anions) in the presence of a molality m of a dissolved electrolyte Y (ionizing to give ν'_+ cations and ν'_- anions). Suppose that X and Y have a common cation. Then the standard equilibrium constant of the reaction:

$$X(s) = X(\text{solute}),$$

$$(20.11.1)$$

is

$$K^\ominus = (\nu_+ s + \nu'_+ m)^{\nu_+}(\nu_- s)^{\nu_-}\{\gamma_\pm(s, m)/m^\ominus\}^{(\nu_+ + \nu_-)},$$

$$(20.11.2)$$

† King, E. J. *Acid-Base Equilibria*. Pergamon: Oxford. **1965**.
   Bell, R. P. *The Proton in Chemistry*. Cornell University Press: New York. **1959**.
   ‡Kortüm, G.; Vogel, W.; Andrussow, K. *Dissociation Constants of Organic Acids in Aqueous Solution*. Butterworths: London. **1961**.
   Perrin, D. D. *Dissociation Constants of Organic Bases in Aqueous Solution*. Butterworths: London. **1965**. Supplement, **1972**.

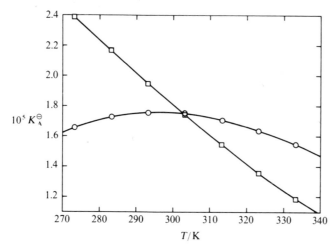

Figure 20.2. Plot of the standard acidity constant against the temperature for two carboxylic acids. ○, acetic acid; □, diethylacetic acid.

where we have again neglected the effect of any difference between p and $p^\ominus$. When there is no common ion, equation (2) simplifies to

$$K^\ominus = \nu_+^{\nu_+} \nu_-^{\nu_-} (s\gamma_\pm/m^\ominus)^{(\nu_+ + \nu_-)}, \qquad (20.11.3)$$

so that if $X = AgCl$ then $K^\ominus = (s\gamma_\pm/m^\ominus)^2$, if $X = PbI_2$ then $K^\ominus = 4(s\gamma_\pm/m^\ominus)^3$, if $X = Ca_3(PO_4)_2$ then $K^\ominus = 108(s\gamma_\pm/m^\ominus)^5$, and so on.

For a reaction like (1) the standard equilibrium constant $K^\ominus$ is called the *standard solubility product*. We have already dealt in effect (on p. 308) with one method for the measurement of such $K^\ominus$'s. The remaining important method, especially for electrolytes of very low solubility for which direct measurements of solubility are not practicable, is by means of a galvanic cell.

The galvanic cell:

$$\text{Ag}|\text{AgCl(s)}\left|\begin{matrix}\text{KCl}(xm)\\ \text{KNO}_3(m-xm)\end{matrix}\right|\text{KNO}_3(m)\left|\begin{matrix}\text{AgNO}_3(xm)\\ \text{KNO}_3(m-xm)\end{matrix}\right|\text{Ag}, \quad (20.11.4)$$

where m denotes molality and x is a fraction, has an e.m.f. E given by

$$-FE/RT = \ln\{m(\text{Ag}^+, \text{L})m(\text{NO}_3^-, \text{L})/m(\text{Ag}^+, \text{R})m(\text{NO}_3^-, \text{R})\}$$

$$+ 2\ln\{\gamma(\text{Ag}^+, \text{NO}_3^-, \text{L})/\gamma(\text{Ag}^+, \text{NO}_3^-, \text{R})\}$$

$$+ \sum_{i \neq \text{NO}_3^-} \int_{\text{L}}^{\text{R}} (t_i/z_i RT)\{d\mu_i + z_i\, d\mu\,(\text{NO}_3^-)\}. \qquad (20.11.5)$$

The standard solubility product of AgCl is determined near the left-hand electrode by the relation:

$$m(\text{Ag}^+, \text{L}) = K^\ominus(m^\ominus)^2/m(\text{Cl}^-, \text{L})\{\gamma(\text{Ag}^+, \text{Cl}^-, \text{L})\}^2. \qquad (20.11.6)$$

Substituting (6) into (5) we obtain

$$\Psi \overset{\text{def}}{=} -FE/RT + \ln\{(xm/m^\ominus)^2/(1-x)\} - 2\alpha m^{1/2}/\{1 + (m/\text{mol kg}^{-1})^{1/2}\}$$

$$= \ln K^\ominus + 2\ln\{\gamma(\text{Ag}^+, \text{NO}_3^-, \text{L})/\gamma(\text{Ag}^+, \text{NO}_3^-, \text{R})\}$$

$$-2[\ln\{\gamma(\text{Ag}^+, \text{Cl}^-, \text{L})\} + \alpha m^{1/2}/\{1 + (m/\text{mol kg}^{-1})^{1/2}\}]$$

$$+ \sum_{i \neq \text{NO}_3^-} \int_L^R (t_i/z_i RT)\{d\mu_i + z_i \, d\mu(\text{NO}_3^-)\} . \qquad (20.11.7)$$

If x is varied for given m then

$$\lim_{x \to 0} \Psi = \ln K^\ominus + 2\ln\{\gamma(\text{Ag}^+, \text{NO}_3^-, \text{L})/\gamma(\text{Ag}^+, \text{NO}_3^-, \text{R})\}$$

$$-2[\ln\{\gamma(\text{Ag}^+, \text{Cl}^-, \text{L})\} + \alpha m^{1/2}/\{1 + (m/\text{mol kg}^{-1})^{1/2}\}], \qquad (20.11.8)$$

and if that limit is found for various values of m, then

$$\lim_{m \to 0} \left\{ \lim_{x \to 0} \Psi \right\} = \ln K^\ominus. \qquad (20.11.9)$$

§ 20.12　Other ionic reactions

The methods already described can similarly be used to determine the standard equilibrium constants, called *standard stability constants*, of reactions like

$$\text{I}_2 + \text{I}^- = \text{I}_3^- ,$$

$$\text{Ag}^+ + \text{NH}_3 = \text{Ag}(\text{NH}_3)^+ ,$$

$$\text{Ag}^+ + 2\text{NH}_3 = \text{Ag}(\text{NH}_3)_2^+ ,$$

$$\text{Ag}^+ + 2\text{CN}^- = \text{Ag}(\text{CN})_2^- ,$$

in which complex ions are formed. Reactions like

$$\text{Pb}^{2+} + \text{Cl}^- = \text{PbCl}^+ ,$$

$$\text{Mg}^{2+} + \text{SO}_4^{2-} = \text{MgSO}_4 ,$$

are no different in principle, but they are usually written in the other direction and described as examples of incomplete dissociation of electrolytes.

Compilations of stability constants can be found in the literature.[†]

[†] Sillén, L. G.; Martell, A. E. *Stability Constants of Metal-Ion Complexes.* Chemical Society: London. **1964**. Supplement, **1971**.

Elementary accounts of all kinds of ionic equilibria are available,[†] as are more specialized treatments of the properties of electrolyte solutions.[‡]

Problems for chapter 20

Problem 20.1

The following values of the activity coefficient γ of HCl in aqueous solutions of molality m were calculated from measurements of the e.m.f. of a galvanic cell at 298.15 K. Calculate the osmotic coefficient ϕ of water in a solution containing 0.1 mol kg^{-1} of HCl.

$m/\text{mol kg}^{-1}$	0.001	0.002	0.005	0.010	0.020	0.050	0.100
γ	0.966	0.952	0.929	0.905	0.876	0.830	0.796

Problem 20.2

Given at 298.15 K that the relative permittivity of water is 78.54 and the density of water is 0.9971 g cm^{-3}, calculate the Debye–Hückel coefficients α and β. Thence calculate the activity coefficient of a 1–1 electrolyte of molality 0.1 mol kg^{-1} in water at 298.15 K, and the osmotic coefficient of the water, (a), according to the Debye–Hückel limiting law and (b), according to the Debye-Hückel equation with $d = 0.3$ nm.

Problem 20.3

The depressions ΔT of the freezing temperature are given for aqueous solutions of HCl of molality m. The cryoscopic constants of H_2O are given in equation (20.6.4).

$m/\text{mol kg}^{-1}$	0.001474	0.003430	0.006879	0.01326	0.02744	0.05006
$\Delta T/\text{K}$	0.00542	0.01253	0.02494	0.04763	0.09746	0.1774

Calculate the osmotic coefficient for each solution. Plot the osmotic coefficients against $m^{1/2}$ and compare them with those calculated according to the Debye–Hückel limiting law and according to the Debye–Hückel equation with $d = 0.3$ nm.

[†] Prue, J. E. *Ionic Equilibria*. Pergamon: Oxford. **1966**.
  Rossotti, H. *The Study of Ionic Equilibria*. Longman: London. **1978**.
[‡] MacInnes, D. A. *The Principles of Electrochemistry*. Reinhold: New York. **1939**. (Dover edition. Dover: New York. **1961**.)
  Harned, H. A.; Owen, B. B. *The Physical Chemistry of Electrolytic Solutions*. Reinhold: New York. **1950**.
  Robinson, R. A.; Stokes, R. H. *Electrolyte Solutions*. Butterworths: London. Second edition. **1959**.

Problem 20.4

The following are values of the e.m.f. E at 298.15 K of the galvanic cell:

$$\text{Pt}|\text{H}_2(\text{g})|\text{HCl}(\text{aq}, m)|\text{HgCl}(\text{s})|\text{Hg}(\text{l})|\text{Pt},$$

corrected to $f(\text{H}_2, \text{g}) = p^{\ominus}$.

$m/\text{mol kg}^{-1}$	0.0016077	0.0050403	0.013968	0.037690	0.075081	0.119304
E/V	0.60080	0.54366	0.49339	0.44516	0.41187	0.38948

Calculate the standard equilibrium constant $K^{\ominus}$ of the reaction: $\text{HgCl}(\text{s}) + \frac{1}{2}\text{H}_2(\text{g}) = \text{Hg}(\text{l}) + \text{HCl}(\text{aq})$, and the activity coefficient of HCl in each of the solutions.

Problem 20.5

The following are molalities s of TlIO_3 in aqueous solutions saturated with TlIO_3 in the presence of molalities m of KCl at 298.15 K.

$m/\text{mol kg}^{-1}$	0	0.004900	0.012570	0.025650	0.040810
$s/\text{mol kg}^{-1}$	0.001843	0.001930	0.002025	0.002158	0.002266

Calculate the standard solubility product of TlIO_3, and the activity coefficient of TlIO_3 in a solution saturated with TlIO_3 in the absence of KCl.

Problem 20.6

Values of the Napierian absorbance $B = \ln(1/T)$ of solutions of 2,4-dinitrophenol, $\text{C}_6\text{H}_3(\text{NO}_2)_2\text{OH}$, of molality m were measured spectrophotometrically at $T = 298.15$ K for light of wavelength 436 nm (a), in water: $B(\alpha)$; and (b), in excess of alkali: $B(1)$.

$10^4 m/\text{mol kg}^{-1}$	0.9245	1.3636	2.3945	3.5827
$\alpha = B(\alpha)/B(1)$	0.5993	0.5334	0.4410	0.3804

Assuming that the light is absorbed by the 2,4-dinitrophenolate ion but not by 2,4-dinitrophenol, calculate the standard acidity constant of 2,4-dinitrophenol.

Problem 20.7

Values are given of the e.m.f. E at 298.15 K of the galvanic cell:

$$\text{Pt}|\text{H}_2(\text{g})|\text{CH}_3\text{CO}_2\text{H}(m_1), \text{CH}_3\text{CO}_2\text{Na}(m_2), \text{NaCl}(m_3)|\text{AgCl}(\text{s})|\text{Ag}(\text{s})|\text{Pt},$$

corrected to $f(\text{H}_2, \text{g}) = p^{\ominus}$.

$m_1/\text{mol kg}^{-1}$	0.004779	0.012035	0.021006	0.04922
$m_2/\text{mol kg}^{-1}$	0.004599	0.011582	0.020216	0.04737
$m_3/\text{mol kg}^{-1}$	0.004896	0.012328	0.021516	0.05042
E/V	0.63981	0.61604	0.60174	0.57997

Given that $E^{\ominus}\{AgCl(s) + \frac{1}{2}H_2(g) = Ag(s) + H^+ + Cl^-, \ 298.15 \ \text{K}\} = 0.22239 \ \text{V}$, calculate the standard acidity constant of CH_3CO_2H at 298.15 K.

Problem 20.8

Values are given of the e.m.f. E at 298.15 K of the galvanic cell:

$$Pt|H_2(g)|KOH(m_1), \ KCl(m_2)|AgCl(s)|Ag(s)|Pt \ ,$$

corrected to $f(H_2, g) = p^{\ominus}$.

$m_1/\text{mol kg}^{-1}$	0.01	0.01	0.01	0.01
$m_2/\text{mol kg}^{-1}$	0.01	0.02	0.03	0.04
E/V	1.05069	1.03295	1.02258	1.01521

Given that $E^{\ominus}\{AgCl(s) + \frac{1}{2}H_2(g) = Ag(s) + H^+ + Cl^-, \ 298.15 \ \text{K}\} = 0.22239 \ \text{V}$, calculate the standard ionization product of H_2O at 298.15 K.

Problem 20.9

Use the following values of the e.m.f. E at 298.15 K of the galvanic cell:

$$Pt|Ag(s)|AgCl(s)\left|\begin{array}{c} KCl(xm) \\ KNO_3(m-xm) \end{array}\right|KNO_3(m)\left|\begin{array}{c} AgNO_3(xm) \\ KNO_3(m-xm) \end{array}\right|Ag(s)|Pt \ ,$$

to calculate the standard solubility product of AgCl at 298.15 K.

$m/\text{mol kg}^{-1}$	0.03	0.03	0.03	0.02	0.02	0.02
x	0.3	0.2	0.1	0.4	0.3	0.2
E/V	0.32643	0.30553	0.26971	0.32196	0.30705	0.28603

Chapter 21

Thermodynamics of fluid surfaces

§ 21.1 Work of stretching a surface

We shall deal only with planar interfaces like that shown in figure 21.1. The boundary planes A and B are so placed that the *surface phase* σ contains the whole of the inhomogeneous region between the homogeneous bulk phases α and β. Provided that this condition is satisfied it does not matter if A is placed further down so as to include some, or more, of the bulk phase α, or if B is placed further up. We shall assume that the interface is rectangular, as in figure 21.1, but our results will be independent of that assumption, though we shall not prove it. Let the rectangular interface have sides of lengths x and y and area $\mathscr{A} = xy$, and let the interfacial layer have thickness τ and volume $V^{\sigma} = \mathscr{A}\tau$.

In the bulk phases the force acting across any unit area is the same in all directions and is of course the pressure p of the phase. In the surface phase the force acting across any unit area *parallel* to the interface is the same as in the bulk phases and equal to the pressure p. The situation is, however, quite different for the force acting across an area *normal* to the interface. Consider the rectangular plane C with its lower and upper sides, of length l, lying in the planes A and B respectively so that the height of the plane is τ. Then the force acting across this plane is

$$p\tau l - \gamma l, \tag{21.1.1}$$

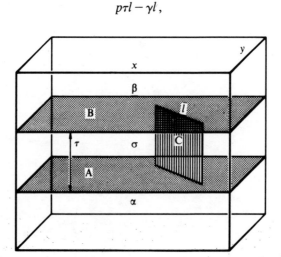

Figure 21.1. Planar interfacial layer σ between two bulk phases α and β.

where γ is a force divided by a length and is called the *interfacial tension*. For the particular case of an air-to-liquid interface γ is called the *surface tension*.

Suppose now that the surface phase be increased infinitesimally in size without change in its material content. In particular let τ be increased to $(\tau + d\tau)$, x to $(x + dx)$, and y to $(y + dy)$. The work done when τ is increased to $(\tau + d\tau)$ is that done against the force $p\tau$ normal to the planes A and B and so is

$$-p\mathscr{A}\,d\tau. \tag{21.1.2}$$

The work done when x is increased to $(x + dx)$ is that done against the force $(p\tau y - \gamma y)$ parallel to the planes A and B and so is

$$-(p\tau y - \gamma y)\,dx. \tag{21.1.3}$$

Similarly the work done when y is increased to $(y + dy)$ is

$$-(p\tau x - \gamma x)\,dy. \tag{21.1.4}$$

The total work W done on the surface phase is therefore

$$\begin{aligned}
W &= -p\mathscr{A}\,d\tau - (p\tau y - \gamma y)\,dx - (p\tau x - \gamma x)\,dy \\
&= -p\mathscr{A}\,d\tau - p\tau\,d(xy) + \gamma\,d(xy) \\
&= -p\mathscr{A}\,d\tau - p\tau\,d\mathscr{A} + \gamma\,d\mathscr{A} \\
&= -p\,d(\mathscr{A}\tau) + \gamma\,d\mathscr{A} \\
&= -p\,dV^{\sigma} + \gamma\,d\mathscr{A}. \tag{21.1.5}
\end{aligned}$$

We note that the work consists of two parts, one having the same form as for the compression of a bulk phase, and the other being an extra term for the work needed to increase the area of the interface. We also note that the two terms in equation (5) have opposite signs: whereas work must be done on a bulk phase to *compress* it, work must be done on a surface phase to *stretch* it.

An apparatus like that shown in figure 21.2 can be used to measure the surface tension whenever a surface can be suspended between the vertical frame A and the

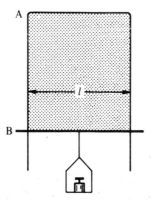

Figure 21.2. An apparatus for the determination of surface tension by measurement of the downward force needed to keep a liquid film stationary.

free horizontal rod B. The downward force on B is mg where m is the mass of the rod and of the weights hanging from it. If the mass m is adjusted so that the rod B is stationary then, in the absence of friction, we have

$$\gamma = mg/2l, \qquad (21.1.6)$$

the factor 2 allowing for there being two air-to-liquid surfaces one in front of the frame and the other at the back of the frame. The experiment is easy if the liquid is for example a soap solution.

§ 21.2   Curved surfaces

Most of the methods commonly used for the measurement of interfacial tension depend on the use of the relation:

$$p^{\alpha} - p^{\beta} = 2\gamma/r, \qquad (21.2.1)$$

for the difference of pressures between the inside α and the outside β of a spherical surface of radius r. Thus, a small spherical drop of oil immersed in water has inside it an excess pressure $2\gamma/r$ where r is the radius of the drop and γ is the oil-to-water interfacial tension. The excess pressure within a ('hollow') bubble is $4\gamma/r$, where γ is now the surface tension; there are two air-to-fluid surfaces having virtually the same radius.

The thermodynamics of curved surfaces has not yet been completely worked out except when the radii of curvature are large compared with the thickness of the surface phase. Accounts of the field can be found in the literature.†

§ 21.3   Practical measurement of interfacial tension

The most commonly used method of measurement of interfacial tension is the 'capillary-rise' method illustrated in figure 21.3. The height h to which the fluid α rises in the capillary tube above the level of the bulk fluid β is related to the interfacial tension γ by the equation:

$$\gamma = (\rho^{\alpha} - \rho^{\beta})(r/\cos \theta)gh/2, \qquad (21.3.1)$$

where r is the radius of the capillary, θ is the angle of contact between the liquid α and the walls of the tube, and ρ^{α} and ρ^{β} are the densities of the fluids α and β.

If the fluid α is an ordinary liquid that 'wets' the tube so that $\theta = 0$ and $\cos \theta = 1$, and if the fluid β is air, then to a good approximation equation (1) takes the simpler form:

$$\gamma = \rho^{\alpha}rgh/2. \qquad (21.3.2)$$

If the surface in the tube is convex upwards rather than downwards, as it would be for example if α were mercury and β were air or water, then $\theta > \pi/2$ or $\cos \theta < 0$ so

† Guggenheim, E. A. *Thermodynamics*. North-Holland: Amsterdam. Fifth edition. **1967**, p. 50.
Defay, R.; Prigogine, I.; Bellemans, A. *Surface Tension and Adsorption*. Everett, D. H.: translator. Longmans: London. **1966**.

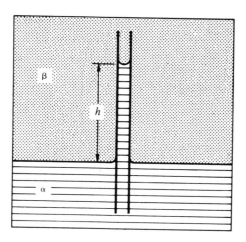

Figure 21.3. Measurement of interfacial tension by the capillary-rise method.

that $h < 0$; the surface in the capillary will lie below the plane surface between the bulk phases.

Other methods used for the measurement of interfacial tension include measurement of the mass or volume of a drop (if the drop that hangs from the tip of a tube of internal radius r were cylindrical and had the same radius r then $\gamma = mg/2\pi r$; in practice several corrections must be made), measurement of the excess pressure needed to blow a bubble (several corrections must be made to equation 21.2.1), determination of the shape of a sessile drop (or of a bubble or pendent drop or rotating drop), and measurement of the force F needed to detach a circular ring of radius r from a surface (ideally, $\gamma = F/4\pi r$; an instrument suitable for rough measurements is marketed under the name "du Nouy tensimeter"). Details of these and other methods can be found in the literature.[†]

§ 21.4 Thermodynamics of a plane surface phase

For any infinitesimal change in the state of a plane surface phase σ we use (21.1.5) to extend equation (5.2.1) by writing

$$dU^\sigma = T\, dS^\sigma - p\, dV^\sigma + \gamma\, d\mathscr{A} + \sum_B \mu_B\, dn_B^\sigma, \qquad (21.4.1)$$

where we restrict ourselves to systems in complete thermal, hydrostatic, and diffusive equilibrium and so do not label T, p, or the μ_B's.

By operating on equation (1) as we did on (5.2.1) we obtain

$$0 = S^\sigma\, dT - V^\sigma\, dp + \mathscr{A}\, d\gamma + \sum_B n_B^\sigma\, d\mu_B, \qquad (21.4.2)$$

† Alexander, A. E.; Hayter, J. B. *Techniques of Chemistry*, Volume I, Part V. Weissberger, A.; Rossiter, B. W.: editors. Wiley-Interscience: New York. **1971**, chapter 9.
  Pugachevich, P. P. *ETd(II)*, chapter 20.

which is the extension of the general Gibbs–Duhem equation (5.5.3) to a plane surface phase.

§ 21.5   Gibbs's adsorption equation

At constant temperature and pressure for a surface phase σ we have according to (21.4.2)

$$0 = \mathscr{A}\, d\gamma + \sum_B n_B^\sigma\, d\mu_B , \tag{21.5.1}$$

whereas for the bulk phase α we have according to (5.5.3)

$$0 = \sum_B n_B^\alpha\, d\mu_B . \tag{21.5.2}$$

Any one of the μ_B's may be eliminated from (1) and (2). In particular, for a binary mixture of A and B we obtain

$$\mathscr{A}\, d\gamma = (n_A^\sigma n_B^\alpha / n_A^\alpha - n_B^\sigma)\, d\mu_B , \quad (T, p \text{ constant}) , \tag{21.5.3}$$

which is Gibbs's adsorption equation. We may rewrite it in the form:

$$(\partial\gamma / \partial\mu_B)_{T,p} = \Gamma_A n_B^\alpha / n_A^\alpha - \Gamma_B , \tag{21.5.4}$$

where $\Gamma_C \overset{\text{def}}{=} n_C^\sigma / \mathscr{A}$ is called the *surface concentration* of C.

For an ideal-dilute solution of B in the solvent A we may write in place of (4)

$$(RT)^{-1}(\partial\gamma / \partial \ln m_B^\alpha)_{T,p} = M_A \Gamma_A m_B^\alpha - \Gamma_B . \tag{21.5.5}$$

According to (5) the surface tension will be lowered by positive adsorption of a solute into the surface: $\partial\gamma / \partial \ln m_B^\alpha < 0$ so that $n_B^\sigma / n_A^\sigma > n_B^\alpha / n_A^\alpha$ or $m_B^\sigma > m_B^\alpha$. For example the surface tension of water is lowered by the addition of a small quantity of alcohol. Conversely the surface tension will be raised by negative adsorption of a solute from a surface: $\partial\gamma / \partial \ln m_B^\alpha > 0$ so that $m_B^\sigma < m_B^\alpha$. For example the surface tension of water is raised by the addition of a small amount of sodium chloride.

Equation (5) has been quantitatively tested by blowing bubbles of air, of known size and number and thence of known total area, through a solution and removing the bubbles for analysis, and better by removing a surface phase (from a dish with flat and greased edges so that the liquid surface was proud) by scooping it into a container attached behind a fast-moving microtome blade, and then analysing it interferometrically.

One application of equation (5) must suffice. Szyszkowski, a Polish schoolmaster, found in 1908 that the surface tensions of not too dilute aqueous solutions of the soluble carboxylic acids from ethanoic to hexanoic were of the form: $\gamma = A + B \ln(m_B^\alpha / m^\ominus)$, and that the coefficient B was independent of the length of the carbon chain of the acid. Application of those results to equation (5) suggests strongly, in view of the non-dependence of the left-hand side of (5) on the molality

m_B^α of solute, that Γ_A must be zero, that is to say that there is no water in the surface phase; and that $\Gamma_B = n_B^\sigma/\mathcal{A}$ is independent of chain length. The results are plausible only if we conclude that molecules of carboxylic acid occupy the whole of the surface and that they are aligned vertically in the surface (presumably with the carboxyl group hydrogen-bonded to the water and the paraffin chain as far as possible from contact with water). The failure of the relation: $\gamma = A + B \ln(m_B^\alpha/m^\ominus)$ at high dilutions is also consistent with that interpretation, as are the results of experiments with insoluble films of longer-chain carboxylic acids spread on water in a surface balance.

Problems for chapter 21

Problem 21.1

Measurements of the surface tension γ of dilute aqueous solutions of molality m_B^α of $B = C_6H_5OH$ at 293 K can be fitted by the empirical formula: $\gamma/N\,m^{-1} = 0.0357 - 0.0136 \ln(m_B^\alpha/mol\,kg^{-1}) - 0.0009\{\ln(m_B^\alpha/mol\,kg^{-1})\}^2$.

By rapid removal of the surface of area $0.0310\,m^2$ from aqueous solutions of C_6H_5OH at 293 K followed by weighing and analysis the following mean products were found of the mass w^σ ($\approx w_A^\sigma$) of the surface phase and the difference $(m_B^\sigma - m_B^\alpha)$ of the molalities m_B^σ in the surface phase and m_B^α in the bulk phase.·

$m_B^\alpha/mol\,kg^{-1}$	0.0532	0.213	0.319
$10^7 w^\sigma (m_B^\sigma - m_B^\alpha)/mol$	0.7 ± 0.2	1.8 ± 0.2	1.8 ± 0.4

Use these two sets of results to test Gibbs's adsorption equation.

Problem 21.2

In Szyszkowski's experiment (described in the last paragraph of § 21.5) the coefficient B had the value $0.013\,N\,m^{-1}$ at 291 K. Calculate the cross-sectional area of a carboxylic acid.

Appendix I

Rules of partial differentiation

If $u = u(x, y, z, \cdots)$ is a function of x, y, z, $\cdots$, then

$$du = (\partial u/\partial x)_{y,z,\ldots} \, dx + (\partial u/\partial y)_{x,z,\ldots} \, dy + (\partial u/\partial z)_{x,y,\ldots} \, dz + \cdots,$$

is called the total or complete differential of u.

The following three theorems relate to a function $u(x, y)$ for which

$$du = (\partial u/\partial x)_y \, dx + (\partial u/\partial y)_x \, dy.$$

Theorem 1: 'the -1 rule':

$$(\partial u/\partial x)_y (\partial x/\partial y)_u (\partial y/\partial u)_x = -1,$$

or

$$(\partial u/\partial x)_y = -(\partial u/\partial y)_x (\partial y/\partial x)_u.$$

Theorem 2: 'the cross-differentiation rule':

$$\{\partial(\partial u/\partial x)_y/\partial y\}_x = \{\partial(\partial u/\partial y)_x/\partial x\}_y.$$

Theorem 3: 'the rule for change of the variable held constant':

$$(\partial u/\partial x)_z = (\partial u/\partial x)_y + (\partial u/\partial y)_x (\partial y/\partial x)_z.$$

Appendix II

Euler's theorem on homogeneous functions

If a function $u(x, y, z, \cdots)$ can be written in the form:

$$u = x^n U(y/x, z/x, \cdots),$$

where n is integral or zero, then the function u is said to be homogeneous of the n'th degree in the variables $x, y, z, \cdots$.

According to Euler's theorem, if u is a homogeneous function of the n'th degree in the variables $x, y, z, \cdots$ then

$$x(\partial u/\partial x)_{y,z,\ldots} + y(\partial u/\partial y)_{x,z,\ldots} + z(\partial u/\partial z)_{x,y,\ldots} + \cdots = nu.$$

Proof of Euler's theorem

Since

$$(\partial U/\partial x)_{y,z} = -yx^{-2}\{\partial U/\partial(y/x)\}_{z/x} - zx^{-2}\{\partial U/\partial(z/x)\}_{y/x},$$

$$(\partial U/\partial y)_{x,z} = x^{-1}\{\partial U/\partial(y/x)\}_{z/x},$$

$$(\partial U/\partial z)_{x,y} = x^{-1}\{\partial U/\partial(z/x)\}_{y/x},$$

it follows that

$$x(\partial U/\partial x)_{y,z} + y(\partial U/\partial y)_{x,z} + z(\partial U/\partial z)_{x,y} = 0.$$

In view of the equation $u = x^n U$ we then obtain

$$x(\partial u/\partial x)_{y,z} + y(\partial u/\partial y)_{x,z} + z(\partial u/\partial z)_{x,y} = xnx^{n-1}U$$

$$+ x^n\{x(\partial U/\partial x)_{y,z} + y(\partial U/\partial y)_{x,z}$$

$$+ z(\partial U/\partial z)_{x,y}\}$$

$$= nx^n U$$

$$= nu.$$

Example of the use of Euler's theorem

At given temperature T and pressure p the volume V of a phase containing amounts of substance n_A of A and n_B of B is a function of n_A and n_B:

$$V = V(n_A, n_B).$$

Since the function V can be expressed in the form:

$$V = n_A \Phi(n_B/n_A),$$

it is a homogeneous function of the first degree in the variables n_A and n_B. It follows from Euler's theorem that

$$n_A(\partial V/\partial n_A)_{T,p,n_B} + n_B(\partial V/\partial n_B)_{T,p,n_A} = V.$$

Appendix III

Three useful mathematical formulae

Taylor's theorem

Taylor's expansion of an analytic function $f(x)$ round the point $x = a$ is

$$f(x) = f(a) + (x - a)f'(a) + (x - a)^2 f''(a)/2! + (x - a)^3 f'''(a)/3! + \cdots,$$

where $f''(a)$ denotes $\{\partial^n f(x)/\partial x^n\}_{x=a}$.

Euler–Maclaurin theorem

The sum of a function $f(n)$ over all integral values of n from 0 to ∞ is related to the infinite integral over all values of n from 0 to ∞ by the formula:

$$\sum_{n=0}^{\infty} f(n) = \int_0^{\infty} f(n)\, dn + \tfrac{1}{2}f(0) - \tfrac{1}{12}(\partial f/\partial n)_{n=0} + \tfrac{1}{720}(\partial^3 f/\partial n^3)_{n=0} - \cdots.$$

A non-elementary infinite integral

The only non-elementary integral used in this book (in chapter 14) is

$$\int_0^{\infty} \exp(-ax^2)\, dx = (\pi/4a)^{1/2}.$$

The integral I can be derived as follows. Write

$$I^2 = \int_0^{\infty} \exp(-ax^2)\, dx \int_0^{\infty} \exp(-ay^2)\, dy = \int_0^{\infty} \int_0^{\infty} \exp\{-a(x^2 + y^2)\}\, dx\, dy.$$

Change the variables from x and y to r and θ where $x = r \sin \theta$ and $y = r \cos \theta$.

$$I^2 = \int_0^{\pi/2} \int_0^{\infty} \exp(-ar^2)\{(\partial x/\partial \theta)(\partial y/\partial r) - (\partial x/\partial r)(\partial y/\partial \theta)\}\, dr\, d\theta$$

$$= \int_0^{\pi/2} \int_0^{\infty} \exp(-ar^2) r\, dr\, d\theta$$

$$= (\pi/4) \int_0^{\infty} \exp(-ar^2)\, dr^2$$

$$= \pi/4a.$$

Appendix IV

Solutions of the problems

1.1: $T(\theta_2)/T(\theta_1) = 53.394$ Pa m^3/45.423 Pa m^3 = 1.17548.
$T(\theta_2) = 1.17548 \times 273.16$ K = 321.095 K.

1.2: $R = 53.272$ Pa m^3/(0.023456 mol 273.16 K) = 8.3143 J K^{-1} mol^{-1}.

1.3: (a), $\{U(T_f, V, \xi_f) - U(T_i, V, \xi_i)\} = 0$. (b), $\{U(T'_f, V, \xi_f) - U(T_i, V, \xi_f)\} =$ (2.13476 A)^2 125.686 Ω 75.2613 s = 43108 J. (c), By simple proportion $\{U(T_f, V, \xi_f) - U(T_i, V, \xi_f)\} = 43108$ J 2.12807 K/2.78304 K = 32963 J; from that result with (a): $\{U(T_i, V, \xi_f) - U(T_i, V, \xi_i)\} = -32963$ J.
(d), $\Delta U_m = -32963$ J 88.106 g mol^{-1}/1.29848 g = -2236.6 kJ mol^{-1}.

1.4: (a), $\Delta_{mix}H = (0.068450$ A)^2 28.132 Ω 14.193 s/1.3011 = 1.4378 J. (b), $n_A = 1.0763$ g/88.225 g mol^{-1} = 0.012200 mol; $n_B = 1.7211$ g/178.829 g mol^{-1} = 0.009624 mol; $(n_A + n_B) = 0.021824$ mol; $x = 0.4410$; $\Delta_{mix}H_m = 1.4378$ J/0.021824 mol = 65.88 J mol^{-1}.

1.5: (a), $\{H(301.60$ K$) - H(291.73$ K$)\} = (104.42 - 37.84)$ J = 66.58 J.
$\{H(311.38$ K$) - H(291.73$ K$)\} = (219.12 - 79.41)$ J = 139.71 J.
(b), Fit the results to a quadratic: $\{H(T) - H(291.73$ K$)\} = 6.3756(T - 291.73$ K$)$ J K^{-1} + 0.037349$(T - 291.73$ K$)^2$ J K^{-2}. $C_p = (\partial H/\partial T)_p = 6.3756$ J K^{-1} + 0.074698$(T - 291.73$ K$)$ J K^{-2}. $C_p(300.00$ K$) = \{6.3756 + 0.074698(300.00 - 291.73)\}$ J K^{-1} = 6.993 J K^{-1}.
(c), $C_{p,m}(300.00$ K$) = 6.993$ J K^{-1} 263.61 g mol^{-1}/76.1415 g = 24.21 J K^{-1} mol^{-1}.

1.6: (a), $\{H_m(383.26$ K, 2.904 kPa$) - H_m(383.26$ K, 33.003 kPa$)\} = 0.029459$ W/ 0.00050211 mol s^{-1} = 58.670 J mol^{-1}. (b), $(\partial H_m/\partial p)_T = \phi_{JT} \approx$ 58.670 J mol^{-1}/(2904 - 33003) Pa = -0.001949 m^3 mol^{-1} = -1949 cm^3 mol^{-1}.

2.1: (a), $V_m = 24.947$ cm^3 mol^{-1}, $\Delta_{mix}V_m = -0.907$ cm^3 mol^{-1}; $V_m = 31.017$ cm^3 mol^{-1}, $\Delta_{mix}V_m = -0.939$ cm^3 mol^{-1}. (b), $V_A = 17.45$ cm^3 mol^{-1}, $V_B = 39.25$ cm^3 mol^{-1}.
(c), $(V_A - V_A^*) = -0.665$ cm^3 mol^{-1}, $(V_B - V_B^*) = -1.400$ cm^3 mol^{-1}; $V_A = 17.40$ cm^3 mol^{-1}, $V_B = 39.34$ cm^3 mol^{-1}. (d), It is easier to obtain results of given accuracy by method (c) than by method (b).

2.2: For $x = 0, 0.25, 0.5, 0.75$, and 1: $\Delta_{mix}H_m$/J mol^{-1} = 0, 190.94, 216.17, 142.60, and 0; $(H_A - H_A^*)$/J mol^{-1} = 0, 98.7, 278.5, 477.1, and 681.0; $(H_B - H_B^*)$/J mol^{-1} = 1246.5, 467.8, 153.8, 31.1, and 0.

2.3: For m/mol kg^{-1} = 0, 0.1, 0.5, and 1.0: V_A/cm^3 mol^{-1} = 18.067, 18.067, 18.061, and 18.045; V_B/cm^3 mol^{-1} = 27.297, 27.642, 28.924, and 30.077;

$(V_A - V_A^*)/\text{cm}^3\,\text{mol}^{-1} = 0,\ -0.000,\ -0.006,$ and $-0.022;$ $(V_B - V_B^\infty)/\text{cm}^3\,\text{mol}^{-1} = 0,\ +0.345,\ +1.627,$ and $+2.780.$

6.1: (a), $\{S_m(400\ \text{K},\ 2\ \text{MPa}) - S_m(300\ \text{K},\ 0.1\ \text{MPa})\} = -18.9\ \text{J K}^{-1}\,\text{mol}^{-1}.$
(b), $\{H_m(400\ \text{K},\ 2\ \text{MPa}) - H_m(300\ \text{K},\ 0.1\ \text{MPa})\} = +2080\ \text{J mol}^{-1}.$

6.2: (a), $\{S_m(450\ \text{K},\ 9\ \text{MPa}) - S_m(350\ \text{K},\ 9\ \text{MPa})\} = +14.70\ \text{J K}^{-1}\,\text{mol}^{-1}.$
(b), $\{H_m(450\ \text{K},\ 9\ \text{MPa}) - H_m(350\ \text{K},\ 9\ \text{MPa})\} = +5814\ \text{J mol}^{-1}.$

6.3: (a), $\{G_m(350\ \text{K},\ 9\ \text{MPa}) - G_m(350\ \text{K},\ 1\ \text{MPa})\} = +5661\ \text{J mol}^{-1}.$
(b), $\{S_m(350\ \text{K},\ 9\ \text{MPa}) - S_m(350\ \text{K},\ 1\ \text{MPa})\} = -24.84\ \text{J K}^{-1}\,\text{mol}^{-1}.$
(c), $\{H_m(350\ \text{K},\ 9\ \text{MPa}) - H_m(350\ \text{K},\ 1\ \text{MPa})\} = -3033\ \text{J mol}^{-1}.$

6.4: (a), $\{S_m(450\ \text{K},\ 9\ \text{MPa}) - S_m(350\ \text{K},\ 1\ \text{MPa})\} = -10.14\ \text{J K}^{-1}\,\text{mol}^{-1}.$
(b), $\{H_m(450\ \text{K},\ 9\ \text{MPa}) - H_m(350\ \text{K},\ 1\ \text{MPa})\} = +2781\ \text{J mol}^{-1}.$
$\{G_m(T_2,\ p_2) - G_m(T_1,\ p_1)\}$ can be determined only when $T_1 = T_2.$

6.5: $c_V = 1751\ \text{J K}^{-1}\,\text{kg}^{-1}.$

6.6: $\kappa_S = 1/\{(1253 \pm 1)^2\ \text{m}^2\,\text{s}^{-2}\ (773.87 \pm 0.01)\ \text{kg m}^{-3}\} = (0.8231 \pm 0.0013) \times 10^{-3}\ \text{MPa}^{-1}.$ $T\alpha^2/\rho c_p = 298.15\ \text{K}\ (1.215 \pm 0.001)^2 \times 10^{-6}\ \text{K}^{-2}/\{(773.87 \pm 0.01)\ \text{kg m}^{-3}\ (1861 \pm 1)\ \text{J K}^{-1}\,\text{kg}^{-1}\} = (0.3056 \pm 0.0005) \times 10^{-3}\ \text{MPa}^{-1}.$ The value $(\kappa_S + T\alpha^2/\rho c_p) = (1.129 \pm 0.001) \times 10^{-3}\ \text{MPa}^{-1}$ of κ_T agrees within the experimental errors with the value $(1.130 \pm 0.002) \times 10^{-3}\ \text{MPa}^{-1}.$

6.7: For the air alone $\delta T \approx \{T(\text{d}B/\text{d}T)/C_{V,m}\}\,\delta p \approx \{288\ \text{K} \times 0.2 \times 10^{-6}\ \text{m}^3\,\text{mol}^{-1}\,\text{K}^{-1}/(2.5 \times 8.3\ \text{J K}^{-1}\,\text{mol}^{-1})\}\ 1.1 \times 10^6\ \text{Pa} \approx 3\ \text{K}.$ For the whole apparatus $\delta T \approx \{n(\text{air})C_{V,m}(\text{air})/37000\ \text{J mol}^{-1}\}\ 3\ \text{K} \approx \{2.2 \times 10^6\ \text{Pa}\ 2.2 \times 10^{-3}\ \text{m}^3/(8.3\ \text{J K}^{-1}\,\text{mol}^{-1}\ 288\ \text{K})\}\ (2.5 \times 8.3\ \text{J K}^{-1}\,\text{mol}^{-1}/37000\ \text{J mol}^{-1})\ 3\ \text{K} \approx 0.003\ \text{K}.$

6.8: The equation fitted to B gives $(B - T\ \text{d}B/\text{d}T) = -(2651 \pm 28)\ \text{cm}^3\,\text{mol}^{-1}$ at $T = 373.5\ \text{K},$ and $-T\ \text{d}^2B/\text{d}T^2 = (15.0 \pm 0.1)\ \text{cm}^3\,\text{K}^{-1}\,\text{mol}^{-1}$ at $T = 402.3\ \text{K}.$ The first of these values agrees within the experimental errors with $(\partial H_m/\partial p)_T = -(2630 \pm 20)\ \text{cm}^3\,\text{mol}^{-1}.$ The second agrees within the experimental errors with $(\partial C_{p,m}/\partial p)_T = (15.3 \pm 0.4)\ \text{cm}^3\,\text{K}^{-1}\,\text{mol}^{-1}.$ The results are thermodynamically consistent within the (rather large) experimental errors.

8.1: (a), $\mu^\alpha(\text{N}_2) + 3\mu^\alpha(\text{H}_2) = 2\mu^\alpha(\text{NH}_3);$ $\mu^\beta(\text{N}_2) + 3\mu^\beta(\text{H}_2) = 2\mu^\beta(\text{NH}_3).$ (b), Since $\mu^\alpha(\text{H}_2) = \mu^\beta(\text{H}_2):$ $\mu^\alpha(\text{N}_2) - \mu^\beta(\text{N}_2) = 2\mu^\alpha(\text{NH}_3) - 2\mu^\beta(\text{NH}_3).$ If the membrane were permeable to two of the substances then $\mu^\alpha = \mu^\beta$ for two of the components, and therefore also for the third; the membrane now has no effect.

8.2: For equilibrium of the two phases we have the Gibbs–Duhem relations: $S_m^\alpha\ \text{d}T - V_m^\alpha\ \text{d}p + (1 - x_B^\alpha)\ \text{d}\mu_A + x_B^\alpha\ \text{d}\mu_B = 0,$ and $S_m^\beta\ \text{d}T - V_m^\beta\ \text{d}p + (1 - x_B^\beta)\ \text{d}\mu_A + x_B^\beta\ \text{d}\mu_B = 0.$ If $x_B^\alpha = x_B^\beta = x$ then $(S_m^\alpha - S_m^\beta)\ \text{d}T - (V_m^\alpha - V_m^\beta)\ \text{d}p = 0,$ so that $(\partial T/\partial x)_p = 0$ and $(\partial p/\partial x)_T = 0.$ If $x_B^\alpha = x_B^\beta$ then plots of the equilibrium (e.g. 'boiling') temperature T at given pressure against x_B^α and against x_B^β must kiss at an extremum. Similarly, plots of the equilibrium (e.g. 'vapour') pressure p at given temperature against x_B^α and against x_B^β must kiss at an extremum. Such a mixture, called an azeotrope, distils or melts without change of composition; see § 16.21 and chapter 17.

8.3: See chapter 10.

9.1: (a), $\Delta S_m = 92.59 \text{ J K}^{-1} \text{ mol}^{-1}$; $\Delta H_m = 17551 \text{ J mol}^{-1}$. (b), $\Delta S_m = 92.53 \text{ J K}^{-1} \text{ mol}^{-1}$; $\Delta H_m = 17543 \text{ J mol}^{-1}$. The two pathways, one through a 'liquid'-to-gas transition and the other from 'gas' to gas, give the same results within the experimental errors. The 'liquid' at 230 K and 2 MPa is identical to the 'gas' at 230 K and 2 MPa. The values of ΔS_m and ΔH_m can be obtained: for a change of pressure at constant temperature from $\Delta S_m = -\int_{p_1}^{p_2} (\partial V_m/\partial T)_p \, dp$ and $\Delta H_m = \int_{p_1}^{p_2} \{V_m - T(\partial V_m/\partial T)_p\} \, dp$ by use of measured values of $V_m(T, p)$; for a change of temperature at constant pressure from $\Delta S_m = \int_{T_1}^{T_2} C_{p,m} \, d\ln T$ and $\Delta H_m = \int_{T_1}^{T_2} C_{p,m} \, dT$ by use of measured values of $C_{p,m}(T, p)$; and for a phase change at the equilibrium temperature T^{l+g} from $\Delta_l^g S_m = \Delta_l^g H_m / T^{l+g}$ where $\Delta_l^g H_m$ is the calorimetrically measurable molar enthalpy of evaporation.

9.2: $dp^{s+l}/dT = \Delta_s^l H_m / T \Delta_s^l V_m = \Delta_s^l H_m \rho^l \rho^g / T M (\rho^l - \rho^g) = 6007 \text{ J mol}^{-1} 0.9999 \text{ g cm}^{-3} 0.9168 \text{ g cm}^{-3} / \{274.15 \text{ K } 18.015 \text{ g mol}^{-1} (0.9168 - 0.9999) \text{ g cm}^{-3}\} = -13.47 \text{ J cm}^{-3} \text{ K}^{-1} = -13.47 \text{ MPa K}^{-1}$. For a change of pressure $\delta p = (0.61 - 101.32) \times 10^{-3} \text{ MPa}$, $\delta T = +100.7 \times 10^{-3} \text{ MPa}/13.47 \text{ MPa K}^{-1} = +0.0075 \text{ K}$. Thus the freezing temperature T^{s+l} of pure water at atmospheric pressure exceeds the triple-point temperature T^{s+l+g} of pure water by 0.0075 K. The difference $(T^{s+l} - T_{ice}) = +0.0025 \text{ K}$, between the freezing temperature T^{s+l} of pure water and the 'normal ice temperature' T_{ice}, arises from air dissolved in water at the 'normal ice temperature'; see problem 18.4.

9.3: (a), According to equation (9.4.3), $\ln\{p^{l+g}(340.08 \text{ K})/p^{l+g}(298.15 \text{ K})\} = 33606$ $\text{J mol}^{-1} (340.08 - 298.15) \text{ K}/8.3144 \text{ J K}^{-1} \text{ mol}^{-1} 298.15 \text{ K } 340.08 \text{ K} = 1.6715$, so that $p^{l+g}(340.08 \text{ K}) = 10.62 \text{ kPa e}^{1.6715} = 56.50 \text{ kPa}$. (b), According to equation (9.4.4), $p^{l+g}(340.1 \text{ K}) \approx 100 \exp\{10 \times (340.1 - 357.3)/340.1\} \text{ kPa} = 60.3 \text{ kPa}$. The errors are (a) an underestimate by 2.2 per cent, and (b) an overestimate by 4.3 per cent.

9.4: Rearranging equation (9.4.5) we obtain $B = \Delta_l^g H_m / T(dp^{l+g}/dT) - RT/p^{l+g} + V_m^l$ so that $B = (21840 \text{ J mol}^{-1}/298.15 \text{ K } 5471 \text{ Pa K}^{-1}) - (8.3144 \text{ J K}^{-1} \text{ mol}^{-1} 298.15 \text{ K}/171418 \text{ Pa}) + 0.000123 \text{ m}^3 \text{ mol}^{-1} = (0.013389 - 0.014661 + 0.000123) \text{ m}^3 \text{ mol}^{-1} = -949 \text{ cm}^3 \text{ mol}^{-1}$. The value of B differs from that obtained by extrapolation of the directly measured values by twice the experimental error of the latter.

9.5: $C_{l+g,m}^g = C_{p,m}^g - (\partial V_m^g/\partial T)_p \Delta_l^g H_m / \Delta_l^g V_m = (36.6 - 86.8 \times 40660/30140)$ $\text{J K}^{-1} \text{ mol}^{-1} = -80.5 \text{ J K}^{-1} \text{ mol}^{-1}$. $C_{l+g,m}^l = C_{p,m}^l - (\partial V_m^l/\partial T)_p \Delta_l^g H_m / \Delta_l^g V_m = (76.0 - 0.012 \times 40660/30140) \text{ J K}^{-1} \text{ mol}^{-1} = 76.0 \text{ J K}^{-1} \text{ mol}^{-1}$.

9.6: The experimental mean densities $(\rho^l + \rho^g)_{obs}/2$ are compared below with those calculated from the equation: $(\rho^l + \rho^g)_{calc}/2 = \{0.8666 - 0.0013400(T/K)\}$ g cm^{-3}, fitted to the results by use of the principle of least squares.

T/K	293.15	295.15	297.15	299.15	300.15
$(\rho^l+\rho^g)_{obs}/2$ g cm^{-3}	0.474	0.473	0.468	0.464	0.464
$(\rho^l+\rho^g)_{calc}/2$ g cm^{-3}	0.474	0.471	0.468	0.466	0.464

T/K	303.15	304.15	305.15	306.15
$(\rho^l+\rho^g)_{obs}/2$ g cm^{-3}	0.462	0.458	0.458	0.458
$(\rho^l+\rho^g)_{calc}/2$ g cm^{-3}	0.460	0.459	0.457	0.456

From the absence of any systematic trend in the discrepancies we conclude that N_2O obeys the law of the rectilinear diameter within the experimental errors. From the given value of the critical temperature we obtain $\rho^c = 0.452$ g cm^{-3}.

9.7: A plot of $\ln(\rho^l/\rho^c - 1)$ and of $\ln(1 - \rho^g/\rho^c)$ on the same diagram against $\ln(1 - T/T^c)$ gives a straight line having a slope $\beta = 0.350 \pm 0.005$. Analysis of the results by the principle of least squares gives $\beta = 0.351$; see figure 9.6.

10.1: Suppose that the gaseous mixture is $\{(1-x^\alpha)H_2 + x^\alpha Ar\}$. Use a chemical potentiometer with a palladium membrane to study the mixture (α) against pure hydrogen (β). Measure p^α and p^β at osmotic equilibrium. Measure the partial molar volumes of H_2 in the mixture at pressures between p^α and p, and the molar volumes of H_2 at pressures between p^β and p. Use equation (10.3.3) to evaluate $\{\mu^*(H_2, T, p) - \mu(H_2, T, p, x^\alpha)\}$. Either, find another membrane permeable to Ar but impermeable to H_2 and so obtain $\{\mu^*(Ar, T, p) - \mu(Ar, T, p, x^\alpha)\}$, or (more likely) repeat the measurements with the palladium membrane for values of x between 1 and x^α, and use the Gibbs–Duhem relation to evaluate $\{\mu^*(Ar, T, p) - \mu(Ar, T, p, x^\alpha)\}$ as $\int_{x=1}^{x=x^\alpha}\{(1-x)/x\}\,d\mu(H_2, T, p, x)$. Then $\Delta_{mix}G_m(T, p, x^\alpha) = -(1-x^\alpha)\{\mu^*(H_2, T, p) - \mu(H_2, T, p, x^\alpha)\} - x^\alpha\{\mu^*(Ar, T, p) - \mu(Ar, T, p, x^\alpha)\}$. Either, repeat all this at other temperatures around T and use the relation $\Delta_{mix}S_m = -(\partial\Delta_{mix}G_m/\partial T)_p$, or measure $\Delta_{mix}H_m$ calorimetrically and use the relation $\Delta_{mix}S_m = (\Delta_{mix}H_m - \Delta_{mix}G_m)/T$.

10.2: If the gases are perfect then $V_B = RT/p$ and $(1-x^\alpha)p^\alpha = p^\beta$, where α and β have the same meanings as in the solution of problem 10.1. Equation (10.3.3) then becomes $\{\mu_A^*(T, p) - \mu_A(T, p, x^\alpha)\} = RT\ln(p^\alpha/p^\beta) = -RT\ln(1-x^\alpha)$. Use the Gibbs–Duhem relation to evaluate $\{\mu_B^*(T, p) - \mu_B(T, p, x^\alpha)\}$ as $-RT\int_1^{x^\alpha}dx/x = -RT\ln x^\alpha$. Then $\Delta_{mix}G_m = RT\{(1-x^\alpha)\ln(1-x^\alpha) + x^\alpha\ln x^\alpha\}$ and $\Delta_{mix}S_m = -R\{(1-x^\alpha)\ln(1-x^\alpha) + x^\alpha\ln x^\alpha\}$. When $x^\alpha = 0.5$, $\Delta_{mix}S_m = -8.314$ J K^{-1} mol^{-1} $\ln 0.5 = +5.763$ J K^{-1} mol^{-1}.

10.3: See § 16.7.

10.4: $\{\mu_A - \mu_A^*\} = -\int_p^{p+\Pi}V_A\,dp \approx -V_A^*\Pi = -18$ cm^3 mol^{-1} 0.139 MPa $= -2.5$ J mol^{-1}.

10.5: $\Pi \approx \Delta_s^l H(T^* - T)/TV_A = 6000 \text{ J mol}^{-1} (273.150 - 271.240) \text{ K}/271.24 \text{ K}$ $18 \text{ cm}^3 \text{ mol}^{-1} = 2.3_5 \text{ MPa}$.

10.6: Let $A = H_2O$. Then $\ln(\lambda_A^*/\lambda_A) = (\Delta_s^l H_A^*/RT_A^*)(1 - T_A/T_A^*) + (\Delta_s^l H_A^*/RT_A^* - \Delta_s^l C_{p,A}^*/2R)(1 - T_A/T_A^*)^2 = 0.003568 + 0.000001 = 0.003569$. If $\Delta_s^l C_{p,A}^*$ were neglected the second term would be 0.000005 instead of 0.000001.

10.7: $\{\mu(\text{HCl}, m = 0.119304 \text{ mol kg}^{-1}) - \mu(\text{HCl}, m = 0.0109474 \text{ mol kg}^{-1})\} = -F(E_1 - E_2) = -96485 \text{ C mol}^{-1} (0.38948 - 0.50532) \text{ V} = +11.176 \text{ J mol}^{-1}$.

10.8: By use of the Gibbs–Duhem relation, $\{\mu_A - \mu_A^*\} = -\int_0^{x^l} \{x/(1-x)\} \times (\partial\mu_B/\partial x)_{T,p} \, dx = -\int_0^{x^l} [\{x/(1-x)\}RT/x - 2A(1-x)] \, dx = RT\ln(1-x^l) + A(x^l)^2$. No, the proposed formulae do not satisfy the Gibbs–Duhem relation.

11.1: $K^\ominus(400 \text{ K}) = \exp(44.04 - 67.33) = 7.68 \times 10^{-11}$; $K^\ominus(1000 \text{ K}) = \exp(24.13 - 23.17) = 2.61$.

11.2: $\ln\{K^\ominus(1000 \text{ K})\} = (25.634 + 17.493 - 1.521 - 24.854) - (-13295 + 29086) \text{ K}/1000 \text{ K} = 0.961$; $K^\ominus(1000 \text{ K}) = 2.61$.

11.3: At 298.15 K, $\Delta H_m^\ominus/R = +15791 \text{ K}$, $\Delta S_m^\ominus/R = +16.123$, $\Delta C_{p,m}^\ominus/R = +1.894$, and $K^\ominus = \exp(-15791/298.15 + 16.123) = 1.00 \times 10^{-16}$. According to equation (11.4.3), $\ln\{K^\ominus(1000 \text{ K})\} \approx -36.84 + 15791 \times (1000 - 298.15)/1000 \times 298.15 = +0.332$ so that $K^\ominus(1000 \text{ K}) \approx 1.39$. According to equation (11.4.5), $\ln\{K^\ominus(1000 \text{ K})\} \approx +0.332 + 1.894\{\ln(1000/298.15) - (1000 - 298.15)/1000\} = 1.295$ so that $K^\ominus(1000 \text{ K}) \approx 3.65$. (The temperature interval is too great to allow these approximations to lead to more accurate results.)

11.4: We use the relation $\Delta H_m^\ominus = -R \, d\ln\{K^\ominus(T)\}/d(1/T)$ and obtain $\Delta H_m^\ominus(400 \text{ K}) \approx +34 \text{ kJ mol}^{-1}$ and $\Delta H_m^\ominus(800 \text{ K}) \approx -13 \text{ kJ mol}^{-1}$.

11.5: $K^\ominus = \exp(-31.11 + 52.95 - 44.30) = 1.76 \times 10^{-10}$. (Thus the solubility, expressed as a molality, of AgCl in pure water at 298.15 K is $1.33 \times 10^{-5} \text{ mol kg}^{-1}$.)

11.6: The mole fractions are $(1 - 2x - 3y)$ of $C_6H_5 \cdot C_2H_5$, x of $C_6H_5 \cdot CH{:}CH_2$, y of $C_6H_5 \cdot C{:}CH$, and $(x + 2y)$ of H_2. Then at 1000 K, $K_1^\ominus = \exp(-47.26 + 47.95) = 1.994 = (x + 2y)x(p/p^\ominus)/(1 - 2x - 3y)$, and $K_2^\ominus = \exp(-55.02 + 47.95) = 8.50 \times 10^{-4} = (x + 2y)y(p/p^\ominus)/(1 - 2x - 3y)$. Thus $y = xK_2^\ominus/K_1^\ominus$, and for $p = p^\ominus$, $(1 - 2x - 3xK_2^\ominus/K_1^\ominus)K_1^\ominus = x^2(1 + 2K_2^\ominus/K_1^\ominus)$. Solving the quadratic for x we obtain $x = 0.4491$. Then $y = 0.4491K_2^\ominus/K_1^\ominus = 0.0002$.

11.7: For $H_2O(l) = H_2O(g)$, $K^\ominus(298.15 \text{ K}) = \exp(92.22 - 95.68) = 0.03143$ so that $p^{l+g}(298.15 \text{ K})/p^\ominus \approx 0.03143$ and $p^{l+g}(298.15 \text{ K}) \approx 3.185 \text{ kPa}$; $\Delta_l^g H_m^\ominus(298.15 \text{ K}) = 5292 \text{ K} \cdot 8.3144 \text{ J K}^{-1} \text{ mol}^{-1} = 44000 \text{ J mol}^{-1}$.

12.1: We have $r(p)/r(0) = [1 + \{B(C_3H_8) - B(N_2)\}p/RT]$, and find by the method of least squares $\{B(C_3H_8) - B(N_2)\}/RT = -0.1631 \text{ MPa}^{-1}$ so that $\{B(C_3H_8) - B(N_2)\} = -400 \text{ cm}^3 \text{ mol}^{-1}$ and $B(C_3H_8, C' = 0) = -405 \text{ cm}^3 \text{ mol}^{-1}$. The ratio $B(C_3H_8, C = 0)/B(C_3H_8, C' = 0) \approx 1 + Bp/RT$, which for the mean pressure is 0.992 so that $B(C_3H_8, C = 0) \approx -402 \text{ cm}^3 \text{ mol}^{-1}$.

12.2: From the relation $\rho_A = \rho_B$ we obtain $B_B = (M_B/M_A)(RT/p_A + B_A) - RT/p_B$. Let A be N_2 and B be C_6H_6. Then $B(C_6H_6)$ has the values -1528, -1524, and $-1526 \text{ cm}^3 \text{ mol}^{-1}$, or $B(C_6H_6, C' = 0) = -(1526 \pm 2) \text{ cm}^3 \text{ mol}^{-1}$. The ratio $B(C_6H_6, C = 0)/B(C_6H_6, C' = 0) \approx 1 + Bp/RT$, which is 0.995 so that $B(C_6H_6, C = 0) \approx -1518 \text{ cm}^3 \text{ mol}^{-1}$.

12.3: At $T = 100 \text{ K}$ we have $B = -187.8 \text{ cm}^3 \text{ mol}^{-1}$, $dB/dT = +3.687 \text{ cm}^3 \text{ mol}^{-1} \text{ K}^{-1}$, and $d^2B/dT^2 = -0.1154 \text{ cm}^3 \text{ mol}^{-1} \text{ K}^{-2}$. We use the formula: $u^2 = (RT/M) \times (1 + Bp/RT)^2/[1 - \{1 + (p/R)(dB/dT)\}^2/\{C_{p,m}(p \to 0)/R - (pT/R)(d^2B/dT^2)\}]$, to calculate $u(p = 30 \text{ kPa}) = 185.60 \text{ m s}^{-1}$ and $u(p \to 0) = 186.25 \text{ m s}^{-1}$.

12.4: The criticality conditions, obtained by solving $\partial^2 \Delta_{mix}G_m/\partial x^2 = 0$, and $\partial^3 \Delta_{mix}G_m/\partial x^3 = 0$, are $RT^c/A = 0.4695$ and $x^c = 0.7384$. At $T/T^c = 0.7$ the compositions of the coexisting phases obtained by drawing a common tangent to a plot of $\Delta_{mix}G_m$ against x are $x' \approx 0.40$ and $x'' \approx 0.95$. Detailed calculations of $(\mu_A - \mu_A^*)$ and $(\mu_B - \mu_B^*)$ around those values of x gave $x' = 0.4043$ and $x'' = 0.9581$ as the compositions at which $\mu_A' = \mu_A''$ and $\mu_B' = \mu_B''$.

12.5: The fugacity $f_{Ar} = p \exp(B_{Ar}p/RT) = 0.5 \text{ MPa} \exp(-187.8 \times 0.5/831.44) = 0.4466 \text{ MPa}$.

12.6: $\Delta G_m^{\ominus} = +7.8 \text{ kJ mol}^{-1}$. $\{G(T, p, \xi) - G(T, p, 0)\} = 0$, -4378, -5843, -5545, -3214, and $+3795 \text{ kJ}$ for $\xi = 0$, 0.2, 0.4, 0.6, 0.8, and 1 mol.

12.7: $\{S_m^{\ominus}(\text{Ar, g, } 87.28 \text{ K}) - S_m(\text{Ar, g, } 87.28 \text{ K, } 101.325 \text{ kPa})\} = R \ln(p/p^{\ominus}) + p \, dB/dT = p \, dB/dT = 0.101325 \text{ MPa } 105.4 \text{ cm}^3 \text{ mol}^{-1} \times (113.0/87.28^2 \text{ K}) \exp(113.0/87.28) = 0.58 \text{ J K}^{-1} \text{ mol}^{-1}$.

12.8: At equilibrium $\xi^e = 0.00933 \text{ mol}$; $x^e(H_2) = 0.0223$, $x^e(I_2) = 0.3557$, and $x^e(HI) = 0.6220$.

12.9: At equilibrium $(1 + \alpha^e) = pV/nRT = 0.1508 \text{ MPa } 308.6 \text{ cm}^3/(0.008367 \text{ mol } 8.3144 \text{ J K}^{-1} \text{ mol}^{-1} 473 \text{ K}) = 1.414$, so that $\alpha^e = 0.414$, $\xi^e = \alpha^e n = 0.00347 \text{ mol}$, $x^e(PCl_5) = 0.414$, $x^e(PCl_3) = x^e(Cl_2) = 0.293$, and $K^{\ominus} = 0.308$.

12.10: At equilibrium $\xi^e/n = (pV/nRT - 1)$ and $K^{\ominus} = 4(\xi^e/n)^2(p/p^{\ominus})/\{1 - (\xi^e/n)^2\}$ so that $K^{\ominus} = 0.170$, 0.165, 0.167, and 0.168. Given $K^{\ominus}(1274 \text{ K}) = 0.168$ and $K^{\ominus}(1073 \text{ K}) = 0.0109$, $\Delta H_m^{\ominus}(1173 \text{ K}) = (8.3144 \text{ J K}^{-1} \text{ mol}^{-1} 1274 \text{ K } 1073 \text{ K}/201 \text{ K}) \ln(0.168/0.0109) = +155 \text{ kJ mol}^{-1}$.

12.11: Let the amounts of substance of I be $2\alpha n$ and of I_2 be $(1 - \alpha)n$ so that $4\alpha^2(p/p^{\ominus})/(1 - \alpha)^2 = K^{\ominus}$ and $\alpha = \{K^{\ominus}/(K^{\ominus} + 4p/p^{\ominus})\}^{1/2}$. Then $H = (1 - \alpha)nH_m^{\ominus}(I_2) + 2\alpha nH_m^{\ominus}(I)$, so that $C_p/n = (1 - \alpha)C_{p,m}^{\ominus}(I_2) + 2\alpha C_{p,m}^{\ominus}(I) + \{2H_m^{\ominus}(I) - H_m^{\ominus}(I_2)\} d\alpha/dT = (1 - \alpha)C_{p,m}^{\ominus}(I_2) + 2\alpha C_{p,m}^{\ominus}(I) + \Delta H_m^{\ominus} \frac{1}{2}(K^{\ominus})^{1/2} \times 4(p/p^{\ominus})(K^{\ominus} + 4p/p^{\ominus})^{-3/2} d\ln K^{\ominus}/dT$, so that $C_p/nR = C_{p,m}^{\ominus}(I_2)/R + \{2C_{p,m}^{\ominus}(I)/R - C_{p,m}^{\ominus}(I_2)/R\}\{K^{\ominus}/(K^{\ominus} + 4p/p^{\ominus})\}^{1/2} + \frac{1}{2}(\Delta H_m^{\ominus}/RT)^2(K^{\ominus})^{1/2}4(p/p^{\ominus}) \times$

$(K^{\ominus}+4p/p^{\ominus})^{-3/2} = 4.5+0.5 \times (0.048/4.048)^{1/2}+\frac{1}{2} \times 171000\,\mathrm{J\,mol^{-1}}/(8.3144\,\mathrm{J\,K^{-1}}$ $\mathrm{mol^{-1}}\,1173\,\mathrm{K})^2 \times 0.048^{1/2} \times 4/4.048^{3/2} = 4.5+0.054+16.540 = 21.094$. The heat capacity C_p of $0.2538\,\mathrm{kg}$ of equilibrium mixture for which $n = 0.2538\,\mathrm{kg}/0.2538\,\mathrm{kg\,mol^{-1}} = 1\,\mathrm{mol}$, is given by $C_p = 21.094 \times 8.3144\,\mathrm{J\,K^{-1}}$ $\mathrm{mol^{-1}}\,1\,\mathrm{mol} = 175.4\,\mathrm{J\,K^{-1}}$. Compare the heat capacity of $0.2538\,\mathrm{kg}$ of 'non-dissociating' I_2 which is $37.4\,\mathrm{J\,K^{-1}}$.

12.12: Given the material from problem 12.11, $V = \{(1-\alpha)n+2\alpha n\}RT/p$ and $T(\partial V/\partial T)_p = (1+\alpha)nRT/p+(nRT^2/p)\,d\alpha/dT$ so that $\{V-T(\partial V/\partial T)_p\} = -(nRT^2/p)(K^{\ominus})^{1/2}2(p/p^{\ominus})(\Delta H_m^{\ominus}/RT^2)/(K^{\ominus}+4p/p^{\ominus})^{3/2} = -2n(K^{\ominus})^{1/2}(\Delta H_m^{\ominus}/p^{\ominus})/$ $(K^{\ominus}+4p/p^{\ominus})^{3/2}$. Then $\{H(T,p_2)-H(T,p_1)\}/n = +(K^{\ominus})^{1/2}\Delta H_m^{\ominus} \times$ $\{(K^{\ominus}+4p_2/p^{\ominus})^{-1/2}-(K^{\ominus}+4p_1/p^{\ominus})^{-1/2}\} = P/f_m$ where P denotes power and f_m molar flow rate of 'undissociated' I_2; $f_m = 0.05\,\mathrm{g\,s^{-1}}/253.8\,\mathrm{g\,mol^{-1}}$. Thus $P = f_m(K^{\ominus})^{1/2}\Delta H_m^{\ominus}\{(K^{\ominus}+4p_2/p^{\ominus})^{-1/2}-(K^{\ominus}+4p_1/p^{\ominus})^{-1/2}\} = (0.05/253.8)$ mol $\mathrm{s^{-1}}$ $0.048^{1/2} \times 171000\,\mathrm{J\,mol^{-1}}\,(0.4428^{-1/2}-1.0349^{-1/2}) = 3.84\,\mathrm{W}$.

12.13: (a), $(y^e)^2/(1-y^e)^4(p/p^{\ominus})^2 = 27K^{\ominus}/256$ where y^e is the mole-fraction yield of NH_3 at equilibrium. Thus $y^e/(1-y^e)^2 = (p/p^{\ominus})(27K^{\ominus}/256)^{1/2} = k_0$, say, so that $y^e = \{2k_0+1-(4k_0+1)^{1/2}\}/2k_0$. When $p = 100p^{\ominus}$, $k_0 = 0.4184$ and $y^e = 0.241$; when $p = 1000p^{\ominus}$, $k_0 = 4.184$ and $y^e = 0.616$. (b), $y^e/(1-y^e)^2 = k_0 \exp[-\frac{1}{2}\{2B(NH_3)-B(N_2)-3B(H_2)\}p/RT] = k_1$, say. When $p = 100p^{\ominus}$, $k_1 = 0.4686$ and $y^e = 0.258$; when $p = 1000p^{\ominus}$, $k_1 = 12.98$ and $y^e = 0.755$. Lewis and Randall's rule gives a considerable improvement at the higher pressure.

13.1: From $Z^c = 0.290$, $V_m^c = 0.290 \times 8.31 \times 377\,\mathrm{J\,mol^{-1}}/6.28\,\mathrm{MPa} = 145\,\mathrm{cm^3}$ $\mathrm{mol^{-1}}$; from figure 13.1, $Z(471.2\,\mathrm{K} = 1.25T^c, 12.56\,\mathrm{MPa} = 2p^c) \approx 0.67$; from equation (13.2.8), $B(565.5\,\mathrm{K} = 1.5T^c) = -70\,\mathrm{cm^3\,mol^{-1}}$; from equation (13.3.7), $T^b(0.1013\,\mathrm{MPa}) = 5.512 \times 377\,\mathrm{K}/\{5.522-\ln(0.1013/6.28)\} = 215\,\mathrm{K}$ (from the approximate relation $T^b/T^c \approx 0.6$, $T^b \approx 226\,\mathrm{K}$); from equation (13.3.7), $p^{l+g}(300\,\mathrm{K} = 0.796T^c) = 6.28\,\mathrm{MPa}\exp(5.522-5.512/0.796) = 1.54\,\mathrm{MPa}$; from equations (13.3.10) and (13.3.11), $V_m^{l,l+g}(300\,\mathrm{K} = 0.796T^c) = 66\,\mathrm{cm^3\,mol^{-1}}$ and $V_m^{g,l+g}(300\,\mathrm{K} = 0.796T^c) = 1280\,\mathrm{cm^3\,mol^{-1}}$.

14.1: $N(\varepsilon_1)/N = 2\exp(-180\,\mathrm{K}/T)/\{2+2\exp(-180\,\mathrm{K}/T)\}$. (a), At $90\,\mathrm{K}$, $N(\varepsilon_1)/N = 0.119$; (b), at $360\,\mathrm{K}$, $N(\varepsilon_1)/N = 0.378$.

14.2: $C_{p,m}/R = 3.500+\{(\varepsilon_1-\varepsilon_0)/kT\}^2\exp\{-(\varepsilon_1-\varepsilon_0)/kT\}/[1+\exp\{-(\varepsilon_1-\varepsilon_0)/kT\}]^2 = 3.836, 3.807, 3.780, 3.756, 3.734, 3.714$, and 3.697 at $120, 130, 140, 150, 160, 170$, and $180\,\mathrm{K}$. The results agree with the measured values within the probable experimental errors.

14.3: $C_{V,m}/R-5/2 = (h\nu/kT)^2\exp(h\nu/kT)/\{\exp(h\nu/kT)-1\}^2$. We construct the following table.

T/K	0	25	50	100	250	500	1000	2500	5000
$C_{V,m}(O_2)/R$ $-5/2$	0	0.00	0.00	0.00	0.01	0.23	0.67	0.94	0.98
$C_{V,m}(Br_2)/R$ $-5/2$	0	0.00	0.01	0.22	0.76	0.93	0.98	1.00	1.00

For O_2 the vibrational mode is effectively unexcited at temperatures below about 250 K and is effectively classical only at temperatures above about 5000 K. For Br_2 the vibrational mode is effectively unexcited at temperatures below about 50 K and is effectively classical at temperatures above about 1000 K.

14.4: $C_{p,m}/R = 4.00 + \Sigma_{i=1}^{i=6} (\theta_i/T)^2 \exp(\theta_i/T)/\{\exp(\theta_i/T) - 1\}^2$. (a), $C_{p,m}(300 \text{ K})/R = 4.00 + 0.11_7 + 0.08_2 + 0.03_9 + 0.01_6 + 0.00_0 + 0.00_0 = 4.25$; $C_{p,m}(300 \text{ K}) = 35.4 \text{ J K}^{-1} \text{ mol}^{-1}$. (b), $C_{p,m}(600 \text{ K})/R = 4.00 + 0.54_0 + 0.48_2 + 0.37_4 + 0.27_5 + 0.05_7 + 0.04_9 = 5.78$; $C_{p,m}(600 \text{ K}) = 48.0 \text{ J K}^{-1} \text{ mol}^{-1}$.

14.5: $\langle \varepsilon_x \rangle = \frac{1}{2}RT/L = 0.5 \times 8.31 \text{ J K}^{-1} \text{ mol}^{-1} 10 \text{ K}/6.02 \times 10^{23} \text{ mol}^{-1} = 6.9 \times 10^{-23} \text{ J}$. $\langle n_x^2 \rangle = 4l_x^2 MRT/L^2h^2 = 4 \times 10^{-6} \text{ m}^2 \ 0.004 \text{ kg mol}^{-1} \ 8.3 \text{ J K}^{-1} \text{ mol}^{-1} 10 \text{ K}/(6 \times 10^{23} \text{ mol}^{-1} 6.6 \times 10^{-34} \text{ J s})^2 \approx 8 \times 10^{12}$, so that $\langle n_x \rangle \approx \langle n_x^2 \rangle^{1/2} \approx 3 \times 10^6$. $\Delta \varepsilon_x = \{n_x^2 - (n_x^2 - 2n_x + 1)\}Lh^2/8l_x^2 M \approx 6 \times 10^6 \times 6 \times 10^{23} \text{ mol}^{-1} (6.6 \times 10^{-34} \text{ J s})^2/(8 \times 10^{-6} \text{ m}^2 \ 0.004 \text{ kg mol}^{-1}) \approx 5 \times 10^{-29} \text{ J}$, so that $\Delta \varepsilon_x/\langle \varepsilon_x \rangle \approx 7 \times 10^{-7}$.

14.6: $I = \Sigma_i m_i r_i^2 = 2 \times 0.0010079 \text{ kg mol}^{-1} \ 0.03756 \times 0.03756 \times 10^{-18} \text{ m}^2/6.022 \times 10^{23} \text{ mol}^{-1} = 4.722 \times 10^{-48} \text{ m}^2 \text{ kg}$. $\theta_R = Lh^2/8\pi^2 IR = 6.022 \times 10^{23} \text{ mol}^{-1} 6.626 \times \times 6.626 \times 10^{-68} \text{ J}^2 \text{ s}^2/(8\pi^2 \times 4.722 \times 10^{-48} \text{ m}^2 \text{ kg } 8.3144 \text{ J K}^{-1} \text{ mol}^{-1}) = 85.3 \text{ K}$. H_2 will fail to act as a classical rotator unless $T/0.42\theta_R \gg 1$ (equation 14.6.7). At 100 K, $T/0.42\theta_R = 2.8$; at 300 K, $T/0.42\theta_R = 8.4$. H_2 can be expected to behave more like a classical rotator at 300 K than at 100 K.

14.7: Some calculated values of $K^{\ominus}(T)$ are $K^{\ominus}(1174 \text{ K}) = 0.049$, $K^{\ominus}(1224 \text{ K}) = 0.093$, $K^{\ominus}(1274 \text{ K}) = 0.170$, $K^{\ominus}(1324 \text{ K}) = 0.296$, and $K^{\ominus}(1374 \text{ K}) = 0.495$. The calculated value of $K^{\ominus}(1274 \text{ K})$ agrees excellently with the experimental value.

14.8: Some calculated values of $K^{\ominus}(T)$ are $K^{\ominus}(600 \text{ K}) = 78.0$, $K^{\ominus}(650 \text{ K}) = 65.0$, $K^{\ominus}(700 \text{ K}) = 55.5$, $K^{\ominus}(750 \text{ K}) = 48.2$, and $K^{\ominus}(800 \text{ K}) = 42.5$. The agreement of the calculated values with the experimental ones is excellent, especially in view of the uncertainty of the value of $\Delta H_m^{\ominus}(298.15 \text{ K})$.

15.1: $\{\Sigma_B \nu_B S_B^{\ominus}(500 \text{ K}) - \Sigma_B \nu_B S_B(s, T \rightarrow 0)\} = 122.4 \text{ J K}^{-1} \text{ mol}^{-1}$. $\Sigma_B \nu_B H_B^{\ominus}(500 \text{ K}) = 73.0 \text{ kJ mol}^{-1}$. $-RT \ln\{K^{\ominus}(500 \text{ K})\} = 11470 \text{ J mol}^{-1}$. Thus, $\Sigma_B \nu_B S_B(s, T \rightarrow 0) = \Sigma_B \nu_B H_B^{\ominus}(500 \text{ K})/500 \text{ K} - \{\Sigma_B \nu_B S_B^{\ominus}(500 \text{ K}) - \Sigma_B \nu_B S_B(s, T \rightarrow 0)\} + R \ln K^{\ominus}(500 \text{ K}) = (73000 - 500 \times 122.4 - 11470) \text{ J mol}^{-1}/500 \text{ K} = 0.66 \text{ J K}^{-1} \text{ mol}^{-1}$, which is zero within experimental error.

15.2: $\Sigma_B \nu_B S_B^{\ominus}(500 \text{ K}) = \text{d}(RT \ln K^{\ominus})/\text{d}T = (122.8 \pm 0.5) \text{ J K}^{-1} \text{ mol}^{-1}$. $\{\Sigma_B \nu_B S_B^{\ominus}(500 \text{ K}) - \Sigma_B \nu_B S_B(s, T \rightarrow 0)\} = 122.4 \text{ J K}^{-1} \text{ mol}^{-1}$. Thus, $\Sigma_B \nu_B S_B(s, T \rightarrow 0) = (0.4 \pm 0.5) \text{ J K}^{-1} \text{ mol}^{-1}$, which is zero within experimental error.

15.3: The difference $[\{S_m^{\ominus}(M, 368.5\text{ K}) - S_m(M, T \to 0)\} - \{S_m^{\ominus}(R, 368.5\text{ K}) - S_m(R, T \to 0)\}]$ is given by the integral $\int_0^{368.5\text{ K}} \{C_{p,m}^{\ominus}(M) - C_{p,m}^{\ominus}(R)\}\,d\ln T = (0.91 \pm 0.09)\text{ J K}^{-1}\text{ mol}^{-1}$. The standard molar entropy of transition is given by $\{S_m^{\ominus}(M, 368.5\text{ K}) - S_m^{\ominus}(R, 368.5\text{ K})\} = (397 \pm 40)\text{ J mol}^{-1}/368.5\text{ K} = (1.08 \pm 0.11)\text{ J K}^{-1}\text{ mol}^{-1}$. Thus, $\{S_m(M, T \to 0) - S_m(R, T \to 0)\} = (0.17 \pm 0.14)$ $\text{J K}^{-1}\text{ mol}^{-1}$, which is probably indistinguishable from zero.

15.4: For the reaction: $Hg(l) + \frac{1}{2}Cl_2(g) = HgCl(s)$, $\Sigma_B \nu_B S_B^{\ominus}(298.15\ K) = F\,dE^{\ominus}/dT = -0.000945\text{ V K}^{-1}\,96485\text{ C mol}^{-1} = -91.2\text{ J K}^{-1}\text{ mol}^{-1}$. $\{\Sigma_B \nu_B S_B^{\ominus}(298.15\text{ K}) - \Sigma_B \nu_B S_B(s, T \to 0)\} = (96.2 - 111.5 - 76.0)\text{ J K}^{-1}\text{ mol}^{-1} = -91.3\text{ J K}^{-1}\text{ mol}^{-1}$. Thus, $\Sigma_B \nu_B S_B(s, T \to 0)$ is zero within the experimental errors.

15.5: $S_m^{\ominus}(H_2O, g, \text{'spectroscopic'}, 298.15\text{ K}) = 22.69 \times 8.3144\text{ J K}^{-1}\text{ mol}^{-1} = 188.6$ $\text{J K}^{-1}\text{ mol}^{-1}$. Thus, $S_m(H_2O, s, T \to 0) = 3.3\text{ J K}^{-1}\text{ mol}^{-1}$, which is not zero within the combined experimental errors. $R \ln(3/2) = 3.4\text{ J K}^{-1}\text{ mol}^{-1}$.

16.1: By use of equations (16.7.8), (16.7.9), and (16.7.10) we obtain $f_A = 1.0538e^{-0.007086} = 1.0464$; $\mu_A^E = RT \ln f_A = 118.1\text{ J mol}^{-1}$; $f_B = 1.03116e^{0.003372} = 1.0346$; $\mu_B^E = RT \ln f_B = 88.7\text{ J mol}^{-1}$; $G_m^E = (1 - x_B^l)\mu_A^E + x_B^l\mu_B^E = (56.2 + 46.5)$ $\text{J mol}^{-1} = 102.7\text{ J mol}^{-1}$.

16.2: According to equation (16.9.3), $x_B^l = 0.10660 \times 10372.7/2334.2 = 0.4737$. According to equation (16.7.10), $G_m^E = 68.9\text{ J mol}^{-1}$.

16.3: According to equation (16.9.5), $f_A = 0.9677e^{0.007139} = 0.9746$; $\mu_A^E = RT \ln f_A = (-80.0 + 17.4)\text{ J mol}^{-1} = -62.6\text{ J mol}^{-1}$.

16.4: We put $\beta(0) = 0.13045 V_A^c = 11.675\text{ cm}^3\text{ mol}^{-1}$, $\beta(1) = 0.13045 V_B^c = 12.915\text{ cm}^3\text{ mol}^{-1}$, and $\beta(\frac{1}{2}) = \{\beta(0) + \beta(1)\}/4 + \{\beta(0)^{1/3} + \beta(1)^{1/3}\}^3/16 = 12.290$ $\text{cm}^3\text{ mol}^{-1}$; $a(0) = 1.3829 RT_A^c V_A^c = 0.12987\text{ J m}^3\text{ mol}^{-2}$; $a(1) = 1.3829 RT_B^c V_B^c = 0.21696\text{ J m}^3\text{ mol}^{-2}$, and $a(\frac{1}{2}) = \{a(0) + a(1)\}/4 + 0.98\{a(0)a(1)/\beta(0)\beta(1)\}^{1/2} \times \{\beta(0)^{1/3} + \beta(1)^{1/3}\}^3/16 = 0.16899\text{ J m}^3\text{ mol}^{-2}$. Solving equation (16.12.13) we obtain $V_m(0) = 33.778\text{ cm}^3\text{ mol}^{-1}$, $V_m(1) = 27.771\text{ cm}^3\text{ mol}^{-1}$, and $V_m(\frac{1}{2}) = 29.691\text{ cm}^3\text{ mol}^{-1}$. Then $G_m^E(x = 0.5) = 169\text{ J mol}^{-1}$, $H_m^E(x = 0.5) = 137\text{ J mol}^{-1}$, and $V_m^E(x = 0.5) = -1.08\text{ cm}^3\text{ mol}^{-1}$.

16.5: According to equations (16.13.1) and (16.13.2) with $z = 8$, $r_A = 4$, $r_B = 15/2$, and $x = 0.5$, $G_m^E \approx A_m^E = RT\,\text{ath}(0.5) + \text{th}(0.5) \times 1290\text{ J mol}^{-1} = \{8.314 \times 293.15 \times (-0.04581) + 0.06056 \times 1290\}\text{ J mol}^{-1} = -34\text{ J mol}^{-1}$; $U_m^E = \text{th}(0.5) \times 2860\text{ J mol}^{-1} = 173\text{ J mol}^{-1}$. According to equations (16.13.3) and (16.13.4), $H_m^E \approx U_m^E + 0.5(T\alpha_6^*/\kappa_{T,6}^* + T\alpha_{13}^*/\kappa_{T,13}^*)V_m^E = (173 - 278 \times 0.370)\text{ J mol}^{-1} = 70\text{ J mol}^{-1}$. The experimental value of $H_m^E(x = 0.5)$ is 64 J mol^{-1}; G_m^E has not been measured.

16.6: The average value of w/kT is 0.436. Using that value we obtain calculated values of $p^{l+g}/p_A^{l+g} = 0.981, 0.946, 0.906, 0.735,$ and 0.542.

16.7: The average value of $K^{\ominus}$ is 3.66. Using that value we obtain calculated values of $\lambda_A/\lambda_A^* = 0.6954, 0.3165,$ and 0.0859; of $\lambda_B/\lambda_B^* = 0.0859, 0.3165,$ and 0.6954;

and of $G_m^E/\text{J mol}^{-1} = -762, -1076,$ and $-762,$ which we compare with the experimental values: $-771, -1069,$ and $-757.$

16.8: For the equilibrium of $C_6H_6(s)$ and liquid mixture we have $T_A = 1193\ \text{K}/\{4.283 - \ln(1-x)\}$, and similarly for the equilibrium of $(C_6H_5)_2(s)$ and liquid mixture we have $T_B = 2023/\{5.888 - \ln x\}$. We plot T_A and T_B against x and find the point of intersection: $T_E = 266\ \text{K}$ and $x_E = 0.175.$

16.9: The cooling curves are steep when no phase is separating, less steep when there are two phases present and one phase is separating, and horizontal when three phases are present. In the last example, the mixture of the azeotropic composition behaves like a pure substance; there is an arrest at the azeotropic temperature.

18.1: We find $\lim_{(w_B/V)\to 0} (w_B/Vh) = 147\ \text{kg m}^{-4}$. Then $\langle M \rangle = (RT/g\rho) \times \lim_{(w_B/V)\to 0} (w_B/Vh) = 147\ \text{kg m}^{-4}\ 8.31 \times 310.2\ \text{J mol}^{-1}/9.81\ \text{m s}^{-2}\ 1566\ \text{kg m}^{-3} = 24.7\ \text{kg mol}^{-1}$. The molar mass obtained is called the 'number-average' molar mass $\langle M \rangle_n = \Sigma_B n_B M_B/\Sigma_B n_B$ where n_B is the amount of the particular polymer B. Other kinds of average, such as the mass average, are obtained by other methods. The molar mass found corresponds to an average of about $24.7/0.219 = 113$ of the amino acids coupled together in the polypeptide.

18.2: At equilibrium $m_P = m(B_2) + m(B) = (1-\alpha)m/2 + \alpha m = (1+\alpha)m/2$ where α is the fraction of $B_2 = (C_6H_5CO_2H)_2$ dissociated. Thus $\alpha = 2m_P/m - 1 = 0.075, 0.081, 0.081, 0.088,$ and $0.148.$ For the reaction: $B_2(\text{solute}) = 2B(\text{solute})$ and $m^\ominus = 1\ \text{mol kg}^{-1}$; $10^3 K^\ominus = 10^3 \times 2\alpha^2(m/m^\ominus)/(1-\alpha) = 4.6, 4.8, 3.7, 4.4,$ and $4.4,$ giving a value of $K^\ominus \approx 4.4 \times 10^{-3}.$

18.3: We have $(1+\alpha) = 2\Delta T_A/k_f m$ where k_f is the cryoscopic constant. Thence, $\alpha = 0.1970, 0.1483, 0.1066, 0.0840, 0.0628,$ and $0.0558,$ and $10^4 K^\ominus = 10^4 \times 2\alpha^2 m/(1-\alpha)m^\ominus = 6.0, 7.3, 6.6, 6.7, 6.0,$ and $6.2,$ giving a mean value: $\langle K^\ominus(278.55\ \text{K}) \rangle = 6.5 \times 10^{-4}.$ Given that $\langle K^\ominus(317.05\ \text{K}) \rangle = 4.4 \times 10^{-3}$ from problem 18.2, $\Delta H_m^\ominus = \{RT_2 T_1/(T_2 - T_1)\} \ln\{K^\ominus(T_2)/K^\ominus(T_1)\} \approx (8.31 \times 317 \times 279/38.5) \ln(44/6.5)\ \text{J mol}^{-1} \approx 36\ \text{kJ mol}^{-1}.$

18.4: Let A denote H_2O. Then $(T_A^* - T_A) = -(RT_A^{*2}/\Delta_s^l H_A^*) \ln(1-x_B) = (8.314 \times 273.15 \times 273.15/6007)\ \text{K}\ 2.4 \times 10^{-5} = 0.0025\ \text{K}.$ According to problem 9.2, $(T_A^{s+l+g} - T_A^{s+l}) = +0.0075\ \text{K}$, where T_A^{s+l} denotes the melting temperature at atmospheric pressure. According to this problem, $(T_A^{s+l} - T_{ice}) = +0.0025\ \text{K}.$ Thus, $(T_A^{s+l+g} - T_{ice}) = 0.0100\ \text{K}.$

18.5: By substitution in (18.7.2) we obtain $x = 0.01542\ \text{K}^{-1}\ 2160\ \text{s}\ 0.4130\ \text{K}$ $1.2132\ \text{K}\ 0.8002\ \text{K}/\{(3548 \times 0.8002 - 1388 \times 0.4130)\ \text{K s}\ (0.8002 - 0.4130)\ \text{K}\} = 0.0152.$ The mole-fraction purity is $(1-x) = 0.9848.$

20.1: The relation between ϕ and γ is $(1-\phi) = -\ln\gamma + (m^\ominus/m)\int_0^{m/m^\ominus} \ln\gamma\ d(m/m^\ominus)$. The area of a plot of $\ln\gamma$ against $m/m^\ominus$ from $m/m^\ominus = 0$ to $m/m^\ominus = 0.1$ is -0.0173_6. Thus $(1-\phi) = +0.228_2 - 10 \times 0.0173_6 = 0.054_6$, so that $\phi = 0.945_4.$

20.2: $\alpha = (2\pi L\rho_A^*)^{1/2}(e^2/4\pi\varepsilon_r\varepsilon_0 kT)^{3/2} = 1.171 \text{ kg}^{1/2} \text{ mol}^{-1/2}$; $\beta = 2(2\pi L\rho_A^*)^{1/2} \times (e^2/4\pi\varepsilon_r\varepsilon_0 kT)^{1/2} = 3.282 \text{ nm}^{-1} \text{ kg}^{1/2} \text{ mol}^{-1/2}$. (a), $-\ln\gamma = \alpha m^{1/2} = 0.3703$; $\gamma = 0.6905$; $(1-\phi) = \frac{1}{3}\alpha m^{1/2} = 0.1234$; $\phi = 0.8766$. (b), $-\ln\gamma = \alpha m^{1/2}/(1+\beta dm^{1/2}) = 0.2824$; $\gamma = 0.7540$; $(1-\phi) = \frac{1}{3}\alpha m^{1/2}\sigma(\beta dm^{1/2})$; $\beta dm^{1/2} = 0.3114$; $\sigma(\beta dm^{1/2}) = 0.6623$; $(1-\phi) = 0.0817$; $\phi = 0.9183$.

20.3: We have $\phi = 0.5376(\Delta T/\text{K})\{1 + 4.8\times10^{-4}\Delta T/\text{K}\}$ mol kg^{-1}/$2m = 0.9884$, 0.9819, 0.9746, 0.9656, 0.9548, and 0.9526. According to the limiting law the corresponding values are 0.9856, 0.9781, 0.9690, 0.9569, 0.9380, and 0.9162. According to the Debye–Hückel equation with $d = 0.3$ nm the corresponding values are 0.9864, 0.9798, 0.9724, 0.9633, 0.9505, and 0.9378.

20.4: We have $E = E^\ominus - (2RT/F)\ln(m/m^\ominus) - (2RT/F)\ln\gamma$. $E' \overset{\text{def}}{=} E + (2RT/F) \times \ln(m/m^\ominus) - (2RT/F)2\alpha m^{1/2}/\{1+(m/m^\ominus)^{1/2}\} = E^\ominus - (2RT/F)\ln\gamma - 2\alpha m^{1/2}/\{1+(m/m^\ominus)^{1/2}\}$. $E'/\text{V} = 0.26792$, 0.26784, 0.26756, 0.26692, 0.26588, and 0.26481. By plotting E' against m we find $E^\ominus = 0.26796$ V. $K^\ominus = \exp(E^\ominus F/RT) = 3.384\times10^4$. The activity coefficients are given by $\gamma = \exp\{(E^\ominus - E)(F/2RT) - \ln(m/m^\ominus)\} = 0.9566$, 0.9277, 0.8904, 0.8436, 0.8094, and 0.7876.

20.5: We have $s\gamma/m^\ominus = (K^\ominus)^{1/2}$, and thence $\ln(s'/m^\ominus) \overset{\text{def}}{=} \ln(s/m^\ominus) - \alpha(s+m)^{1/2}/[1+\{(s+m)/m^\ominus\}^{1/2}] = \frac{1}{2}\ln K^\ominus - \ln\gamma - \alpha(s+m)^{1/2}/[1+\{(s+m)/m^\ominus\}^{1/2}]$. Then $\ln(s'/m^\ominus) = -6.3446$, -6.3395, -6.3287, -6.3061, and -6.2911. Plotting $\ln(s'/m^\ominus)$ against m we find $\frac{1}{2}\ln K^\ominus = -6.348$ so that $K^\ominus = 3.062\times10^{-6}$. The activity coefficient for $m = 0$ is $\gamma = (K^\ominus)^{1/2}m^\ominus/s = 0.949$.

20.6: $K_A^\ominus = \alpha^2 m\gamma^2/m^\ominus(1-\alpha)$ so that $\ln K_A^\ominus = \lim_{\alpha m\to 0}[2\ln\alpha - \ln(1-\alpha) + \ln(m/m^\ominus) - \alpha_{\text{DH}}(\alpha m)^{1/2}/\{1+(\alpha m/m^\ominus)^{1/2}\}]$, where $\alpha = B(\alpha)/B(1)$ is the fraction of $A = C_6H_3(NO_2)_2OH$ that has reacted with H_2O, so that the ionic strength is αm. The values of $[\cdots]$ are

$10^4 m/\text{mol kg}^{-1}$	0.9245	1.3636	2.3945	3.5827
$[\cdots]$	-9.415	-9.415	-9.418	-9.415

so that $\ln K_A^\ominus = -9.415$ and $K_A^\ominus = 8.15\times10^{-5}$.

20.7: We have $E = E^\ominus - (RT/F)\ln\{m_H m_3/(m^\ominus)^2\} - (2RT/F)\ln\{\gamma(H^+, Cl^-)\} = E^\ominus - (RT/F)\ln\{(m_1 - m_H)K_A^\ominus m_3/(m_2 + m_H)m^\ominus\} - (2RT/F)\ln\{\gamma(H^+, Cl^-)/\gamma(H^+, A^-)\}$ so that $E' \overset{\text{def}}{=} E - E^\ominus + (RT/F)\ln\{(m_1 - m_H)m_3/(m_2 + m_H)m^\ominus\} = -(RT/F)\ln K_A^\ominus - (2RT/F)\ln\{\gamma(H^+, Cl^-)/\gamma(H^+, A^-)\}$ and $\lim_{I\to 0} E' = -(RT/F)\ln K_A^\ominus$, where m_H is the molality of H_3O^+ and $I \approx (m_2 + m_3)$. We begin by evaluating E' for $m_H = 0$:

$E'(m_H = 0)/\text{V}$	0.28174	0.28169	0.28170	0.28181

so that $\lim_{I\to 0}\{E'(m_H = 0)\} \approx -(RT/F)\ln K_A^\ominus \approx 0.282$ V and $K_A^\ominus \approx 1.71\times10^{-5}$. We now use the approximation $m_H \approx m^\ominus K_A^\ominus m_1/m_2$ to re-evaluate E':

$E'(m_H \approx m^\ominus K_A^\ominus m_1/m_2)/V$ 0.28154   0.28161   0.28166   0.28179

so that $\lim_{I\to 0}\{E'(m_H \approx m^\ominus K_A^\ominus m_1/m_2)\} = -(RT/F)\ln K_A^\ominus = 0.2815_5$ V and $K_A^\ominus = 1.74_1\times 10^{-5}$. If the acid were stronger we might need to continue, and to use the accurate expression for m_H in the two factors $(m_1 - m_H)$ and $(m_2 + m_H)$, to find a more accurate value of $K_A^\ominus$; in our case the value of $K_A^\ominus$ has converged.

20.8: We have $E = E^\ominus - (RT/F)\ln\{m_H m_2/(m^\ominus)^2\} - (2RT/F)\ln\{\gamma(H^+, Cl^-)\} = E^\ominus - (RT/F)\ln(K_w^\ominus m_2/m_1) - (2RT/F)\ln\{\gamma(H^+, Cl^-)/\gamma(H^+, OH^-)\}$, where we safely neglect the small difference between $m(OH^-)$ and m_1. Thus, $E' \overset{\text{def}}{=} E - E^\ominus + (RT/F)\ln(m_2/m_1) = -(RT/F)\ln K_w^\ominus - (2RT/F)\ln\{\gamma(H^+, Cl^-)/\gamma(H^+, OH^-)\}$, and $\lim_{I\to 0} E' = -(RT/F)\ln K_w^\ominus$ where $I = (m_1 + m_2)$. We evaluate E':

E'/V 0.82830   0.82837   0.82842   0.82844

so that $\lim_{I\to 0} E' = -(RT/F)\ln K_w^\ominus = 0.82821$ V and $K_w^\ominus = 1.001\times 10^{-14}$.

20.9: If $E' \overset{\text{def}}{=} E - (2RT/F)\ln(xm/m^\ominus) + (2\alpha RT/F)m^{1/2}/\{1 + (m/m^\ominus)^{1/2}\}$ then $\lim_{m\to 0}(\lim_{x\to 0} E') = -(RT/F)\ln\{K^\ominus(AgCl)\}$. We evaluate E':

$m/$mol kg^{-1}	0.03	0.03	0.03	0.02	0.02	0.02
x	0.3	0.2	0.1	0.4	0.3	0.2
E'/V	0.57737	0.57729	0.57709	0.57752	0.57739	0.57721

For $m = 0.03$ mol kg^{-1}, $\lim_{x\to 0} E' = 0.576_8$ V and for $m = 0.02$ mol kg^{-1}, $\lim_{x\to 0} E' = 0.576_8$ V. There being no discernible trend with m, $-(RT/F)\ln\{K^\ominus(AgCl)\} = 0.576_8$ V and $K^\ominus(AgCl) = 1.78\times 10^{-10}$.

21.1: We calculate values of $\partial\gamma/\partial\ln m_B^\alpha$ from the empirical equation, and values of $(RT/\mathscr{A})w^\sigma(m_B^\alpha - m_B^\sigma)$. Those two quantities should be equal if Gibbs's adsorption equation is obeyed.

$(\partial\gamma/\partial\ln m_B^\alpha)/$N m^{-1}	-0.008	-0.011	-0.012
$(RT/\mathscr{A})w^\sigma(m_B^\alpha - m_B^\sigma)/$N m^{-1}	-0.006 ± 0.002	-0.014 ± 0.002	-0.014 ± 0.003

The two quantities agree almost if not quite within the experimental errors.

21.2: Given the interpretation of B in the text, the cross-sectional area a of a solute molecule is given by $a = \mathscr{A}/Ln_B^\sigma = RT/LB = 8.314\times 291$ J mol$^{-1}/6.02\times 10^{23}$ mol^{-1} 0.013 J m$^{-2} = 0.309\times 10^{-18}$ m$^2 = 0.309$ nm^2. (This value is rather greater than that found for insoluble films by use of a surface balance.)

Index